DIVINE VISITOR

The Gaia Method Series

DIVINE VISITOR

The Science Behind Why We Choose This Life

MJ MANCINI

Founder, Gaia Institute for Coherent Prosperity

Staten House

Published by Staten House
for The Gaia Institute for Coherent Prosperity

First Edition

ISBN 979-8-90417-704-1
Printed in the United States of America

To my late mother Barbara
and her mother Rose.

The two strongest women
I ever was blessed
to have in my life.

I can't wait to see you two again.

We are participators in bringing into being not only the near and here

but the far away and long ago.

— John Archibald Wheeler

CONTENTS

CHAPTER 1

The Anthropic Architecture

One cool September morning in 2002, a Tibetan monk arrived at the Dane County Regional Airport in Madison, Wisconsin. His journey had begun seven thousand miles away at a monastery on a hill at the edge of Kathmandu. Eighteen hours in the air over three days. Ten time zones. He was the only shaven-headed man in gold-and-crimson robes in the airport, and the customs officer, after a moment of hesitation, smiled at him the way the officer might have smiled at a small child whose parents had walked off without him.

He was twenty-seven years old. He wore rimless glasses with gold frames. He was unfailingly polite and clean-shaven and quick to smile, with a somewhat reserved demeanor that softened in the presence of children and strangers in equal measure. He spoke English with the careful enunciation of a man who had learned it as his fourth or fifth language, after Tibetan and Nepali and the liturgical Sanskrit of his tradition and the fragments of Hindi he had picked up across the years he had lived in northern India. By that morning he had spent nine of the most recent ten years of his life in formal meditation retreats. The lifetime hours of formal practice he had accumulated by his late twenties exceeded what most professional surgeons accumulate in their hands across their entire careers, only the practice was internal, and the instrument was the substrate of his own attention.

It was not, looking at the polite young man with the rimless glasses, an obvious thing to have spent a life on.

His name was Yongey Mingyur Rinpoche, and he had been a frightened child. The fact does not usually appear in the literature about him because the literature about him is preoccupied with the gamma synchrony and the fMRI activations and the brain-age numbers that came later, and a child's panic does not survive translation into those vocabularies. But the panic was the seed of the practice, and the practice was the seed of the substrate that came to be measured. He had been born in a village called Nubri, on the Nepal-Tibet border, in 1975. His father was Tulku Urgyen Rinpoche, one of the most respected meditation masters of his generation. His mother was a descendant of two Tibetan kings whose names mean nothing to most readers and everything to the lineage Mingyur was born into. From early childhood — earlier than memory should reach — he had been afflicted with what would today be called panic disorder. The episodes seized him without warning. His heart would pound. He could not breathe. The walls of whatever room he was in would seem to lean inward. He was, by his own later account, terrified most of the time of nothing in particular and everything at once.[1]

He learned to run away to the caves around the village to meditate. Eight years old, ten years old. He sat in the caves the local hermits had used for centuries, sometimes for hours at a time, and what he found in the sitting was that the panic could not maintain itself inside the kind of attention the caves drew out of him. The panic was not gone. It was a thing he was sitting next to, not a thing he was being. The difference was the foothold.

---

[1]Mingyur Rinpoche's biographical details and his own account of his childhood panic disorder are drawn from his memoir, Mingyur Rinpoche, Y., and Tworkov, H. (2019). In Love with the World: A Monk's Journey Through the Bardos of Living and Dying. New York: Spiegel & Grau, pp. 9-22; from his earlier book, Mingyur Rinpoche, Y., and Swanson, E. (2007). The Joy of Living: Unlocking the Secret and Science of Happiness. New York: Harmony Books, pp. 3-19; and from the family memoir of his father: Tulku Urgyen Rinpoche (2005). Blazing Splendor: The Memoirs of Tulku Urgyen Rinpoche. Kathmandu: Rangjung Yeshe Publications. The biographical chronology of his enthronement, retreats, and monastic positions follows the official Tergar lineage account at tergarsangha.org and the parallel account at the Sherab Ling and Palpung sources, which are in close agreement on dates and events.

At nine he was sent to study with his father at a small hermitage called Nagi Gonpa, on the outskirts of the Kathmandu valley. For nearly three years his father guided him through Mahamudra and Dzogchen, the high practices of the Karma Kagyu and Nyingma traditions, which by their own accounting are not given to children. The exception was made because the child had the temperament for them and because the father had the rare authority to recognize this and the rarer willingness to act on the recognition. The instruction proceeded by what is called pith — a few sentences, sometimes only a few words, after which the student was sent to a cushion to find out what the sentences meant from the inside. The father would call him back when enough days had passed. The student would report. The father would adjust the next instruction by what the report contained. This is not the way the educational systems of most countries are organized. It is, in lineages of this kind, the way the practice is preserved.

At eleven he was requested to take up residence at Sherab Ling, the monastic seat of Tai Situ Rinpoche in northern India. He left his father and his mother and the village where the moon was said to have appeared on the roof during the month before his birth, and he went to a monastery in the Himalayan foothills where a thousand-year-old lineage was being preserved by men who had walked out of Tibet ahead of the Chinese army with the lineage in their saddlebags. He studied the rituals of the Karma Kamtsang school with the retreat master, Lama Tsultrim. He learned the tantric texts that had been brought into Tibet eight hundred years earlier by the great translator Marpa, who had crossed the Himalayas on foot to retrieve them from India.[2]

[2]On the Tibetan lineage transmission and the history of Sherab Ling, see Mingyur Rinpoche, Y. (2009). Joyful Wisdom: Embracing Change and Finding Freedom. New York: Harmony, esp. pp. 17-41. On the political-religious history of the Tibetan diaspora into India and the establishment of monasteries like Sherab Ling, see Goldstein, M. C. (1997). The Snow Lion and the Dragon: China, Tibet, and the Dalai Lama. Berkeley:

At twelve, in 1987, he was formally enthroned at Sherab Ling as the seventh incarnation of Yongey Mingyur Rinpoche — a teaching lineage of the Karma Kagyu and Nyingma traditions that extends back through six prior recognized incarnations. The ceremony was conducted by Tai Situ Rinpoche, then in his thirties, one of the highest-ranking lamas of the Karma Kagyu and one of the four heart sons of the sixteenth Karmapa. The recognition itself had been earlier — Mingyur had been identified as the seventh incarnation in early childhood by the sixteenth Karmapa and by Dilgo Khyentse Rinpoche, two of the most respected figures in twentieth-century Tibetan Buddhism — but the enthronement was the public ceremony at which the recognition was formalized and the twelve-year-old child took on the role the recognition entailed. He wore the robes the role required. He accepted the offerings the role accepted. He bowed to the teachers the role bowed to. He was, by the structure of the institution he was born into, no longer entirely a child by the time the ceremony ended, and he would not be entirely a child again.

At thirteen he begged both his father and Tai Situ Rinpoche for special permission to enter the traditional three-year retreat that was about to begin at Tendzin Gephel Ling, the retreat center of Sherab Ling. The retreat was not designed for thirteen-year-olds. It was designed for adults who had completed extensive preliminary practices and were ready to undertake what the tradition calls the path of accomplishment — three years, three months, and three days of formal meditation, eight to ten hours a day, conducted in a small cell-like room from which the practitioner emerges only to bathe, to receive food, and, on occasion, to take instruction from the retreat master. The retreat master that year was Saljey Rinpoche, then in his late seventies, who had spent half of his own life in strict retreat in Tibet before walking out into India after the Chinese invasion. He

---

University of California Press.

had given the three-year retreat instructions in their full form to a great many students. He had not given them to a thirteen-year-old.[3]

Mingyur was thirteen and a half when the door of his retreat cell closed behind him. The cell was small. The bed was a meditation cushion thicker than most meditation cushions. The window was small enough to be inconsequential. The practices, which he had begged to be allowed to do, included the preliminary practices of the Ocean of Kagyu Siddhas — the foundational disciplines that an adult practitioner is expected to complete before approaching the development stage — and then the development stages for Jinasagara and Chakrasamvara and Vajravarahi, the deity practices through which the practitioner trains the substrate to recognize itself as a configured signal embedded inside a larger configured field. After the development stages came the completion stages: the Six Doctrines of Naropa, which include the practices of inner heat and dream yoga and clear light, and the practice of Mahamudra, which the tradition considers its deepest. The retreat ran for three years, three months, and three days. The boy went in at thirteen and a half. He came out at sixteen and three quarters. Saljey Rinpoche died shortly after the retreat ended, and Tai Situ Rinpoche, finding the retreat master's position vacant and the seventeen-year-old Mingyur the most accomplished available successor, appointed him retreat master for the next three-year retreat at Sherab Ling. He was, by some accounts, the youngest lama in living memory to hold the position.

---

[3]On Saljey Rinpoche (1910-1991): Rigpa Wiki entry, drawn from Mingyur Rinpoche's memorial account 'The Life and Teachings of Saljey Rinpoche.' Saljey Rinpoche was recognized in his own youth by the fifteenth Karmapa, Khakhyap Dorje, and had completed a three-year retreat in Tibet under the oral instructions of the second Jamgön Kongtrul and the Palpung retreat master Lama Tashi Tsering. On the structure of the three-year retreat in the Karma Kagyu tradition, see Kongtrul, J. (1994). Jamgön Kongtrul's Retreat Manual. Translated by N. Zangpo. Ithaca, NY: Snow Lion Publications.

By the time he stepped off the plane in Madison that September morning, he had spent more than nine thousand hours in formal practice across multiple retreats, and a great many additional hours in informal practice during the years in between. The number that would become widely cited in the academic literature about him a decade and a half later — over sixty thousand hours of lifetime meditation — was already well underway. He was, by any sober reckoning, one of the most heavily trained substrates of human attention then alive on the planet. He had also been, until age eight or nine, the child who could not breathe in small rooms.

He had come to Madison at the invitation of Richard Davidson. The invitation had been mediated through Francisco Varela, the Chilean-French neuroscientist who had co-founded the Mind and Life Institute with the Dalai Lama in the 1980s and who, in the years before his death in 2001, had traveled regularly to Nepal to receive instruction from Mingyur's father Tulku Urgyen Rinpoche. Mingyur, in his twenties, had sat in on a number of conversations between his father and Varela in which the questions traveled freely between phenomenology and physiology and the philosophy of mind. The conversations had given the young lama an early curiosity about what neuroscience could and could not say about the substrate his lineage had spent a thousand years training, and the curiosity had survived Varela's death and ripened, by 2002, into a willingness to be measured.[4]

---

[4]On the Varela connection and the formation of the Mind and Life Institute's research agenda: Varela, F. J., Thompson, E., and Rosch, E. (1991). The Embodied Mind: Cognitive Science and Human Experience. Cambridge, MA: MIT Press, esp. introduction and Ch. 11; and Houshmand, Z., Livingston, R. B., and Wallace, B. A. (eds.) (1999). Consciousness at the Crossroads: Conversations with the Dalai Lama on Brain Science and Buddhism. Ithaca, NY: Snow Lion. On Davidson's own arc from cognitive psychology into the neuroscience of meditation, see Davidson, R. J., & Begley, S. (2012). The Emotional Life of Your Brain. New York: Hudson Street Press, esp. Ch. 9.

Davidson had met him briefly in 1995 at a Mind and Life conference in Dharamsala on the subject of destructive emotions, but had forgotten what Mingyur looked like in the intervening seven years. The gold-and-crimson robes resolved that problem on arrival. The two men recognized each other across the baggage claim. They shook hands. The drive from the airport into Madison took half an hour. Mingyur asked questions about Wisconsin in his careful English. He looked out the window at the suburbs, at the trees beginning to color, at the late summer light. The country was the third he had ever set foot in. The light fell differently from the way it fell on the Himalayan foothills. He noticed but did not comment.

The lab the next morning was a small set of rooms at the Waisman Center for Brain Imaging and Behavior. The lead investigator running the EEG protocol that week was Antoine Lutz, a French postdoctoral researcher in Davidson's group who had studied with Francisco Varela before Varela's death and who had inherited from Varela the conviction that the right substrates, examined with the right instruments, would disclose phenomenology in the language of physiology. Lutz had spent the preceding months developing a meditation protocol that could be reliably presented to expert practitioners and to novices in matched conditions. The protocol called for the subject to enter, on a researcher's cue, a state of compassion meditation — a practice in which the practitioner generates an unbounded wish for the welfare of all beings, holding the wish steadily without object, without effort, without reaction. Lutz had run the protocol on novice meditators in the months preceding Mingyur's arrival. The novices had produced, during the compassion sequence, an increase of ten to fifteen percent in the activity of the neural circuits associated with empathy and positive emotion.

The EEG headgear had 256 thin wires, each connected to a sensor capable of registering the electrical activity of the cortex beneath it. The application took twenty minutes. Mingyur sat patiently in the chair while the technician parted his shaven scalp into sections and applied conductive gel and pressed each sensor home. He looked, in the photographs the lab would later publish, like a man wearing a wired wig. The wires were gathered into a thick cable that ran from the back of the cap to the EEG console behind the chair, where the signal would be displayed in real time on a screen the technicians could see and the subject could not.

On the researcher's cue, Mingyur entered the practice. The console registered what the room had not seen before. The gamma-frequency synchrony in the cortex — the high-frequency, broadband neural coherence that the electroencephalographic literature had previously documented only in fragments, of much shorter duration and much smaller amplitude — appeared at levels and durations that exceeded everything then in the published literature on human cognition. Lutz checked the readings against the calibration. The calibration was correct. He checked the connections at the back of the cap. The connections were sound. He looked at Davidson. Davidson looked back. The empathy circuits that the novices had elevated by ten percent were elevated, in Mingyur's substrate, by seven hundred percent. In some sequences they reached eight hundred percent. The researchers' first response was to suspect the equipment. The equipment was working.[5]

---

[5]The initial published results on the gamma-synchrony findings: Lutz, A., Greischar, L. L., Rawlings, N. B., Ricard, M., & Davidson, R. J. (2004). "Long-term meditators self-induce high-amplitude gamma synchrony during mental practice." Proceedings of the National Academy of Sciences 101(46): 16369-16373. The fMRI compassion-meditation findings are reported in Lutz, A., Brefczynski-Lewis, J., Johnstone, T., & Davidson, R. J. (2008). "Regulation of the neural circuitry of emotion by compassion meditation: Effects of meditative expertise." PLoS ONE 3(3): e1897. Davidson's own retrospective account of the moment in the lab appears in Davidson, R. J., & Begley, S. (2012). The Emotional Life of Your Brain. New York: Hudson Street Press, pp. 215-225.

Mingyur sat in the chair the entire time. He had been told, before the protocol began, that the researchers would give him a cue to begin a sequence and a cue to end it, and that between cues he was to remain in whatever state the previous cue had asked him to enter. He did this. The cues came. The cues went. The screens behind him registered values the laboratory's published baselines had no slot for. The session lasted several hours. When it was over, the technician helped him out of the headgear. The gel was wiped from his scalp with a damp towel. He thanked everyone in the room individually, by name. He bowed slightly. He left.

Over the next fourteen years Davidson's lab would scan him four times — in 2002, 2005, 2007, and 2016 — and the longitudinal record would, by 2020, document a substrate that had structurally aged at roughly two-thirds the rate of the matched-age control population, so that at the age of forty-one Mingyur's brain resembled that of a thirty-three-year-old. The aging differential was not localized to any specific brain region. It was distributed across the gray matter, as if the entire substrate had been kept, by the practice running on top of it, in a configuration the substrate would otherwise have begun to lose. The longitudinal data would establish, with the level of rigor the literature of contemporary neuroscience demands, that what the man could do at age twenty-seven was not a fluke of measurement. It was a property of the substrate the practice had built. And the substrate the practice had built was the substrate that had once belonged, before the practice took it up, to the small boy who could not breathe in the small room.

None of that was on the screens the first morning. What was on the screens were values that exceeded the published literature, and Antoine Lutz and Richard Davidson, looking at the values, agreeing in silence that they were looking at a configured signal of a kind their field had never measured. The session ended. Mingyur thanked everyone. He went back to the guesthouse the university had arranged for him. He slept. The next morning the protocol was run again. The values held.

✦ ✦ ✦

This book is not, in the main, about Mingyur Rinpoche. The chapters that follow examine the framework his life happens to demonstrate, drawing on the data from many lives and many investigative literatures. But the morning he stepped off the plane in Madison is the right place to begin, because it discloses, in one image, the question that organizes everything else in the book.

The question is not whether the substrate of an individual human being can produce configured signals of remarkable specificity. The lab established that it can, and the published record has since extended the finding through twenty-one additional yogis with lifetime practice hours ranging from twelve thousand to sixty-two thousand, all of whom produced configured signatures in the same general range Mingyur did. The configured substrate is real. The configured signal is real. The question that follows is a different one, and it is the question this chapter takes up. If the substrate is configured, what about the universe the substrate was assembled in? If the signal carries the marks of a long preparation, what about the conditions that made the preparation possible at all?

The question pulls in the opposite direction from where the lab pulled. The lab pulled inward, toward the cortex of one man. The question pulls outward, toward the architecture in which that cortex came to exist. The lab found a configured signal. The question is whether the configured signal had to arise inside an equally configured field. The arithmetic in the rest of this chapter argues that it did.

The Cosmological Constant

The number is 10^{-122}, and it represents the ratio between the cosmological constant predicted by quantum field theory and the cosmological constant actually measured in the expansion of the universe. The prediction overshoots the observation by 122 orders of magnitude — the worst mismatch between theory and observation in the history of physics, and the figure that one of the architects of the Standard Model, Steven Weinberg, once called the bone in our throat.[6] He meant it the way the phrase is normally used: as something one cannot swallow and cannot ignore.

The mismatch matters because the cosmological constant is the term in Einstein's field equations that determines whether the universe expands fast enough to keep forming stars or so fast that matter never coheres at all. If the constant were larger by even a small fraction of its predicted value, the universe would have flown apart in microseconds, with no chance for hydrogen to clump, no chance for gravity to assemble galaxies, no chance for any of the structures on which complex chemistry depends. If it were negative by the same fraction, the universe would have collapsed before a single atom assembled. Neither outcome leaves room for a planet, a body, or a reader. You are inside the surviving fraction. The constant is small enough, and stable enough, and positive enough, that the universe expanded slowly enough to assemble galaxies and not so slowly that those galaxies fell back into themselves. The window was narrow. The arrow flew through it. You are downstream of that flight.

That observation — narrow window, surviving arrow — is the seed of everything that follows. Before you can speak intelligibly about anyone choosing to be here, you have to establish that the *here*

[6]Weinberg, S. (1989). "The cosmological constant problem." Reviews of Modern Physics 61(1): 1-23. The phrase recurs in Weinberg's later popular work, including Dreams of a Final Theory (Pantheon, 1992).

in question is the kind of place that could have been chosen. A chaotic universe with no stable physics, no stable chemistry, no stable matter, is not a destination. It is noise. The universe Mingyur stepped into in 2002, the universe his substrate had spent twenty-seven years cultivating itself inside, the universe these sentences are reaching you across, is not noise. It is configured. And the configuration is the subject of this chapter.

Where the Discrepancy Comes From

The cosmological constant problem is not a numerical curiosity. It is the cleanest existing example of the structural tension at the heart of modern physics, and it deserves a few paragraphs of its own before we move on to the other numbers that have to be tuned.

In quantum field theory, the vacuum is not empty. It is a sea of fluctuations — pairs of particles and antiparticles that appear and annihilate in intervals so short they cannot be directly observed, but whose collective energy has measurable effects on the world. The Casimir effect, demonstrated experimentally in 1948 and refined many times since, shows that two metal plates placed very close together in a vacuum will be pushed together by the pressure differential between the fluctuations inside the gap and those outside it.[7] The vacuum has energy. The energy is real. It can move plates.

When physicists try to calculate the total energy of the vacuum, they sum the contributions of all the quantum fields up to some cutoff — typically the Planck scale, the smallest length at which our current physics is thought to apply. The number they get back is astronomical: roughly 10^{113} joules per cubic centimeter. The number they should get, judging from the observed expansion of the

[7]Casimir, H. B. G. (1948). "On the attraction between two perfectly conducting plates." Proc. K. Ned. Akad. Wet. 51: 793-795. The first high-precision experimental confirmation is Lamoreaux, S. K. (1997). "Demonstration of the Casimir force in the 0.6 to 6 μm range." Physical Review Letters 78: 5-8.

universe, is about 10^{-9} joules per cubic centimeter. The ratio between prediction and observation is 10^{122}, written out as a one followed by 122 zeros.

There is no working physicist who calls this an acceptable state of affairs. It is the largest discrepancy between theory and experiment in any branch of physics, and after sixty years of effort, no one has produced a mechanism that explains it. The standard responses fall into three categories. The first proposes that an unknown symmetry cancels the vacuum contributions almost exactly, leaving the small residual we observe; no such symmetry has been found. The second proposes that the calculation is being done incorrectly, and that some future framework will yield the right answer from first principles; decades of work has not identified the error. The third proposes that the value is what it is for selection reasons, which is where the multiverse arguments enter the literature in their most serious form. Each response is taken seriously in the field. None is settled.

The discrepancy is laid out here for one reason: it is the first and largest of the fine-tunings, and it is real. It is not metaphysics. It is the measured offset between what physics predicts and what physics observes. The other tunings follow.

Six Numbers

The cosmological constant is the most extreme example, but it is not isolated. The British astronomer Martin Rees, who served as Astronomer Royal and as president of the Royal Society, set out in his short book *Just Six Numbers* (1999) the case that there are six independent dials in the underlying physics of the universe, each of which has to be set within a narrow window for the universe to be capable of producing complex matter.[8] The book is widely cited in

---

[8]Rees, M. (1999). Just Six Numbers: The Deep Forces That Shape the Universe. London: Weidenfeld & Nicolson; New

cosmology and is one of the most accessible summaries of what the field calls the fine-tuning problem.

The first number is **N**, the ratio of the electromagnetic force to the gravitational force between two protons. Its value is approximately 10^{36} — a one followed by thirty-six zeros. The electromagnetic force is enormously stronger than gravity at the scale of individual particles, which is why a small magnet can lift a paperclip against the gravitational pull of the entire Earth. But gravity scales with mass, and electromagnetic forces between neutral matter cancel out at large scales, which is why galaxies are gravitationally bound rather than electromagnetically bound. If N were significantly smaller — say, 10^{36} instead of 10^{3}□— gravity would dominate more aggressively at smaller scales. Stars would be smaller. They would burn through their fuel in a few million years rather than a few billion, and the evolutionary runway required to assemble complex multicellular organisms would not exist. N is set for the long burn.

The second number is **ε** (epsilon), the efficiency with which hydrogen converts to helium when fused under stellar pressure. Its measured value is 0.007, and that number reports the fraction of mass converted to energy each time four hydrogen nuclei fuse into one helium nucleus. At 0.006, the strong nuclear force would not be quite strong enough to bind two protons against their mutual electrical repulsion, and hydrogen could not fuse at all. The universe would remain a hydrogen cloud forever, cooling and thinning and producing no chemistry. At 0.008, the binding energy would be high enough that all the free protons in the early universe would have fused into helium and heavier elements in the first few minutes after the Big Bang, leaving no hydrogen for stars to burn and no water for chemistry. The window for stellar chemistry is the narrow corridor between those two values, and epsilon sits inside the corridor.

York: Basic Books.

The third number is **Ω** (omega), the density parameter of the universe — the ratio between the actual density of matter and energy and the critical density required for a flat geometry. If Ω is much greater than one, the universe collapses gravitationally before structures form. If Ω is much less than one, matter disperses so quickly that gravitational binding cannot produce galaxies. Ω is observed to be very close to one, which means the universe is geometrically flat. The flatness problem — the fact that Ω cannot have been close to one at the Big Bang unless it was tuned to absurd precision at that moment — was one of the motivations for the theory of cosmic inflation, which proposes a brief period of exponential expansion in the first 10^{-36} seconds. Inflation explains how Ω could be flat now without being unnaturally tuned at the beginning. It does not eliminate the need for tuning. It pushes the tuning to a different parameter (the inflaton field) and to a different stage.

The fourth number is **Λ** (lambda), the cosmological constant already discussed at length. It is the most extreme of the six tunings and the one that drove the framing of this chapter. It appears here again because Rees lists it among the six, and its inclusion is not redundancy — it is the most dramatic instance of the pattern, the tuning so severe that it forced cosmology as a discipline to admit it had an unsolved problem at the foundations of its own model.

The fifth number is **Q**, the amplitude of the irregularities in the early universe — the unevennesses in the cosmic microwave background that grew, under gravity, into galaxies and clusters of galaxies. Its measured value is approximately 10^{-5}. If Q were lower than about 10^{-6}, the unevennesses would not be deep enough to clump matter against the universe's overall expansion, and no galaxies would form. The universe would be a uniform thinning gas with no centers of gravitational organization. If Q were higher than about 10^{-4}, the unevennesses would be so deep that matter would clump into supermassive black holes long before stars could light. The window of Q that produces galaxies of the kind we see —

galaxies inside which stars burn long enough for planets to form, cool, and host chemistry — is roughly one part in 100,000.

The sixth number is **D**, the number of large spatial dimensions. The universe has three, and the value matters more than it appears. In 1917, the Austrian physicist Paul Ehrenfest published a short paper asking what would change if the number were different.[9] He proved that in spaces of more than three dimensions, no stable orbits are possible — a planet would either spiral into its star or spiral away from it, depending on the smallest perturbation. He also showed that the hydrogen atom would not have stable bound states in spaces with fewer than three dimensions, which means no chemistry. The result has been refined many times in the century since, but the basic conclusion has held: three dimensions are the unique value at which both atoms and planetary orbits can be stable. Lower and chemistry breaks. Higher and orbits break. Three is the only working setting.

Each of these six is independent of the others. Each one had to land within its narrow window. The probability of all six landing in their respective windows by chance, on the standard accounting, is small enough that the field has spent the better part of three decades looking for a non-chance explanation. Several have been proposed. None has been confirmed.

The History of Noticing

The fine-tuning was not always recognized as a problem. For most of the twentieth century, physicists treated the constants of nature as brute facts — measurements to be made, parameters to be

[9]Ehrenfest, P. (1917). "In what way does it become manifest in the fundamental laws of physics that space has three dimensions?" Proceedings of the Amsterdam Academy 20: 200-209. The argument has been refined and generalized many times since; see Tegmark, M. (1997). "On the dimensionality of spacetime." Classical and Quantum Gravity 14: L69-L75.

fed into equations, but not subjects for explanation in their own right. The shift began in the early 1970s, with a presentation given by the Australian physicist Brandon Carter at a 1973 symposium in Krakow held to mark the 500th anniversary of the birth of Copernicus.[10]

Carter introduced what he called the *Anthropic Principle*, in two formulations. The weak version stated that the observed values of physical constants must be compatible with the existence of observers capable of measuring them — a statement that sounds trivial but has real force when applied to questions about the early universe. The strong version went further: the universe, and the parameters on which it depends, must be such as to admit the creation of observers within it at some stage. The strong version was controversial then and remains so. The weak version is now widely accepted.

The Princeton physicist John Wheeler, who had coined the term "black hole" a few years earlier and who would later coin "it from bit," found the anthropic question generative rather than embarrassing. He pushed his students to take seriously the question of why the universe was the way it was, and not some other way. One of those students was John Barrow, who, with the mathematical physicist Frank Tipler, published *The Anthropic Cosmological Principle* in 1986 — a 700-page survey of the fine-tunings then known, which remains the standard reference work on the subject and which collected dozens of additional tunings beyond the original six.[11]

The physicist Paul Davies, in books such as *The Goldilocks Enigma* (2006), summarized the position to which much of the field

---

[10]Carter, B. (1974). "Large number coincidences and the anthropic principle in cosmology." In M. S. Longair (ed.), Confrontation of Cosmological Theories with Observational Data (IAU Symposium 63), Dordrecht: Reidel, pp. 291-298.
[11]Barrow, J. D., and Tipler, F. J. (1986). The Anthropic Cosmological Principle. Oxford: Oxford University Press.

had drifted by the early 2000s.[12] The constants of nature appear to be specifically set to permit the formation of complex structures, the magnitude of the fine-tuning is too large to be plausibly attributed to chance, and physics has no current mechanism within its existing framework to derive the values from first principles. The field is, in Davies' phrase, in a state of explanatory crisis. The crisis is not concealed. It is the explicit subject of textbooks, conferences, and the latest generation of multiverse proposals.

What changed between Carter's 1973 talk and the present is not the physics. The numbers in 1973 were already the numbers we observe today. What changed is the field's willingness to admit that the numbers required explanation. That admission is the foundation on which any further argument depends, because the third interpretation — the interpretation this book pursues — cannot even be stated coherently in a field that refuses to acknowledge the question.

The Hoyle Prediction

The most famous of the fine-tunings, and the one with the cleanest experimental story, was identified twenty years before Carter's talk by the British astrophysicist Fred Hoyle. Hoyle was an unlikely candidate for the role of accidental theologian. He was a working-class Yorkshireman, a public atheist, and one of the most combative figures in twentieth-century physics. He opposed the Big Bang model on philosophical grounds for most of his career, preferring the steady-state cosmology he co-developed with Hermann Bondi and Thomas Gold in 1948, and he coined the term "Big Bang" itself as a dismissive joke during a BBC radio broadcast. The name stuck. Hoyle did not.

---

[12]Davies, P. (2006). The Goldilocks Enigma: Why Is the Universe Just Right for Life? Boston: Houghton Mifflin.

In 1953, Hoyle was working on the problem of stellar nucleosynthesis — the question of how the heavier elements in the periodic table are forged inside stars. The problem had a sharp difficulty at carbon. Carbon is essential to all known organic chemistry, and any plausible account of how the universe came to contain carbon-based life had to explain how carbon-12 was produced in stellar interiors at a sufficient rate. The route had to go through helium: three helium-4 nuclei combining to form carbon-12. But the calculation, when done with the energy levels then known, gave a rate so slow that almost no carbon should exist in the universe at all.[13]

Hoyle reasoned in the reverse direction. He was made of carbon. The stars must therefore be making carbon at a rate compatible with his own existence. Therefore there must be a missing piece in the calculation — an energy state in the carbon-12 nucleus that no one had yet observed, at approximately 7.6 million electron volts, which would allow three helium nuclei to combine resonantly and form carbon at a rate compatible with the observed abundance.

He took the prediction to William Fowler, the nuclear physicist at the California Institute of Technology who ran the experimental facility most likely to find such a state if it existed. Fowler's later account is that he initially thought Hoyle was wasting his time, but agreed to look out of collegial courtesy. Within a few months, Fowler's team had found the resonance at 7.6549 MeV — within fractions of a percent of the value Hoyle had named, with no theoretical input other than the demand that the resonance had to exist for Hoyle to exist.[14] Fowler later won a share of the 1983 Nobel Prize in Physics for the larger body of work on stellar

[13]Hoyle, F. (1954). "On nuclear reactions occurring in very hot stars. I. The synthesis of elements from carbon to nickel." Astrophysical Journal Supplement Series 1: 121-146. The prediction itself dates from 1953 work circulated before publication.

nucleosynthesis to which this discovery belonged. Hoyle was controversially passed over for the same prize, which many in the field considered a serious slight.

The Hoyle resonance is the cleanest instance in the fine-tuning literature of a prediction made on anthropic grounds — made, that is, from the bare fact of the predictor's existence — that turned out to be correct. The prediction was not vague. It named a number. The number was found. And the number was found at a level of precision that no random tuning of the underlying physics would have produced.

Hoyle's response to his own discovery, written years later in an article for the Caltech publication *Engineering and Science* (November 1981), is the source of the most often-quoted passage in the fine-tuning literature:[15]

> *A common sense interpretation of the facts suggests that a superintellect has monkeyed with physics, as well as with chemistry and biology, and that there are no blind forces worth speaking about in nature. The numbers one calculates from the facts seem to me so overwhelming as to put this conclusion almost beyond question.*

Hoyle was not converting to anything. He spent the rest of his life resisting both conventional religion and conventional cosmology in roughly equal measure, and he died in 2001 an atheist by his own statement. He was reporting, in the register of a working physicist, what the arithmetic forced him to conclude. The quote is worth carrying with the same weight Hoyle carried it — not as proof of any

[14]Dunbar, D. N. F., Pixley, R. E., Wenzel, W. A., and Whaling, W. (1953). "The 7.68-MeV state in C12." Physical Review 92: 649-650. The value has since been refined to 7.6542 MeV.
[15]Hoyle, F. (1981). "The Universe: Past and Present Reflections." Engineering and Science (Caltech), November 1981, pp. 8-12. The quoted passage appears on p. 12.

particular metaphysical system, but as the recorded testimony of a hostile witness. Hostile witnesses are the most credible kind. They have the strongest possible reason to dismiss the evidence, and when they fail to dismiss it, the evidence has done unusual work.

Three Interpretations

The fine-tuning of the constants is not a contested observation. The numbers are the numbers. What is contested is what to make of them, and the field has converged on three live interpretations. Each deserves to be presented in its strongest form before any is set aside.

The first is coincidence. The universe simply happens to have the constants it has. The fact that we are here to notice them does not require explanation, because if the constants had been different, no one would be here to notice the alternative. This is the weakest position in formal terms, but it has serious defenders. The argument is that any time a system can be examined only by observers who depend on its parameters, the parameters will appear tuned to those observers, regardless of whether any tuning occurred. The objection is that the magnitude of the apparent tuning — 10^{122} in the cosmological constant alone — is too steep to be dissolved by the framing alone. The shrug does not scale to numbers of that size. A coincidence one in a hundred is unremarkable. A coincidence one in a trillion is suspicious. A coincidence one in 10^{122} is, in the working judgment of most cosmologists, not a coincidence at all.

The second is multiverse selection. The constants we observe are simply the constants in our universe, and there are very many other universes — perhaps infinitely many — with different constants. We find ourselves in this one because we could not exist in most of the others. The hypothesis comes in several variants. Andrei Linde's eternal inflation model, developed in the 1980s, proposes that the inflationary expansion that began our universe is generic, and that distinct inflationary "bubbles" continually emerge with different physical parameters.[16] Leonard Susskind's string-theory landscape proposes that the equations of string theory admit something on the order of 10^{500} distinct stable configurations of the vacuum, each with its own constants, all realized somewhere.[17] Max Tegmark's mathematical universe hypothesis goes further, proposing that every mathematically consistent structure exists as a physical universe somewhere in the larger ensemble.[18] Lee Smolin's cosmological natural selection proposes that universes reproduce through black holes, with daughter universes inheriting slight variations on parental constants, and that the universes most fit to produce black holes will eventually dominate the population.[19] Each variant is internally consistent. None is currently testable in the standard sense, because no proposed observation could distinguish a multiverse with our constants from a single universe with our constants. The proposals are taken seriously because they are the cleanest available response to the fine-tuning problem within strictly physicalist terms, and physics, by its working method, prefers physicalist responses whenever they remain available.

The third is the hypothesis this book pursues. The constants were not selected at the level of the universe. The universe was selected at the level of the visitor. The visitor — the consciousness now reading this sentence — entered a configured field already optimized for the kind of entity it is. The fine-tuning was not luck, and it was not statistical filtering across infinite parallel universes. It

[16]Linde, A. (1986). "Eternal chaotic inflation." Modern Physics Letters A 1: 81-85.
[17]Susskind, L. (2003). "The anthropic landscape of string theory." In B. Carr (ed.), Universe or Multiverse? Cambridge: Cambridge University Press, pp. 247-266.
[18]Tegmark, M. (2008). "The mathematical universe." Foundations of Physics 38: 101-150.
[19]Smolin, L. (1992). "Did the universe evolve?" Classical and Quantum Gravity 9: 173-191. Full treatment in Smolin, L. (1997). The Life of the Cosmos. Oxford: Oxford University Press.

was the precondition for the visit, and the visit was the operative selection.

These three positions are not equally weighted in the academic literature. The multiverse view dominates serious physics journals. The third is rarely entertained at all. That asymmetry is not a verdict on the third position. It is a verdict on what physics is currently configured to test. Physics, by its working method, cannot test claims about non-physical selection, and so it does not engage with them. The absence of testing is not the absence of evidence. It is the absence of the relevant instrument, and the instrument cannot be built within physics itself because the instrument would have to register a kind of selection that physics, by definition, does not measure.

A reader inclined to push back will already have noticed that the third position carries an argumentative debt. It claims the existence of a visitor whose consciousness predates the body in which the consciousness is now expressed. That claim is not established by the fine-tuning. The fine-tuning, on its own, establishes only the second of the two conditions selection requires: that the field of options has the kind of stability and configuration that selection presupposes. The first condition — the agent — is established elsewhere, in the chapters on consciousness research, on near-death experiences, and on the prior-life case archive maintained for sixty years at the University of Virginia.[20] The argument is cumulative, not single-step. Chapter 1 is the first plank, and a plank

[20]The University of Virginia Division of Perceptual Studies (originally established by Ian Stevenson in 1967) maintains the largest research archive of cases suggestive of reincarnation. Foundational work: Stevenson, I. (1974). Twenty Cases Suggestive of Reincarnation (2nd ed.). Charlottesville: University Press of Virginia. Continuation: Tucker, J. B. (2013). Return to Life: Extraordinary Cases of Children Who Remember Past Lives. New York: St. Martin's Press.

is not a floor. The floor comes by the end of Part 1. The visitor, as such, arrives in Part 2.

What Selection Requires

The argument of this book proceeds from a simple structural observation. The word *selection* presupposes two things. It presupposes an agent capable of selecting, and it presupposes a field of options stable enough to be selected from. Both conditions have to hold. A selection from chaos is not a selection — it is a coincidence given a name. A selectable field with no selector remains a field. Selection is the meeting of the two, and the meeting is what produces the visit.

The fine-tuning data establishes the second condition. The universe is not a chaotic field of arbitrary physics. It is a configured field, sitting on a narrow ridge, with the structural integrity required to host the entities now reading sentences inside it. Whether or not the configuration was chosen by an agent, the configuration is the kind of thing that could be chosen. It carries the signature of selectability. A library of one billion books, organized and indexed and stable on its shelves, is a different object than a billion books on fire in a heap. You can select from the first. You cannot select from the second, no matter how skilled the selector. The selectability is a property of the field, not of the chooser, and it has to be present before any chooser can act.

The fine-tuning, in this framing, is not the proof of selection. It is the establishment of the prerequisite for selection — the proof that the library exists, is organized, and is intact. The librarian is a separate question, addressed by separate evidence. To establish the agent, you need a different class of evidence: the evidence that consciousness exists in a form that can persist across the boundary the body cannot cross, the evidence that memory of pre-bodily existence is sometimes recoverable, the evidence that the specific pairing of soul to body bears the markings of a choice rather than an accident. That evidence is the subject of Chapters 5 through 11, and it is qualitatively different from the evidence of this chapter. The fine-tuning is measured by instruments. The agency is measured by case files. Both are evidence in their own register, but the registers must be kept distinct, and the case for each must be made on the terms that register accepts.

What Chapter 1 establishes is narrower, and it is worth holding the narrowness clearly. It establishes that the universe is the right shape for selection to be a coherent claim. The body you inhabit was assembled in a chemistry that depends on the carbon resonance Hoyle predicted, inside galaxies that exist because Q sat at 10^{-5}, on a planet stabilized by the gravitational ratio N. None of those numbers had to be what they are. All of them had to be what they are for you to be here. The library is real. The shelves are stocked. The argument from there proceeds by adding the librarian, but the library comes first, and the existence of the library is not in dispute.

✦ ✦ ✦

Mingyur Rinpoche left the Madison lab that September and flew back to Kathmandu. He would return three more times across the following fourteen years, and each return would add another set of measurements to a record that, by 2020, would establish him as one of the most extensively neuroimaged practitioners of contemplative discipline in the scientific literature. What the lab took from him on the first visit was a set of numbers about his cortex. What he took from the lab was a confirmation, in the register of a working neuroscience, that the practice he had inherited from his teachers was producing the configured signal it had always claimed to produce.

He did not take the confirmation as license for triumph. He took it the way the cell takes oxygen, which is to say without comment, because it was what the substrate had been built to receive. His interest in the practice, he would later tell Davidson on a visit nearly a decade afterward, had never been the gamma synchrony or the cortical efficiency. His interest was insight and happiness — the same two things he had been after, eight years old in a Himalayan cave, when he first found that the panic could not maintain itself inside the attention the cave drew out of him. The data the lab produced was a side effect, a confirmation in one language of a project the practice had always been pursuing in another.

The boy who could not breathe in small rooms had grown up, by the bare arithmetic, into the man whose substrate produced the most extensive gamma synchrony in the scientific record. The boy had not arrived at the substrate by chance. The substrate had not arrived at the lab by chance. The lab had not arrived in Madison by chance. The university at which the lab sat had not arrived in Wisconsin by chance. Wisconsin had not arrived on a temperate planet by chance. The planet had not arrived in a galaxy by chance. The galaxy had not arrived in a universe with a flat geometry, a low-amplitude perturbation spectrum, a carbon resonance at 7.6549 MeV, and a cosmological constant 122 orders of magnitude smaller than quantum field theory predicts, by chance. None of these arrivals was guaranteed by the level above it. Each was contingent on what the level above had been configured to permit. The contingency runs all the way up. It also runs all the way down. The arrival at the bottom of the chain is the substrate of attention that the panic-attack child built across nineteen years of practice. The arrival at the top is the configured universe inside which the child was able to do the practice at all.

Whatever you think the universe is, it is the kind of place that produces a Mingyur Rinpoche. Whatever you think the universe was meant to do, it is doing what was necessary to produce, in your own substrate, at the moment you read this sentence, the thought that is now occurring to you about whether any of this is true. The instrument that thinks the thought is configured. The field the instrument was assembled in is configured. The configurations are not independent. The smaller one is downstream of the larger one. The remaining chapters examine the case, but the first observation is simple and ungainsayable: the door was narrow, and you came through it anyway.

Mingyur, on the long flight back to Kathmandu the day after the lab session, would have had hours to consider what he had been measured doing. There is no record of what he thought. There is only the record of what he did. He returned to his monastery. He resumed his teaching. He prepared the practice that would, nine years later, lead him to walk out of that same monastery at eleven o'clock at night with nothing but a watch, a few thousand rupees, and the certainty that the substrate the lab had measured would hold under the conditions of total austerity. He would learn, in those four and a half years of wandering, what the substrate could do without the props the world had wrapped around it. The chapters that follow will arrive at that night, in Bodh Gaya in 2011, in due course. The chapter you are now reading is the chapter about the field the substrate came from. The substrate did not produce the field. The field produced the substrate. The substrate, having been produced by the field, was now able to produce the configured signal the lab had been unable to dismiss. The chain of production runs in one direction only. The chapter that follows examines what has to be true of the next link upstream in the chain — the planet — for any of this to have been possible at all.

Where the Argument Goes Next

The next chapter narrows the question from the universe to the planet. Of two hundred billion stars in this galaxy alone, the visit was routed to a single address. The chapter after that narrows from the planet to the body, examining the four-billion-year DNA library from which the visitor's vehicle was assembled — the carbon framework, the pentadactyl limb, the precision grip, the encephalized brain. The chapter after that examines the ascent of the spirit through successive forms in that library, each form carrying its perspective forward into the next, until the encephalization leap produced a substrate capable of integrating the whole. The chapter after that narrows from the body to the brain, and from the brain to the receiver problem, and the receiver problem to the question that has been deferred since the first paragraph of this chapter: who, exactly, is doing the selecting.

For now, hold the number — 10^{-122}. It is the first piece of evidence in a case that takes a book to make, and on its own it is the most striking single number in modern physics. Whatever you think it means, it is the number you are inside, and that fact alone is enough to start.

CHAPTER 2

The Goldilocks Address

The girl was fifteen, and she was hungry, and she was walking from her grandfather's house on the Rozendaalselaan toward the Velp hospital, where she volunteered most days that the war permitted. The walk was a few minutes. Every step of it was filtered through the small calculations that had become, over the four years of occupation, the only way a person whose name had been changed to hide its English origin could be in the streets at all. Was the soldier on the corner one of the older ones, who tended not to bother schoolgirls, or one of the younger SS replacements who had been transferred in after the Arnhem fighting and who had not yet acquired the older soldiers' weariness with their assignment. Was the bicycle missing from its place by the gate because her aunt had taken it to the farms north of the village to barter for sugar beets, or because someone else had taken it. Was the back of the Rotterdamsche Bank — which had been a regular bank when she was eleven and which the Germans had converted to a prison sometime in 1944 — silent today, or were the screams coming through the bricks the way they sometimes did, the screams of Dutchmen being interrogated for things she was glad she did not know.

Her name on her ration card was Edda van Heemstra. Her mother had changed it from Audrey four years earlier, in the days after the German army had crossed into Holland in May 1940, because Audrey was an English name and English names were dangerous.[21] The people of Velp knew her as Adriaantje van

[21]The definitive biographical source on Audrey Hepburn's wartime life is Matzen, R. (2019). Dutch Girl: Audrey Hepburn and World War II. Pittsburgh: GoodKnight Books, which draws on previously unexamined Dutch archival material and interviews with family members and surviving Velp residents. A companion volume covers the postwar years: Matzen, R.

Heemstra, which sounded comfortably Dutch in the mouths of the older women who watched her grow up. She did not speak English in public. She had not spoken English in public since she was eleven. She had been born in Brussels to an Anglo-Irish banker and a Dutch baroness, and her early childhood had been a thing of country houses and boarding schools and parents who had once attended Nuremberg rallies and now no longer spoke of having attended them. The family had walked away from its earlier sympathies. The earlier sympathies had walked away from the family. The walking had been mutual and bitter. The family was on its third year, by late autumn 1944, of doing what could be done from the inside of an occupation that the family had once, in another life, been disposed to admire.

She was five feet five inches tall and weighed, that autumn, around ninety pounds. The weight would continue downward through the winter. She had been a star pupil at the Arnhem Conservatory of Music until the conservatory had been destroyed by German tanks the previous September, in the days of the battle that the British had given the code name Market Garden and that the Dutch would later call the Battle of Arnhem and that the family at Villa Beukenhof on the Rozendaalselaan would later remember as the week the war became personal. She had performed her last formal recital on January 8, 1944, in the Stadsschouwburg in Arnhem, dancing to music by Debussy and Grieg as the star of her school. There would not be another formal recital. Her ballet shoes, by the late autumn, were made by her mother from scraps of felt because there was no leather to be had and no money to buy what little leather there was. She had stopped growing taller. She had begun to develop the symptoms that the Velp hospital, where she

(2021). Warrior: Audrey Hepburn. Pittsburgh: GoodKnight Books. The standard earlier biography is Walker, A. (1994). Audrey: Her Real Story. New York: St. Martin's Press. Hepburn's own brief autobiographical statements about the war years were given in scattered interviews collected after her death in Ferrer, S. H. (2003). Audrey Hepburn, An Elegant Spirit. New York: Atria Books.

walked most mornings, would diagnose as anemia and edema and the early stages of jaundice.

She was carrying, in her shoe, a folded slip of paper she had been asked to deliver to a contact at the hospital. She did not know what was written on it. She had learned not to know. The local resistance leader in Velp, a physician named Dr. Hendrik Visser 't Hooft, used several couriers for several reasons, and a fifteen-year-old who had been seen volunteering at the hospital and who was small enough to be unremarkable was one of his more reliable ones.[22] The papers she carried were sometimes coded and sometimes not. The contents were sometimes the names of downed British or American airmen being moved through the safe-house network toward the Allied lines, sometimes the locations of weapons caches, sometimes the addresses of Jewish families that had to be moved because someone had been observed observing them. She had volunteered at the hospital long enough that her presence in the corridors did not trigger any closer attention from the German soldiers who occasionally passed through. The hospital was not, in any strict sense, a safe place. It was a less unsafe place than most of the alternatives. She entered it, that morning, the way she had entered it every morning for months, and she left it with what she had been given to leave with, and she walked back the few minutes to the Rozendaalselaan, and her shoe felt the same to her on the walk back as it had on the walk in.

She had performed dance recitals in basements through the previous two years, in homes whose owners had drawn the blinds

[22]On Hepburn's resistance work as a courier for Dr. Hendrik Visser 't Hooft, see Matzen (2019), pp. 165-188. Visser 't Hooft was the leading physician at the Velp hospital and the local coordinator of resistance medical work, including the care of downed Allied airmen and the distribution of the underground newspaper Trouw. The work of teenage couriers in the Dutch resistance is surveyed more broadly in Vinke, H. (2018). Children in the Resistance: Dutch Boys and Girls Fight the Nazis 1940-1945. Berkeley: Penguin Random House.

and locked the doors and turned the lamps as low as they could be turned. The performances were called *zwarte avonden* — black evenings — and they were illegal under the occupation, both because the gatherings were unauthorized and because the music played at them was sometimes the work of composers the Reichskulturkammer had banned and because the audiences paid, into a hat passed at the end of each performance, money that went directly to the resistance. The audiences did not applaud. Applause would have been audible from the street and would have been an answer to a question the street had been instructed to ask.[23] She had said, much later, that the best audience she ever had made not a single sound at the end of her performances. The line had carried, in the public versions of her biography, the gloss of a witty self-deprecation. In its original context it had been the description of a working method that depended, like everything else in the occupation, on silences in places where ordinary life would have had sounds.

Her uncle was the central absence in the family's life that autumn, and his absence was the reason her mother and her aunt and her grandfather had completed the long internal migration from political sympathy with the German cause to active material support of the resistance against it. The uncle's name had been Otto Ernst Gelder, Count van Limburg Stirum, and he had been married to Audrey's aunt Miesje, who was her mother's elder sister. He had been a lawyer and a man of cheerful gravity, and he had loved Audrey in the way of an uncle for a niece who had appeared in his life late and unexpectedly when his marriage had not produced children of its own. He had been arrested by the Germans in the spring of 1942 and held with several hundred other prominent Dutch men at an internment camp at St. Michielsgestel that had been a Catholic

[23]On the zwarte avonden see Matzen (2019), pp. 142-156; and Hepburn's own brief recollection in Ferrer (2003), p. 33, where the famous line appears: 'The best audience I ever had made not a single sound at the end of my performances.'

seminary before the war. On August 15, 1942, in retaliation for a resistance attack on a German military train near Rotterdam, the Germans had taken Otto and four other men from the seminary, driven them to a forest at the edge of the camp, lined them up in front of a shallow grave, and shot them. Audrey had been thirteen. Her mother had received the news at the family table in the Sonsbeek house on the Sickeszlaan in Arnhem, in the form of a letter delivered by hand. The family did not speak of Otto's death afterward except in the most glancing terms, even with each other, even decades later. Some pains were not made smaller by being shared. The five men were known in Velp as De Vijf, the Five, and a small annual commemoration would continue at the killing ground in St. Michielsgestel for as long as anyone who had known any of them was still alive to attend it.

After Otto's death, the family had moved from Arnhem to Velp — a quieter village a few miles east, where Audrey's grandfather Baron Aarnoud van Heemstra had a house called Villa Beukenhof. The grandfather had once been the mayor of Arnhem and later the governor of Suriname, and the house was the family's last property of any substance after the Germans had confiscated the bank accounts and the jewelry and the securities. The grandfather was in his seventies. The aunt Miesje had not remarried. The mother Ella had stopped attending the parties she had once attended. Audrey had been transferred from the Arnhemsche Muziekschool, which would later be destroyed in the September fighting, to a smaller dance teacher in Velp who would also, by the late autumn, stop teaching because there was no fuel to heat the studio and no food to feed the students who had begun to faint at the barre.[24]

[24]On the family's move to Villa Beukenhof and the structure of the household in Velp 1942-1945, see Matzen (2019), pp. 110-128. On the genealogy and previous public roles of Baron Aarnoud van Heemstra, see the entries in the Genealogie van het Geslacht Van Heemstra, Centraal Bureau voor Genealogie, Den Haag.

The Battle of Arnhem had begun in September 1944 with the kind of optimism the family had not allowed itself in years. British paratroopers had been dropped behind German lines as part of an Allied plan to seize the bridges across the Rhine and open a path into Germany. Audrey had watched the parachutes coming down from the upstairs window of Villa Beukenhof — small white circles in the late-afternoon sky over Arnhem, six miles to the west. The plan had failed. The bridges had not been held. The British had been driven back across the river with catastrophic losses. The Germans, recovering Arnhem street by street, had blown up the conservatory she had studied at and a great deal else, and they had ordered the population of Arnhem evacuated as a punitive measure. Ninety thousand people had walked east through the autumn rain. Velp had received a great many of them. The grandfather's house had taken in cousins and second cousins and friends of friends until every room had been a sleeping space and the food, which had already been short, had become acutely short. Audrey had volunteered her bedroom and slept on a sofa in her grandfather's study. The new arrivals had brought with them stories from the city she had known as a child — which buildings had been destroyed, which neighbors had been killed, which streets had been impassable. The Stadsschouwburg, where she had given her last recital nine months earlier, had survived. The conservatory had not.

And in the basement of Villa Beukenhof, through the worst of that autumn, the family had hidden a British paratrooper. He had landed near Oosterbeek during Market Garden, been wounded in the leg, made his way through the back gardens of Velp as the German lines closed in, and arrived at the kitchen door of the van Heemstra house in the small hours of one of the September nights. The family had taken him in. The grandfather, the aunt, the mother, and the fifteen-year-old girl had all known what the penalty would be if the Germans found him, and they had all agreed that the penalty was something they were prepared to accept, although they did not speak of accepting it in those terms. He had stayed for about a week — long enough for his leg to be cleaned and bandaged and long enough for the resistance to arrange his transfer to the network that would, eventually, with luck, get him back across the river to Allied lines. Audrey had brought him food in the basement at the times of day when the soldiers in the village were known to be at their meals. She had spoken to him in English — the only English she had spoken in four years — and he had spoken to her in English, and she had not asked his name and he had not asked hers. He had called her *miss*. She had called him *sir*. When he had left, her mother had burned the bandages and the cloth and the small remnants of his uniform in the kitchen stove, and the smoke had carried up the chimney into the night sky over Velp.[25]

That had been autumn. The winter that followed would be remembered in the Netherlands as the Hongerwinter. By November the Germans, in retaliation for a Dutch rail strike called at the Allies' request, had cut off the food shipments from the agricultural east to the densely populated west. The ration in the western cities had fallen below five hundred calories a day, and by January would fall below four hundred. People had begun to eat tulip bulbs, ground into a flour that produced a bread of remarkable bitterness, and grass and nettles and sugar beets and pets and the bark of certain trees. In Velp, on the German side of the front, the situation was less catastrophic than in the west but it was bad enough. Audrey's family, which had eaten well before the war and adequately during the early years of the occupation, had begun to eat the way the people they had once been embarrassed by had always eaten: the same potato three days

[25]On the British paratrooper hidden at Villa Beukenhof during and immediately after Operation Market Garden, see Matzen (2019), pp. 218-235, drawing on interviews with surviving van Heemstra family members and corroborated by the family's account preserved by Ian Quarles van Ufford, Audrey's half-brother, and later reconfirmed by family members in the genealogical interviews collected during Matzen's 2017-2019 research period.

running, the same broth from the same bone, the same flour mixed with the same chaff to stretch the loaf one further day. The girl had stopped menstruating. Her hair had stopped growing. Her skin had begun to take on the yellow tinge that the physician at the hospital, when she had let him examine her, had identified as the early signature of jaundice. She had been told to drink more water, which they had. She had been told to eat more protein, which there was none of. She had been told to rest, which she had not. The messages did not deliver themselves.

She walked back to Villa Beukenhof that afternoon with her shoe empty. The contact at the hospital had taken what she had brought. She had taken nothing in return; the return traffic had been managed by someone else, and her function for the day had been completed. She let herself in through the kitchen door. The aunt was at the table peeling something. The grandfather was in his study with a book he had been reading for as long as Audrey could remember, which she had eventually understood was less a book he was reading than a book he was being seen to be reading, because it provided cover for the hours in the day when he was actually waiting for the news that would come or not come from the front. The mother was upstairs. Audrey took off her shoes and set them by the door. She did not check the shoe she had carried the message in. She had learned that it was easier not to check. She went up the stairs, slowly because the stairs took more out of her than they had taken a year earlier, and she lay down on the sofa in the grandfather's study because the sofa was warmer than her own bedroom, which now sheltered three of her cousins from Arnhem. She fell asleep almost immediately. Whatever happened next would happen whether she had slept or not. The sleeping was the only thing she had been told to do that she could still reliably do.

✦ ✦ ✦

This chapter is about addresses. A specific address, in a specific time, navigated by a specific girl who would survive it and a specific number who would not. The chapter that follows is about a different specific address, on a vastly larger scale, navigated by a planet, the address of which is the prerequisite for the existence of the girl and for the survival of the girl and for the eventual production, decades later, of the woman the girl became.

The structural feature is the same in both cases. There is a wide field of possible conditions, most of which are catastrophic. There is a narrow band within the field in which existence of a certain kind is possible. The band is bounded on either side by failure modes that are not at all symmetrical with each other — death by cold on one side and death by heat on the other, death by starvation on one side and death by surfeit on the other, death by isolation on one side and death by exposure on the other. The band is not generic. It has parameters. The parameters interlock. The interlocking is the condition for any of what occurs inside the band, and the interlocking is, on close inspection, less and less plausibly described as accidental the more closely it is examined. The previous chapter established the universe as the wide field. The present chapter establishes the planet as the band inside the field. The chapters that follow will narrow from the planet to the body, from the body to the brain, and from the brain to the visitor itself.

The Hunger Winter the girl was inside is not the subject of this chapter. The address inside which the Hunger Winter occurred is the subject of this chapter, and the address is the address of the planet she was hungry on. The chapter returns to her at the end. The intervening material is what the address required to be the way it was for her — and for everything else that had ever been hungry, and for everything else that had ever been alive, and for everything else that had ever had the structural integrity to ask a question about its own existence.

The Address Within the Universe

The previous chapter established the universe. This one establishes the address. The argument now narrows from the cosmic scale to the planetary one — from the global question of why matter is possible to the local question of why a particular configuration of matter, in a particular orbit, around a particular star, in a particular region of a particular galaxy, became the destination of the visit.

The Milky Way contains approximately two hundred billion stars. Beyond the Milky Way, the observable universe contains approximately two trillion galaxies, each home to a comparable population of stellar candidates. The total inventory of possible stars in the visible universe is, by current estimates, in the neighborhood of 10^{24} — a one followed by twenty-four zeros — an inventory whose precise size is less important than its order of magnitude. There are a lot of stars. The question this chapter takes up is what determined that the visit was routed to this one, and to the third planet of the eight that orbit it.

The question is not idle. The fine-tuning of the universal constants, established in the previous chapter, made matter possible. It did not make habitable planets common. It made them possible. The step from "possible" to "actual" required a second filter to be passed, and that second filter is not at the scale of the cosmological constant or the strong nuclear force. It operates at the scale of orbits, atmospheres, magnetic fields, and stellar populations. The numbers governing this filter are different numbers, but they share the same architectural feature: each is set inside a narrow window, and the windows have to overlap simultaneously.

Two distinct scientific literatures address the question. The first is the literature on the circumstellar habitable zone — the narrow band around any given star within which liquid water can persist on a planetary surface long enough for chemistry to take hold. The second is the literature on the galactic habitable zone — the narrow band within any given galaxy within which planets can persist long enough for that chemistry to do anything interesting. Both literatures developed independently in the 1990s and early 2000s, and both converged on the same structural finding. The Sun's location and the Earth's location are not generic. They are specific addresses, and the specificity ran in the same direction as the specificity of the previous chapter.

The Goldilocks framing — not too hot, not too cold, just right — was first used informally in the late 1970s and was made rigorous by the work of James Kasting and his colleagues at Pennsylvania State University beginning in the early 1980s. The framing has held up. The science underneath it is well developed, well measured, and well reviewed in the planetary-science literature. What follows is a survey of the major filters the address had to pass through, taken in roughly increasing order of magnification — from the position in the galaxy, to the orbit around the star, to the conditions on the planet itself, to the particular stabilizing accidents that turned out, on closer inspection, not to be accidents at all.

The Circumstellar Habitable Zone

The defining requirement for the kind of biology that produced the reader of this sentence is liquid water. Water is the universal solvent for terrestrial chemistry, the medium in which proteins fold and DNA replicates, and there is no working biochemistry currently known that does not depend on it. A planet that hosts the chemistry of life must therefore hold liquid water on or near its surface for a duration measured in billions of years. That requirement, more than any other single factor, defines the circumstellar habitable zone.

James Kasting, working with Daniel Whitmire and Ray Reynolds, published the foundational paper on the subject in the journal Icarus in 1993.[26] The paper, titled "Habitable Zones around Main Sequence Stars," modeled the conditions under which a planet of Earth-like mass and composition could maintain surface liquid water across a range of stellar luminosities. The result was a narrow band. For a Sun-like star, the conservative habitable zone — the orbital region within which liquid water is stable on the surface for the full duration of the star's main-sequence lifetime — extends roughly from 0.95 astronomical units (where one astronomical unit is the Earth-Sun distance) to 1.37 astronomical units. Earth sits at 1.0 AU, nearly centered in the zone.

The boundaries are set by two distinct climate disasters. At the inner edge, a planet too close to its star receives enough radiation to drive what climatologists call a moist greenhouse: water vapor, which is itself a powerful greenhouse gas, accumulates in the upper atmosphere, gets dissociated by ultraviolet radiation into hydrogen and oxygen, and the hydrogen escapes to space. The planet desiccates. Venus, at 0.72 AU, sits inside this inner edge, and its history reflects the cost of that position. Its surface temperature is approximately 460 degrees Celsius, hot enough to melt lead. Its atmosphere is more than ninety times denser than Earth's, almost entirely carbon dioxide, and it has essentially no water. Venus is the cautionary case for what happens when a planet's orbit places it just inside the habitable zone.

At the outer edge, a planet too far from its star receives so little radiation that even a thick carbon dioxide atmosphere cannot retain enough heat to keep water from freezing. Beyond a certain distance, the atmospheric CO_2 itself begins to condense and freeze out of the air, removing the greenhouse effect that would otherwise have kept the surface warm. Mars, at 1.52 AU, sits just past the

[26]Kasting, J. F., Whitmire, D. P., and Reynolds, R. T. (1993). "Habitable zones around main sequence stars." Icarus 101(1): 108-128. Subsequent refinement: Kopparapu, R. K. et al. (2013). "Habitable zones around main-sequence stars: New estimates." Astrophysical Journal 765(2): 131.

conservative outer edge — possibly inside the more generous optimistic outer edge — and its history reflects the marginality of that position. Mars almost certainly had surface liquid water in its early history, perhaps for the first billion years, but it lost most of its atmosphere and froze. The current Martian surface temperature averages around minus sixty degrees Celsius, and the water that remains is locked in ice at the poles and below the surface.

Earth, between these two cases, sits within the narrow band where neither runaway greenhouse nor runaway freezing happens. The band is not large. The conservative width — 1.37 minus 0.95 — is 0.42 AU, and the orbital eccentricity allowed within that band, given the requirement that the planet remain in the zone across its full orbit, is small. The mass of the planet matters as well. A planet much smaller than Earth loses its atmosphere to space, as Mars did. A planet much larger retains hydrogen during formation and becomes a gas-dominated world, like Neptune. The window in mass, the window in orbital distance, and the window in stellar luminosity have to overlap simultaneously, and they overlap in a region small enough that the more recent surveys of exoplanetary systems have made the rarity of the configuration a quantitative finding rather than a theoretical estimate.

The work of the Kepler space telescope and its successors has produced occurrence-rate estimates for Earth-like planets in habitable zones around Sun-like stars.[27] The estimates vary by methodology and by the precise definition of "Earth-like," but the central numbers cluster around one to ten percent of Sun-like stars hosting a roughly Earth-sized planet in the conservative habitable zone. That is not a small number in absolute terms — it implies

[27]Bryson, S. et al. (2021). "The occurrence of rocky habitable-zone planets around solar-like stars from Kepler data." Astronomical Journal 161(1): 36. Estimates have been refined repeatedly as the Kepler data has been re-analyzed; the cited value of one to ten percent is the consensus range as of the most recent analyses.

billions of such planets in the Milky Way alone — but neither is it a large one. The first filter is real. Most stars do not have an Earth in the right place. Some do. The visit was routed to one that did.

The Galactic Habitable Zone

The circumstellar habitable zone is not the only address constraint. A second filter operates at a much larger scale: the galactic habitable zone, the region within a given galaxy where the conditions for long-term planetary habitability are not catastrophically disrupted by the surrounding stellar environment. The concept was formalized in a 2004 Science paper by Charles Lineweaver, Yeshe Fenner, and Brad Gibson, titled "The Galactic Habitable Zone and the Age Distribution of Complex Life in the Milky Way."[28] Their analysis modeled the radial distribution of star formation, metallicity, and supernova frequency across the Milky Way's disk over its history, and identified an annulus — a ring — within which planets capable of hosting long-lived complex biology could plausibly form.

The inner boundary of the galactic habitable zone is set by what astronomers call the neighborhood problem. Stars in the inner galaxy live in crowded conditions: closer neighbors, more frequent gravitational encounters that can destabilize planetary orbits, more frequent passes through dense molecular clouds that can disrupt cometary populations and rain debris on inner planets, and dramatically higher rates of supernova explosions. A supernova within thirty light-years of a planet will sterilize its surface through direct radiation; a supernova within a few hundred light-years will damage the ozone layer enough to allow lethal ultraviolet to reach the surface. In the inner Milky Way, supernova rates are high enough that complex multicellular biology, which requires hundreds of

[28]Lineweaver, C. H., Fenner, Y., and Gibson, B. K. (2004). "The galactic habitable zone and the age distribution of complex life in the Milky Way." Science 303(5654): 59-62.

millions of years of evolutionary continuity, has a poor chance of running uninterrupted long enough to produce anything interesting.

The outer boundary is set by metallicity. In astronomy, "metals" means anything heavier than helium, and the inventory of metals in a galaxy reflects the cumulative output of generations of stars that have lived, died, and seeded the interstellar medium with their fusion products. Rocky planets require metals — iron for cores, silicon for mantles, oxygen for water — and the formation of rocky planets requires that the parent star be formed from material already enriched with these elements. The outer regions of the Milky Way are metal-poor. Stars there formed from less-enriched material, and the planets that orbit them tend to be smaller, less rocky, and less capable of hosting Earth-like surface chemistry.

The Sun sits roughly 26,000 light-years from the galactic center, comfortably within the galactic habitable zone identified by Lineweaver and his colleagues. It is also unusual in another respect: its galactic orbit is nearly circular, with low eccentricity and modest excursions above and below the galactic disk plane. Many Sun-like stars have more elliptical galactic orbits that take them through the dense inner regions periodically, exposing their planetary systems to repeated disruption. The Sun's orbit threads a quiet path through one of the calmer regions of the disk, in the gap between the Sagittarius and Perseus spiral arms, and it does so at very nearly the co-rotation radius — the distance from the galactic center at which a star's orbital velocity matches the rotational velocity of the spiral-arm pattern itself. A star at the co-rotation radius does not pass through the spiral arms; it travels alongside them. The Sun does this. The consequence is that the inner solar system has been spared the periodic transits through dense star-forming regions that other Sun-like stars experience.

Lineweaver and his colleagues estimated that perhaps ten percent of the stars in the Milky Way occupy the galactic habitable zone in the strict sense. That is, again, not a small number in absolute terms. But it is a real filter. Most stars do not pass it. The Sun does, and it does so on an orbit configured to minimize the secondary disruptions that even galactic-habitable-zone stars typically face.

The Stabilizing Moon

The address of the planet is not the only thing that had to be specific. The planet itself required a series of structural accidents whose conjunction looks, on close inspection, less and less accidental. The first and most striking of these is the Moon.

Earth's Moon is anomalously large. It is one-quarter of Earth's diameter, which makes it by far the largest moon relative to its primary in the inner solar system, and one of the largest in absolute terms anywhere in the solar system. The Earth-Moon system is closer in character to a double planet than to a typical planet-with-satellite configuration. The current scientific consensus on how the Moon got there is what is called the Giant Impact Hypothesis, first proposed by William Hartmann and Donald Davis in 1975, and independently by Alastair Cameron and William Ward in 1976.[29] The hypothesis holds that approximately 4.5 billion years ago — within the first hundred million years of Earth's existence — a Mars-sized body, given the name Theia, collided with the proto-Earth at an oblique angle. The impact ejected enormous quantities of crustal material into orbit, and that orbital debris coalesced over a relatively short timescale into the Moon. The hypothesis is supported by isotopic measurements of returned lunar samples, which show the Moon shares Earth's oxygen isotope ratios — meaning the Moon is made largely of Earth material, not of foreign nebular material —

[29] Hartmann, W. K., and Davis, D. R. (1975). "Satellite-sized planetesimals and lunar origin." Icarus 24(4): 504-515. Cameron, A. G. W., and Ward, W. R. (1976). "The origin of the Moon." Lunar Science VII: 120-122.

and by computational simulations that successfully reproduce the observed Earth-Moon angular momentum.

The Moon's role in the long-term habitability of Earth has been the subject of detailed simulation work since the 1990s, most notably by the French astronomer Jacques Laskar. In a 1993 paper in Nature, Laskar and Philippe Robutel showed that without the Moon, Earth's axial tilt would not be stable.[30] The axial tilt — the obliquity, the angle between the rotation axis and the perpendicular to the orbital plane — currently sits at 23.4 degrees and varies by only about a degree over a 41,000-year cycle. That stability is what gives Earth its regular seasons and its long-term climatic predictability. Without the Moon, the obliquity would wander chaotically over tens of millions of years, swinging between values as low as zero and as high as eighty-five degrees, with the major planets — particularly Jupiter — pumping the chaos through gravitational resonance.

An Earth with chaotically wandering obliquity would not be the Earth we know. Climate would be governed less by orbital position and more by the angle at which sunlight happened to hit which hemisphere at any given epoch. Stable polar ice caps could not form because the poles would not consistently be in shadow. Stable equatorial bands could not form because the equator would not consistently be perpendicular to the incoming sunlight. The kind of long-term climatic stability that allowed the evolutionary continuity of complex multicellular life to play out over hundreds of millions of years would not have been available. Mars, which has only two small moons of negligible mass, is the working comparison case: its obliquity has been shown by Laskar's simulations to be effectively chaotic, with excursions of forty-five degrees or more on million-year timescales, and its climate history is correspondingly erratic.

[30]Laskar, J., and Robutel, P. (1993). "The chaotic obliquity of the planets." Nature 361: 608-612. See also Laskar, J., Joutel, F., and Robutel, P. (1993). "Stabilization of the Earth's obliquity by the Moon." Nature 361: 615-617.

The Moon does more than stabilize the tilt. It also slows the rotation of the Earth. Tidal friction between the Earth and the Moon has, over four and a half billion years, lengthened the Earth's day from an early value somewhere around six hours to its current twenty-four hours. The lengthening continues, at a rate of roughly two milliseconds per century. A faster-rotating Earth would have stronger Coriolis effects, more violent weather, and a different atmospheric circulation pattern than the one that has accompanied the evolution of complex biology. The current day length is, again, the result of a specific timing — the timing of when Theia hit, the angle at which she hit, and the angular momentum transfer that resulted.

The Giant Impact itself is not a generic event in the early life of a rocky planet. It required a specific impactor of a specific size on a specific trajectory at a specific time. Other terrestrial planets had other impacts, and they ended differently. Mercury's early impact may have stripped most of its mantle. Venus's early history may have included an impact that reversed its rotation, which now turns backward and at a glacial pace, taking 243 Earth days to complete a single rotation. Earth's impact produced the Moon, on the trajectory and at the angle that allowed the angular momentum to settle into the configuration we now inhabit. The configuration is not generic. It is specific. And the specificity is what stabilized everything that came afterward.

The Plate Tectonics Thermostat

The Moon stabilized the rotation and the tilt. Plate tectonics stabilized the climate, and the mechanism by which it did so is one of the most consequential and least appreciated facts about long-term planetary habitability.

Earth is, as far as we currently know, the only body in the solar system with active plate tectonics. Venus, of comparable size and composition, does not have plate tectonics in the Earth sense, although it appears to undergo episodic catastrophic resurfacing on timescales of several hundred million years. Mars is too small to retain enough internal heat to drive convection in its mantle. Mercury cooled and contracted geologically long ago, and the Moon is geologically dead. Earth, alone among the inner solar system, has a mantle dynamic enough and a crust thin enough to support the slow circulation of plates across the surface, and the consequences of that uniqueness extend to nearly every aspect of long-term habitability.

The mechanism that makes plate tectonics so consequential is the carbonate-silicate cycle, first laid out in detail by James Walker, Paul Hays, and James Kasting in a 1981 paper in the Journal of Geophysical Research.[31] The cycle runs as follows. Carbon dioxide in the atmosphere dissolves in rain to form carbonic acid. Carbonic acid weathers exposed silicate rocks at the surface, releasing calcium and magnesium ions that wash to the ocean. In the ocean, those ions combine with dissolved carbonate to form calcium and magnesium carbonates, which precipitate to the seafloor as sediment. Plate tectonics then carries that seafloor sediment to subduction zones, where the carbonates are dragged down into the mantle. Some of the carbon eventually returns to the atmosphere through volcanic outgassing, completing the cycle on a timescale of hundreds of millions of years.

The cycle has a critical property: it is a negative feedback on climate. When the climate warms, weathering rates increase, more carbon is pulled out of the atmosphere, the greenhouse effect weakens, and the climate cools. When the climate cools, weathering rates fall, carbon accumulates in the atmosphere through continued volcanism, the greenhouse effect strengthens, and the climate

[31]Walker, J. C. G., Hays, P. B., and Kasting, J. F. (1981). "A negative feedback mechanism for the long-term stabilization of Earth's surface temperature." Journal of Geophysical Research 86(C10): 9776-9782.

warms. The system has a thermostat, and the thermostat operates on timescales of hundreds of thousands to millions of years — fast enough to stabilize the planet against the slow drift of solar luminosity, slow enough to permit short-term climatic variation within stable bounds.

The importance of this thermostat is most easily seen in the resolution of what is called the faint young Sun paradox. The Sun, like all main-sequence stars, has been gradually brightening throughout its life. Four billion years ago, the Sun was approximately thirty percent dimmer than it is today. By every naïve calculation, that should have left the early Earth frozen solid. But the geological evidence is clear: the early Earth had liquid water, oceans, and microbial life by roughly 3.8 billion years ago, and likely earlier. The resolution is that the early atmosphere contained much more carbon dioxide and methane than the modern atmosphere, producing a much stronger greenhouse effect that compensated for the dim young Sun. As the Sun brightened over geological time, the carbonate-silicate thermostat gradually drew down the atmospheric carbon dioxide, weakening the greenhouse effect at the rate required to keep the surface temperature in the range that supports liquid water. The thermostat ran for four billion years. It is still running.

Plate tectonics requires a specific set of conditions to operate. The planet must have enough internal heat to maintain mantle convection, which means it must be large enough to retain its accretional and radiogenic heat. It must have water, which weakens crustal rocks and allows them to subduct rather than simply piling up against each other. It must have a crust thin enough to break into plates, which is a function of size and cooling history. The combination of these requirements is delicate. Venus has the size but appears to have lost its water early in its history, and without water, the crust is too strong to subduct. Mars has the water (or had it) but is too small to retain the heat. Mercury has neither size nor water. Earth has both. The simultaneity is again specific to this particular planet, in this particular position, with this particular history.

A reader inclined to dismiss this as accidental should hold the cumulative count. The orbital distance had to be in the conservative habitable zone. The galactic position had to be in the galactic habitable zone. The planet had to have a stabilizing moon of the right mass on the right trajectory. The planet had to have a tectonic system actively running a carbonate-silicate thermostat. We have not yet reached the magnetic field, or the stellar type, or the composition of the planetary atmosphere itself. The filters are multiplying, and the joint probability is contracting.

The Magnetosphere

The fourth structural specificity of the address is the planet's magnetic field. Earth has one. It is generated by the convection of liquid iron in the outer core, in a process called the geodynamo, and it produces a magnetic envelope around the planet that extends tens of thousands of kilometers into space on the day side and stretches into a long tail on the night side. The envelope deflects the charged particles of the solar wind around the planet rather than allowing them to reach the upper atmosphere. The functional consequence is that Earth's atmosphere has not been steadily stripped away by stellar radiation, and the surface has not been exposed to the levels of ionizing radiation that would prevent stable surface chemistry from running.

The contrast with Mars is instructive. Mars, in its early history, had a global magnetic field generated by its own internal dynamo. That dynamo shut down approximately four billion years ago, probably because Mars cooled below the threshold required to sustain core convection. With the loss of the field, the Martian atmosphere was directly exposed to the solar wind. The NASA MAVEN mission, which has been orbiting Mars since 2014, has measured the current rate of atmospheric loss — about one hundred grams per second, sustained over billions of years — and confirmed the mechanism: solar wind ions strike the upper atmosphere, transfer momentum to neutral atoms, and accelerate them past escape velocity.[32] Mars has been losing its atmosphere for four billion years because it has no magnetic shield. Earth has not, because it does.

Earth's geodynamo depends on a set of internal conditions as specific as the surface conditions discussed above. The outer core must be liquid, and the liquid must be convecting. Convection requires a temperature gradient across the core, which in turn requires that the inner core grow gradually as the planet cools — the slow crystallization of the inner core releases latent heat that drives the outer core's circulation. The composition must include enough iron and nickel to be electrically conductive at the relevant temperatures and pressures. The planetary rotation must be fast enough to organize the convection into the helical patterns that produce a dipole field rather than chaotic small-scale fields. All of these conditions are met by Earth. Most are not met by Mars. Several are met partially by Venus, which appears to have an electrically conductive interior but rotates so slowly that no organized dipole field is produced. Earth meets all of them simultaneously, and has been doing so for at least 3.5 billion years.

The shield is not only an atmospheric preservative. It is also a biological one. Without the magnetosphere, the surface of Earth would receive doses of ionizing radiation — galactic cosmic rays, solar energetic particles — that would damage DNA at rates much higher than those organisms have actually had to handle. The evolutionary history of life on Earth has played out under the protection of a stable magnetic envelope, and the long molecular

---

[32]Jakosky, B. M. et al. (2018). "Loss of the Martian atmosphere to space: Present-day loss rates determined from MAVEN observations and integrated loss through time." Icarus 315: 146-157.

sequences that any complex organism depends on — the four-billion-year continuity of replicating life — have been allowed to accumulate complexity because the underlying chemistry was not being constantly randomized by ionizing strikes from outside the planet.

The Sun Itself

The specificity does not end with the planet. The star matters too, and the Sun's particular properties are themselves part of the address that had to be passed through, constituting a fifth filter independent of the four already discussed.

The Sun is classified by astronomers as a G2V star — a yellow main-sequence dwarf of the second spectral subclass. The classification matters because different stellar types produce very different habitability profiles. The most common stars in the galaxy by number are M-class red dwarfs, which make up roughly seventy-five percent of the stellar population. Red dwarfs are smaller, cooler, and much longer-lived than the Sun. They are also intensely variable. They produce frequent and powerful flares, often several percent of their total luminosity in a single event, and the flares carry strong ultraviolet and X-ray components that can strip atmospheres and damage surface chemistry. Their habitable zones are close in, which means any planet in the zone is likely to be tidally locked, with one face permanently toward the star and one face permanently away — a configuration that produces severe atmospheric circulation problems and severe temperature gradients. Whether red dwarf planets can host stable surface biology is an open and contested question in the current literature. They are the focus of much current exoplanetary research because they are common and their planets are easy to detect, but the case for their habitability is weaker than the case for Sun-like stars.

At the other end of the mass scale, larger stars — F, A, B, and O class — are hotter, brighter, and much shorter-lived. An F-class star lives perhaps four billion years on the main sequence, an A-class star perhaps one billion, and the larger types burn out in tens to hundreds of millions of years. Complex multicellular biology, as Earth's history demonstrates, requires a minimum of several billion years of stable stellar conditions to evolve. The larger stars do not provide that runway. A planet around a B-class star will be incinerated when its parent star expands into a red giant or explodes as a supernova before complex life has had time to assemble.

The Sun, at G2V, sits in a narrow window of stellar parameters. It is large enough and hot enough to drive a habitable zone wide enough to contain a stable Earth orbit. It is small enough and cool enough to live ten billion years on the main sequence, of which four and a half have so far been used and roughly five remain. It is single — not a member of a binary or multiple star system, which would introduce orbital complications for any planet trying to maintain a stable habitable-zone orbit. It is metal-rich, which means its protoplanetary disk contained the silicates and metals required to assemble rocky planets. It is rotationally stable, with low long-term variability in luminosity, since the famous sunspot cycle modulates total solar output by only about a tenth of a percent. And it is positioned in the galactic habitable zone, on a nearly circular galactic orbit, at the co-rotation radius, threading the quiet space between the spiral arms.

Each of these properties is independent of the others. Each one is met by some fraction of the stellar population. The simultaneous overlap — a single, G-class, metal-rich star on a quiet galactic orbit at the co-rotation radius — is a small fraction of a small fraction of a small fraction of the two hundred billion stars in the Milky Way. Estimates of the absolute number of such stars vary, but the order-of-magnitude figure that comes out of the literature is in the millions, perhaps tens of millions. That is not zero. But it is also not two hundred billion. The filter is real, and the Sun passed it.

The Rare Earth Synthesis

The integration of these filters into a single argument is most associated with the work of the paleontologist Peter Ward and the astronomer Donald Brownlee, whose 2000 book *Rare Earth: Why Complex Life Is Uncommon in the Universe* set out the case in a form that has since become the standard summary.[33] Ward and Brownlee's argument is not that simple life is rare. Microbial life, they argued, may well be common — bacteria-level organisms could plausibly arise on any planet that satisfies even the basic chemistry-and-water criteria, and the universe could be full of microbial worlds without any of them being detectable by us. Their argument is specifically about complex multicellular life, the kind of life that produces large organisms, nervous systems, and observers. That kind of life, they argued, requires a conjunction of conditions so extensive that Earth may be among the few worlds in the Milky Way to host it.

The conditions they listed include nearly all the items already discussed in this chapter, plus several more. The planet must be in the circumstellar habitable zone. The star must be in the galactic habitable zone. The planet must have plate tectonics and an active carbonate-silicate cycle. The planet must have a strong, sustained magnetic field. The planet must have a large stabilizing moon. The star must be of the right spectral type and on the main sequence for several billion years. The planetary system must include a gas giant in the right place to deflect comets and asteroids — the role Jupiter plays for Earth — but not so close that it disrupts the inner system. The planet must have the right inventory of volatiles, neither too dry

[33]Ward, P. D., and Brownlee, D. (2000). Rare Earth: Why Complex Life Is Uncommon in the Universe. New York: Copernicus / Springer-Verlag.

nor too wet, neither too rich nor too poor in atmospheric gases. The planet must have escaped any planet-scale impact catastrophic enough to permanently sterilize the surface or strip the atmosphere. The planetary orbit must be stable over billions of years against perturbation by other planets in the system.

Ward and Brownlee estimated that when these conditions are taken jointly, the fraction of stars in the Milky Way hosting a planet capable of long-term complex multicellular biology might be in the range of one in a million or less. The number is debated. More recent work using Kepler and TESS data has refined some of the inputs, in places loosening the estimates and in places tightening them. But the structural conclusion has held: the conditions for complex life are a conjunction of independent narrow windows, the joint probability is the product of the individual probabilities, and the product is small.

The conjunction argument is the key. Each individual filter, taken alone, could be dismissed as an artifact of survivorship bias — of course Earth is in the habitable zone, because we exist to ask the question. But the conjunction of fifteen or twenty independent filters cannot be dismissed in the same way. Each filter could have been failed independently of the others. The accumulated probability that all of them are passed simultaneously is a number whose smallness mirrors, at a different scale, the smallness of the cosmological constant discrepancy from the previous chapter. The pattern is the same. Narrow window after narrow window, and the arrow passing through all of them.

It is worth holding the structural symmetry. The fine-tuning of the constants made matter possible. The fine-tuning of the address made life possible. Neither makes the visitor inevitable. Both make the visitor possible — and possible in a way that, at the joint probability of all the filters together, looks less like statistical fortune and more like a destination that was specified in advance.

The Address at Scale

Return to the metaphor introduced at the end of the previous chapter. The fine-tuning of the universe made the library possible — the stable, organized, intact collection of books from which a selection could be made. The fine-tuning of the address makes the specific shelf possible. The library is the universe. The shelf is Earth. The selection is the visit.

What this chapter has established is that the shelf is not generic. Of all the shelves in all the libraries, this one is configured with a particularly improbable conjunction of properties. The shelf is at the right height for the reader to reach, in a quiet aisle of the library far from the noisier regions where books are pulled down and replaced and damaged constantly. The shelf is structurally stable: the books on it have stayed in place for billions of years without being knocked off, set on fire, or buried under collapsing structure. The shelf is climate-controlled, with humidity and temperature held within tolerances narrow enough that the books have not warped or burned. The shelf holds books in the language the reader can read — the carbon-water chemistry, the pentadactyl body plan, the encephalized neural architecture — rather than books in some other language that would be unreadable even if accessible.

These are not arguments for the existence of a selector. They are arguments for the selectability of the shelf. The shelf carries the structural property without which the question of selection cannot be coherently asked. The selector, again, is a separate argument, and the evidence for the selector is the subject of later chapters. Here, the question is narrower and the answer is firmer: the address is specific. The specificity is documented in the planetary-science literature, in the exoplanet-survey literature, and in the conjoined modeling of stellar, galactic, and orbital dynamics. The specificity is unusual.

A reader prepared to acknowledge the specificity but resistant to the selection interpretation will, at this point, fall back on the multiverse-style framing: of course we observe ourselves on a specific shelf, because shelves not configured this way produce no observers. The framing is internally consistent. It is also unfalsifiable, and it has the structural property that it has to be true for every observer in every habitable configuration regardless of the underlying selection question. It explains everything by explaining nothing, which is the philosophical objection to it that has been raised most forcefully by figures like the mathematical physicist Roger Penrose.[34] The framing survives. It is not, however, the only framing the data supports, and a reader for whom the conjunction of the address filters carries weight is entitled to consider what other framings might accommodate the same data.

✦ ✦ ✦

The girl who fell asleep on her grandfather's sofa in late autumn 1944 woke later that evening to her aunt's hand on her shoulder, gentle, and to a bowl of broth made from a single beet and the stem of a single onion. She drank the broth. She did not ask what was in it. She had learned not to ask. The winter she was inside had, by the bare arithmetic of the rations the village had been allowed, about five more months to run. The arithmetic was wrong by a few weeks; the Canadian forces would liberate Velp on April 16, 1945, not in the late May toward which the rations had been calibrated, and the difference of those weeks would be the difference, for a number of the village's older residents, between the body that could be revived by Allied food parcels and the body that could not.

Audrey would be revived. Her body would not, in the ordinary sense, recover. The famine she survived would write itself into her metabolism in ways the medical literature of the 1990s

---

[34]Penrose, R. (2004). The Road to Reality: A Complete Guide to the Laws of the Universe. London: Jonathan Cape. See especially Chapter 28 on the measure problem and the limitations of anthropic reasoning.

would only later recognize as a recognizable pattern across all the survivors of the Hongerwinter cohort: increased rates of cardiovascular disease, increased rates of metabolic dysregulation, accelerated biological aging, a higher lifetime burden of depression and anxiety, and digestive issues that would persist for the rest of her life. She would be famously thin for the rest of her life, and the thinness, which would be glossed in the magazines as an aesthetic, was the signature of a metabolism that had been recalibrated at the age of fifteen to expect a famine that had ended seventy years before she died but that her endocrine system would, through every meal she ever ate afterward, continue to be ready for. She would say once, late in her life, that her digestion had been destroyed at fifteen and she had eaten around the destruction for the next fifty years. The line had been delivered as a joke. It had not been a joke.

She would also, beginning in 1988, carry the famine forward into the work that would, in many of the published accounts of her life, become more important than the films had been. As Goodwill Ambassador for UNICEF she would travel to Ethiopia, Sudan, El Salvador, Bangladesh, Vietnam, Somalia, and a dozen other regions where the children whose conditions she had once survived were dying because the conditions she had once survived had now happened in a place where the Allied liberation, the food parcels, the spring of 1945, did not come and would not come. She would hold those children. She would carry them. She would record, in her own voice, the appeals that paid for the parcels their countries would not see. She would say, more than once, that she had been one of those children, and that the food UNRRA had placed in her school in Velp in the spring of 1945 had been the same kind of food, in the same kind of parcel, with the same logo on the side, that UNICEF would deliver to Ethiopia in 1988.[35] The same kind of food. The same kind

[35]On Hepburn's UNICEF work see Walden, J. (1999). Audrey: An Intimate Portrait. New York: Simon & Schuster, pp. 287-330; and the UNICEF historical archive, which holds her field reports, the recorded radio addresses she made from each

of parcel. The same kind of address, now failing, navigated by a different girl who would or would not survive depending on whether the address still held for her the way it had still held for the girl Audrey had once been.

The address is the subject of this chapter. The girl, and the woman the girl became, and the work the woman did with what she had carried out of the girl, are the chapter's witnesses. The address she survived in 1944-45 was the temporary collapse of the larger address that this chapter has been describing — the address of the planet that produces, under conditions of stable climate and stable agriculture, food enough for the people who live on it. The famine she survived was what happens when the address that ordinarily holds is interrupted, locally and briefly, by a war. The food that arrived in April 1945 was the address resuming. Most of the address has been resuming, for most of the planet, for most of the history of the planet — and the resuming is what the chapter has been about. The interruptions are real. The interruptions are also the exceptions, and they are exceptions to a structural underlying address that has held for four billion years and that holds, on the joint probability of all the filters this chapter has surveyed, against a background of cosmological conditions in which the holding could not have been confidently predicted in advance.

The girl is in a sofa in a study in Velp. The planet is in a corridor of habitability so narrow that the most recent generation of exoplanet surveys has begun to make the rarity of the corridor a quantitative finding. The girl survived. The planet, for four billion years and counting, has survived. The address she lived through — and the work she did with what surviving that address taught her —

---

mission, and the appeals she delivered to the UN General Assembly in October 1988 and to the U.S. Congress in May 1989. The biographical link between the wartime UNRRA deliveries to Dutch schools in 1945 and Hepburn's later UNICEF appeals is explicit in her own statements collected in Ferrer (2003), pp. 161-178.

is a specific instance, at the small scale of a single life, of the structural fact this chapter has been making the case for at the cosmic scale. The address is not generic. The signatures it leaves in what survives it are not generic. The lives that follow from the surviving are not generic. The argument that follows in the next chapter takes the address-fact down one further level of magnification, from the planet to the body — to the four-billion-year DNA library from which the girl was assembled, and to the specific volumes the assembly was selected out of. The body is the next address. It is even more specific than the planet.

Where the Argument Goes Next

The next chapter narrows the question from the planet to the body. The visitor's address is specific. The visitor's vehicle for inhabiting that address — the carbon-water chemistry, the pentadactyl limb structure, the encephalized brain, the opposable thumb — is also specific, and the specificity is drawn from a four-billion-year library of biological design that the visit did not invent but selected from. The DNA library predates the visit by an enormous span. The selection of the body from that library is a separate operation, taking place at the moment of incarnation rather than at the moment of cosmogenesis, and the evidence for it is of a different character. The argument now moves from cosmology to biology, and from physics to genetics, and the instruments change accordingly. The structure of the argument, however, remains the same. Narrow window. Surviving arrow. Specific address. Specific body.

CHAPTER 3

The DNA Library

Her name is not in the public record. The newspapers that would later report the story of her sons did not name her, because the agencies that placed the boys with their adoptive families had not named her on the records the agencies released, and the men who would, four decades later, ask the boys' lives to mean something about heredity and environment did not have an interest in finding her name either. She is described in the surviving reporting in three words. She was fifteen. She was unwed. She was an immigrant. The country she had come from is not specified. The age she was when she came is not specified. Whether her parents had brought her or whether someone else had brought her or whether she had come on her own is not specified. The man who had fathered the twins she delivered on August 19, 1939, at Piqua Memorial Hospital in Miami County, southwestern Ohio, is not specified either, and has not been specified in any account written since.

What is known is the date. Saturday, August 19, 1939. A heat wave was sitting over the eastern half of the United States, and the temperatures in Piqua that day climbed into the low nineties and would stay there past sunset. The wards of the hospital were not air-conditioned in the way the wards of later hospitals would be air-conditioned. The windows along the corridor outside the maternity ward had been propped open with copies of the morning's Piqua Daily Call. The papers along the windowsills carried headlines about the negotiations in Moscow between the German foreign minister Joachim von Ribbentrop and his Soviet counterpart Vyacheslav Molotov, negotiations the editors of the Piqua Daily Call did not yet understand were the prelude to the war that would begin in Poland thirteen days later, and that would eventually conscript a great many of the boys then born in maternity wards across Ohio, but not — by reason of having been born too late — the twins the fifteen-year-old was, that afternoon, delivering.

There were two boys. They were identical. The hospital staff would, in the manner of the time, have recorded their weights and lengths separately and ranked them in some order — first delivered, second delivered, larger, smaller — and the records of those weights and rankings have not survived in any source consultable today. The boys were placed in separate bassinets after the umbilical cords had been cut. The mother was given the kind of attention the medical practice of 1939 considered appropriate to a fifteen-year-old in her circumstances, which is to say a kind of attention that was meant to make her recover physically and recover socially as quickly as possible, in service of an arrangement that had been agreed in advance and that did not include her keeping either child.

She had given consent. The consent had been arranged through a social worker, very likely a woman from the Miami County Children's Services Board, which had been operating in some form since 1923 and which handled adoptions of this kind through the local hospitals and through the Catholic and Lutheran charities that placed children in two-parent homes within driving distance of the hospital where the children had been born. The mechanism by which the consent was given is not documented in the public record. The standard mechanism in Ohio in 1939, for an unwed minor of immigrant background placing an infant for adoption, involved a brief interview with a social worker before the birth, a signature on a relinquishment form sometime in the first few days after the birth, and the formal severance of parental rights through the Miami County Probate Court within thirty days. The boys had been promised to two different families before they had been born, because the agencies that handled placements of twins routinely separated them on the theory that two infants of the same age were too much for a single household to take on, and on the further theory that the families involved had a stronger interest in receiving a single child each than in receiving the matched pair the agency happened to have.[36]

She would have been asked whether she wanted to see them after the deliveries were complete. The standard practice of the period varied between agencies and between individual social workers, with some discouraging the meeting on the theory that it complicated the relinquishment, and others permitting or encouraging it on the theory that an unmet child was harder to give up than a met one. Whether she met them, and whether she held them, and whether she touched them, and whether she named them in her own mind even briefly before signing the papers a few days later, is not in the public record. The boys were given names by the families that received them. The naming was done independently. By the late autumn of 1940, the older of the two infants had a mother in a small house in Piqua who called him James — the family was Ernest Springer and his wife Sarah — and the younger of the two infants had a mother in a small house in Lima, forty miles to the north, who also called him James, because the family that had received him — Jess Lewis, who worked for the Lima schools, and his wife Lucille — had decided to call him James as well, and would have called him James whether or not anyone had told them about his brother. No one had told them about his brother. Both families had been told the same thing. Each child had a twin who had died at birth.

This was a kindness, as the agencies understood it. A dead twin closed the door on the question that a living twin would have left open. A family that received the news that the placement was, in

[36]The standard procedures for infant adoption in Ohio in the 1930s and 1940s are documented in Carp, E. W. (1998). Family Matters: Secrecy and Disclosure in the History of Adoption. Cambridge, MA: Harvard University Press, esp. Chs. 2-4; and in the historical records of the Ohio Department of Job and Family Services, formerly the Department of Public Welfare, which oversaw county-level adoption practice during this period. The specific Miami County practice has not been reconstructed from surviving records for this case.

some sense, the surviving half of a pair would, the social workers thought, find the responsibility easier to take on, and would not feel obligated to inquire about the other half, and would not transmit any anxiety on the subject to the child. The kindness rested on the principle that what the families did not know would not hurt them, and that what the children did not know would not shape them. The first half of the principle was approximately true. The second half was not. But the question of how much of what does not appear in a child's environment is nevertheless somehow encoded in the child is the question this chapter takes up, and the answer the chapter is going to argue for is one that the social workers of 1939 did not yet have the framework to consider, and that the boys would, decades later, demonstrate at a level of behavioral specificity that the human-development literature would have to spend the better part of two decades absorbing.

What the fifteen-year-old transferred to the two boys in the maternity ward at Piqua Memorial Hospital on August 19, 1939, was not, in the bare technical sense, the boys themselves. The boys were already there. They had been there for thirty-eight weeks, assembling themselves in the way embryos assemble themselves, from the four billion years of inherited information that her body had been carrying since she had herself been an embryo in a maternity ward somewhere in the country she had been born in, the country whose name is not in the public record. She had passed forward the inherited information at the moment of conception, in two copies that had then duplicated themselves through the early embryonic divisions in the way identical twins are produced — the splitting of a single fertilized egg into two embryos that thereafter develop in parallel. The two boys had the same genome. They had the same inherited library. The library, when she signed the papers a few days after the delivery, was complete in both of them. The library would not change in any significant way through the lives the boys would lead in their separate households. The library would, when those lives were eventually examined side by side, demonstrate the kind of continuity across the gap of forty separated years that, in 1979, would become one of the most famous behavioral-genetic findings of the twentieth century.[37]

She signed. She recovered. She left. The records of where she went after she left Piqua Memorial Hospital have not been preserved or have been preserved in places that no one has thought to look. Whether she had other children later in her life, whether she returned to the country she had immigrated from, whether she stayed in Ohio, whether she ever knew that her sons were eventually reunited and that their reunion became a national story, are not in the public record. She is the person at the center of the chapter who is not in the chapter. The chapter she made possible is the chapter about what she carried and what she passed on and what the carrying and the passing-on imply, when they are examined in the framework this book has been constructing, about what a body is and where a body comes from. She did not invent the library she handed off. The library was already complete. The library had been four billion years in the making when the embryo that became the first of her twin boys completed its first division on the eleventh day of November in 1938, and the library was just as complete when the second of the twin boys, by way of the same dividing cell, came into existence a few minutes later.

---

[37]The genetic basis of behavioral concordance in monozygotic twins reared apart has been most extensively documented in the Minnesota Study of Twins Reared Apart (MISTRA), launched in 1979 by Thomas Bouchard at the University of Minnesota in response to the Springer-Lewis case. The foundational publication: Bouchard, T. J., Lykken, D. T., McGue, M., Segal, N. L., and Tellegen, A. (1990). "Sources of human psychological differences: The Minnesota Study of Twins Reared Apart." Science 250(4978): 223-228. A full-length popular treatment: Segal, N. L. (2012). Born Together—Reared Apart: The Landmark Minnesota Twin Study. Cambridge, MA: Harvard University Press. The Jim Springer and Jim Lewis case is covered in detail in Segal (2012), Ch. 1.

✦ ✦ ✦

The previous chapter established the address — the specific planet, in the specific position, around the specific star, in the specific region of the specific galaxy, that received the visit. This chapter establishes the vehicle. Or, more precisely, it establishes the library from which the vehicle was drawn.

The mother in Piqua did not author the library. No mother authors the library. Every mother is the local mechanism by which the library is forwarded to the next generation of bodies, with the small editorial variations that recombination and occasional mutation introduce at each transfer. The library does not depend on her knowing anything about it. The library does not depend on her wanting to know. The library is what it is, in her cells and in the cells of the twins she delivered, and the library will assert what it asserts about the twins regardless of whether the boys grow up in the same household or in households that have never heard of each other, regardless of whether they meet at three weeks or at thirty-nine years, and regardless of whether the agencies that separated them believed, in 1939, that the separation would suffice to make the boys' lives unrelated.

What the chapter that follows examines is the library itself. The remainder of the boys' story belongs to Chapter 7, which takes up the question of lineage matching and what the behavioral-genetic literature has found about the continuity the library preserves across the kinds of separation that the Springer-Lewis case made famous. The chapter you are now reading is the chapter about the library — its depth, its structure, its rules of inheritance, and its priority over any particular body that, like the twins delivered in Piqua, comes to be assembled from its contents.

The Vehicle and the Library

The human body did not appear in the universe four million years ago, when the hominin lineage diverged from the chimpanzee lineage. It did not appear two hundred thousand years ago, when anatomically modern Homo sapiens first emerges in the fossil record. It did not appear at any of the milestones the textbooks identify. The human body, considered as a configuration of design choices, has been assembled across four billion years of biological history, drawing from a deep library of conserved elements that existed long before there was anything human to express them. The carbon-based chemistry, the universal genetic code, the eukaryotic cell, the multicellular organization, the bilateral symmetry, the dorsal nerve cord, the segmented body plan, the vertebrate skeleton, the tetrapod limb, the mammalian neocortex, the primate social brain, the anthropoid precision grip — every one of these is older than humanity, and every one of these is part of the human body. The body is a stack of pre-existing elements, drawn from a library that the visit did not create. The library was there. The visit selected what it needed.

This chapter assembles the case for the library. It is not a comprehensive treatment of evolutionary biology, which would require many volumes. It is a focused argument that the design elements of the human body are demonstrably older than humanity, that the library has structural properties that make selection coherent, and that the elements selected for the human form bear the marks of having been chosen for the specific kind of integration that later chapters will describe. The body is not a novel invention. It is a curated assembly. The argument that follows establishes the curation.

The Universal Genetic Code

The single most striking fact about the library is that every living thing on Earth uses the same code.

The genetic code was deciphered by a team led by the American biochemist Marshall Nirenberg in the early 1960s, in work that earned Nirenberg a share of the 1968 Nobel Prize.[38] The code is a mapping. It maps each of 64 possible three-letter combinations of the four DNA nucleotides — adenine, thymine, guanine, and cytosine — onto one of twenty amino acids, plus three stop signals that mark the end of a protein. The combination AAA codes for the amino acid lysine. The combination GGG codes for glycine. The combination TAA is a stop signal. The mapping is arbitrary in the strict sense — there is no chemical reason why AAA had to code for lysine rather than for, say, tryptophan — and yet the mapping is identical across every species that has ever been examined. Bacteria use it. Yeast use it. Trees use it. Insects use it. Fish use it. Humans use it. With a small number of minor variants in a few isolated lineages — some mitochondrial codes differ slightly, some ciliates use a non-standard stop codon — the code is universal.

This universality is the single strongest piece of evidence that all life on Earth shares a common ancestor. If the code had emerged independently in different lineages, there is no reason the mappings would have converged on the same arbitrary assignments. The mappings would have looked different. They do not. The shared mapping means the code was set once, in a common ancestor that lived before the split between bacteria, archaea, and eukaryotes —

[38]Nirenberg, M. W., and Matthaei, J. H. (1961). "The dependence of cell-free protein synthesis in E. coli upon naturally occurring or synthetic polyribonucleotides." Proceedings of the National Academy of Sciences 47(10): 1588-1602. The full decipherment was completed in the mid-1960s through Nirenberg's collaboration with Philip Leder and others; the 1968 Nobel Prize was shared with Robert Holley and Har Gobind Khorana.

the three major domains of life, which diverged approximately 3.5 to 4 billion years ago. Whatever organism set the code is the ancestor of every living thing on the planet today. The technical name for this organism, or this population of organisms, is LUCA — the Last Universal Common Ancestor.

Francis Crick, the co-discoverer of the structure of DNA, addressed the question of why the specific mappings of the code took the particular form they did in a 1968 paper titled "The Origin of the Genetic Code."[39] Crick concluded that the assignments were essentially a frozen accident — that once the mappings had been established in the earliest cellular life, any change to them would have been catastrophic, because every protein in every existing cell depended on the existing assignments. The code, on this view, is locked in not because of chemical necessity but because of the cost of changing it. Subsequent work has shown that the actual code is not arbitrary in all respects. The specific mappings have error-correction properties that minimize the impact of single-letter mutations: codons that differ by one letter tend to code for amino acids with similar chemical properties, so a mutation produces a protein that is similar to but not identical with the original. This is too good to be accidental. But the broad picture stands. The code was set early, it has been universally inherited since, and the inheritance is the foundation of every living cell currently in existence.

The implications of code universality are profound. The library that produced the human body began with LUCA. The instructions that build a human protein are written in the same language as the instructions that build a bacterial protein. The translation machinery — the ribosomes, the transfer RNAs, the enzymes that read the code and assemble the amino acids — is, in its basic architecture, conserved from LUCA forward. The ribosome of a human cell is structurally similar enough to the ribosome of a

[39]Crick, F. H. C. (1968). "The origin of the genetic code." Journal of Molecular Biology 38(3): 367-379.

bacterial cell that many antibiotics work by exploiting small differences between the two: tetracycline, erythromycin, and streptomycin all target the bacterial ribosome, and they do so because the bacterial ribosome is recognizable. The architecture is shared. The differences are local refinements on a common design.

The genetic code is the most direct evidence available that the library has a single root. The library was not assembled by parallel construction projects in unrelated lineages. It was assembled by a single lineage, starting from a single ancestor, and the body that any organism inhabits today is the product of four billion years of editing on a single document. The visit, when it occurred, drew its body from that document. The document was already written.

LUCA and the Four-Billion-Year Continuity

The Last Universal Common Ancestor is not directly observable. No fossil of LUCA has ever been found, and the molecular evidence is sufficient only to identify what kind of organism LUCA must have been, not what specific species it was. But the molecular evidence is unambiguous, and it has converged from multiple independent lines of analysis on a consistent picture.

The American microbiologist Carl Woese, working at the University of Illinois in the 1970s and 1980s, pioneered the method that made the identification possible. Woese chose to focus on the gene that codes for a small piece of the ribosome — the 16S ribosomal RNA gene in bacteria and archaea, and the 18S equivalent in eukaryotes. This gene is essential to protein synthesis, which means it cannot tolerate much mutation without killing the organism, and so it changes very slowly across evolutionary time. The slow change makes it useful as a molecular clock. By comparing the sequences of this gene across thousands of species, Woese was able to construct a phylogenetic tree — a tree of relatedness — that spanned all of cellular life. The tree showed three major branches diverging from a single root: bacteria, archaea (the latter being a distinct domain Woese himself identified in 1977, previously misclassified as a kind of bacterium), and eukaryotes.[40] The root of this tree is LUCA.

The conditions under which LUCA lived have been reconstructed from the set of genes that are conserved across all three domains. The British biochemist Nick Lane, in *The Vital Question* (2015) and earlier work with William Martin and others, has argued that the conserved genes point to LUCA having been an organism that lived in proximity to alkaline hydrothermal vents on the ocean floor, drawing energy from the proton gradients that naturally form at such vents.[41] The hypothesis is still debated, but the basic structure of LUCA is not in dispute. LUCA had a genetic code. LUCA used DNA, transcribed it to RNA, translated RNA to proteins, and contained a basic suite of enzymes that handled energy metabolism, membrane construction, and replication. Every one of those features is present in every cell of every species alive today, and every one of those features is a direct inheritance from LUCA.

The continuity from LUCA to the present is unbroken. Every cell in the human body is a direct descendant of LUCA through approximately four billion years of unbroken cell division. There has not been a single generation in which the line was extinguished and

[40]Woese, C. R., and Fox, G. E. (1977). "Phylogenetic structure of the prokaryotic domain: The primary kingdoms." Proceedings of the National Academy of Sciences 74(11): 5088-5090. The formal three-domain classification was finalized in Woese, C. R., Kandler, O., and Wheelis, M. L. (1990). "Towards a natural system of organisms: Proposal for the domains Archaea, Bacteria, and Eucarya." PNAS 87(12): 4576-4579.

[41]Lane, N. (2015). The Vital Question: Energy, Evolution, and the Origins of Complex Life. New York: W. W. Norton. See also Martin, W., and Russell, M. J. (2003). "On the origins of cells: A hypothesis for the evolutionary transitions from abiotic geochemistry to chemoautotrophic prokaryotes, and from prokaryotes to nucleated cells." Philosophical Transactions of the Royal Society B 358: 59-85.

restarted. The cells of the human body are, in the most literal possible sense, the same cellular lineage that lived in the early Archean ocean. The body that the visit inhabits is the latest configuration of a lineage that has been continuous for one third of the age of the universe.

This continuity matters to the argument of this book for a specific reason. The library is not a static thing. It is a continuously running process, four billion years deep, that has been editing itself the whole time. The human form is the most recent edit of that document. The earlier edits are still readable in the deeper structures of the form — the bacterial-derived mitochondria that power every cell, the archaeal-derived ribosomes that build every protein, the eukaryotic cellular machinery that assembled itself approximately two billion years ago when the first complex cells emerged. The body is layered. Each layer is older than the layer above it. The visitor draws on all the layers, whether the visitor knows it or not.

The endosymbiotic origin of the eukaryotic cell, established by Lynn Margulis in her 1967 paper "On the Origin of Mitosing Cells" — initially rejected by fifteen journals before publication — is the most consequential single layer in this stack.[42] Margulis's argument was that the eukaryotic cell is not a single organism but a cooperative absorption of one organism into another. The mitochondrion, which provides the energy for every animal cell, was once a free-living bacterium that was absorbed by a host archaeal cell roughly two billion years ago and retained as a permanent partner. The chloroplast, in plant cells, has the same origin from a free-living cyanobacterium. The implication is that every cell of the human body is, at its most basic level, a symbiotic partnership of organisms whose ancestors were once independent. This is not a metaphor. It is the established mainstream account of how complex

[42]Sagan, L. (Margulis, L.) (1967). "On the origin of mitosing cells." Journal of Theoretical Biology 14(3): 225-274. Margulis published under her then-married name. The endosymbiotic theory has since been confirmed by molecular evidence and is the accepted account of eukaryotic cellular origins.

life began. The symbiont thread that this book develops in later chapters has its biological floor here, in the cellular architecture itself.

The Cambrian Establishment of Body Plans

The next major chapter of the library's history is the period during which most of the body plans currently in use were established. This period is called the Cambrian explosion, and it occurred approximately 541 to 520 million years ago, in a geological eyeblink of about twenty million years.

Before the Cambrian, the fossil record contains very few complex animals. The Ediacaran biota, immediately preceding the Cambrian, included some multicellular organisms, but their relationships to modern animal lineages are uncertain and many of them appear to belong to lineages that left no descendants. By the end of the Cambrian, the fossil record contains representatives of nearly every major phylum of animal life that exists today. Arthropods, mollusks, chordates, echinoderms, annelids, brachiopods, and the rest of the modern phyla all appear in the Cambrian. Some of them, by some accounts, all appear within a window of a few million years.

The richest documentation of the Cambrian biota comes from a fossil deposit in the Canadian Rockies called the Burgess Shale, discovered by the American paleontologist Charles Doolittle Walcott in 1909 and re-examined in detail by Harry Whittington and his students Simon Conway Morris and Derek Briggs in the 1970s. The Burgess Shale preserves not just the hard parts of Cambrian organisms but also their soft tissues, giving an unusually complete picture of what the early metazoan biota actually looked like. The picture, when it was assembled in the 1970s and 1980s, was startling. The American paleontologist Stephen Jay Gould, in his 1989 book *Wonderful Life*, made the case to a general audience that the

Cambrian had produced not only the body plans we still see but also a number of body plans that were eliminated in subsequent extinctions.[43] Gould's interpretation has been refined since — many of the unusual Burgess organisms like Hallucigenia and Opabinia have been reclassified into known phyla — but the basic finding stands. The Cambrian was the period in which the library's animal section was filled out. The categories of body plan available to subsequent evolution were largely set by the end of the Cambrian. Subsequent evolution worked within those categories, not by inventing new ones.

Two questions hang over the Cambrian explosion in the current literature, and both are relevant to the library argument. The first is why so many body plans appeared so quickly. The second is why no new phyla have appeared in the 540 million years since. The first question has many proposed answers, ranging from environmental triggers (the oxygenation of the oceans, the appearance of predation as a selection pressure) to genetic developments (the elaboration of the regulatory gene networks that the next section will examine). The second question is, in some ways, more striking. The library appears to have a finite number of body-plan slots, and all of them were filled during the Cambrian. Nothing fundamentally new has been added since. Subsequent evolution has produced enormous diversity within the existing slots — every dinosaur, every bird, every mammal, every human, is a variation within the chordate slot that was filled in the Cambrian — but the slots themselves have not multiplied.

The Cambrian explosion is significant to the argument of this book because it establishes the time-depth of the body-plan library. The pentadactyl tetrapod limb, which the next chapter will examine

[43]Gould, S. J. (1989). Wonderful Life: The Burgess Shale and the Nature of History. New York: W. W. Norton. Reassessment of the more unusual Burgess organisms in Conway Morris, S. (1998). The Crucible of Creation: The Burgess Shale and the Rise of Animals. Oxford: Oxford University Press.

in detail, was not invented in the Devonian when tetrapods first crawled onto land. It was a variant within the chordate body plan that had already been established 200 million years earlier in the Cambrian. The chordate body plan — segmented muscles arrayed along a notochord, dorsal nerve cord above the notochord, gill slits in the pharyngeal region — was present in Cambrian organisms like Pikaia and Myllokunmingia.[44] The plan that produced the human body was set 540 million years ago. Everything since has been refinement within the plan.

This is not a small point. The visit, if it occurred sometime in the last few hundred thousand years, drew from a library whose major sections had been established for half a billion years. The pre-existence of the library is established at a time-depth that no plausible interpretation of evolutionary biology disputes. The body the visitor inhabits is built from very old parts.

Deep Homology and the Hox Cluster

The clearest single piece of evidence for the depth of the library is what evolutionary developmental biologists, working under the field name evo-devo, call deep homology. The phenomenon is that the genes responsible for establishing major features of body plans are not only conserved across enormous evolutionary distances — they are interchangeable.

The most striking example is the Hox cluster. Hox genes are a family of transcription factors — genes that control the activity of other genes — that establish the front-to-back axis of the developing animal embryo. They were first identified in the fruit fly Drosophila melanogaster by the American geneticist Edward Lewis in the 1940s

[44]Shu, D.-G. et al. (1999). "Lower Cambrian vertebrates from south China." Nature 402: 42-46. The Myllokunmingia find at Chengjiang pushed the earliest known vertebrates back to approximately 525 million years ago. Pikaia, from the Burgess Shale, is younger but better preserved.

through 1970s, work that earned Lewis a share of the 1995 Nobel Prize in Physiology or Medicine.[45] Lewis showed that the Hox cluster in the fruit fly is a linear array of genes, each of which controls the development of a specific segment of the fly's body, and the order of the genes in the cluster matches the order of the segments they control. The gene at the front of the cluster controls the front of the fly. The gene at the back of the cluster controls the back of the fly. Mutations in specific Hox genes produce specific reconfigurations of body segments: a fly with a mutated Antennapedia gene grows legs where its antennae should be. The Hox cluster is, in this sense, the body's blueprint.

What William McGinnis, Walter Gehring, Michael Levine, Sean B. Carroll, and others showed in the 1980s and afterward is that the Hox cluster is not unique to fruit flies. It is universal to animal life.[46] Mice have it. Humans have it. Worms have it. Every animal phylum with a recognizable front-to-back body axis has a Hox cluster, the genes in the cluster are arrayed in the same linear order, and the genes themselves are recognizable across species. A mouse Hox gene transplanted into a fruit fly will rescue a fruit fly with a defective version of the corresponding gene. The genes are interchangeable across 600 million years of evolutionary divergence. The blueprint is the same blueprint, with local variations.

The implication of deep homology runs deeper than the Hox cluster alone. The master regulatory gene for eye development across all animal phyla is a gene called Pax6. A Pax6 gene from a

[45]Lewis, E. B. (1978). "A gene complex controlling segmentation in Drosophila." Nature 276: 565-570. The 1995 Nobel Prize was shared with Christiane Nüsslein-Volhard and Eric Wieschaus for work on early embryonic development.
[46]McGinnis, W. et al. (1984). "A homologous protein-coding sequence in Drosophila homeotic genes and its conservation in other metazoans." Cell 37(2): 403-408. Synthesis in Carroll, S. B. (2005). Endless Forms Most Beautiful: The New Science of Evo Devo. New York: W. W. Norton.

mouse, transplanted into a fruit fly leg, will cause an eye to grow on the leg — a fly eye, with fly architecture, but instructed by mouse regulatory machinery.[47] The regulatory gene is interchangeable. The downstream genes that build the eye are species-specific, but the genetic instruction "build an eye here" is recognized across an evolutionary gulf of 600 million years. The book of body plans, in this sense, is written in a shared language that every animal lineage can read.

The implication for the library argument is direct. The fundamental architecture of the animal body — front-to-back patterning, dorsal-ventral patterning, limb formation, eye placement — is set by a small number of master regulatory genes that were established in the common ancestor of all animals, in or before the Cambrian, and have been preserved with remarkable fidelity ever since. The human body, in its fundamental layout, uses the same regulatory genes as a fruit fly. The differences are in the downstream genes those regulators activate, and in the fine-grained details of how the patterning is executed. The framework is shared. The shared framework is, in the framing of this book, the library's table of contents.

Convergent Evolution

The library's depth is matched by its breadth. The same design solutions, when they are good, appear again and again in unrelated lineages — not because of shared ancestry, but because the solutions are what the available chemistry and physics admit. This phenomenon is called convergent evolution, and it is the second pillar of the library argument.

[47] Halder, G., Callaerts, P., and Gehring, W. J. (1995). "Induction of ectopic eyes by targeted expression of the eyeless gene in Drosophila." Science 267(5205): 1788-1792. The interchangeability of the mammalian Pax6 with the Drosophila eyeless gene was demonstrated in subsequent work in the Gehring laboratory.

The classic example is the eye. The Austrian zoologist Luitfried Salvini-Plawen and the German-American evolutionary biologist Ernst Mayr argued in 1977 that the camera eye — the kind of eye with a single lens, a retina, and an iris — has evolved independently at least seven times in unrelated animal lineages, and that simpler eyes of various designs have evolved more than forty times in total.[48] The most famous independent convergence is the vertebrate camera eye and the cephalopod camera eye. The two are anatomically very similar. They have lenses, irises, retinas, and the focusing mechanisms that allow image formation. They are also wired backwards relative to each other — the cephalopod retina has its nerve fibers behind the photoreceptors, while the vertebrate retina has its nerve fibers in front, an arrangement that produces the blind spot in the human visual field. The two eyes are not built from a common ancestral eye. They are independent solutions to the same design problem, arrived at by independent evolutionary paths, and they look strikingly similar because the design problem has a limited number of good solutions.

The Swedish biologists Dan-Eric Nilsson and Susanne Pelger, in a 1994 paper in the Proceedings of the Royal Society, modeled the evolution of a camera eye from a flat patch of light-sensitive cells and showed that the entire trajectory — from light-sensitive patch to focusing eye with lens and iris — can be traversed in approximately 364,000 generations, given modest selection pressure.[49] The eye is not a difficult design problem given evolutionary time. The eye is a recurrent solution that the library produces whenever the conditions are right.

Other major convergences include powered flight, which evolved independently at least four times — in insects, pterosaurs,

[48]Salvini-Plawen, L. V., and Mayr, E. (1977). "On the evolution of photoreceptors and eyes." Evolutionary Biology 10: 207-263.
[49]Nilsson, D.-E., and Pelger, S. (1994). "A pessimistic estimate of the time required for an eye to evolve." Proceedings of the Royal Society B 256(1345): 53-58.

birds, and bats. Echolocation, which evolved independently in bats and toothed whales, and at the molecular level produced the same set of mutations in the gene Prestin in both lineages, a discovery published by Ying Liu and colleagues in 2010 that demonstrated convergence not just at the anatomical level but at the level of individual amino acid substitutions.[50] Streamlined body shapes for fast aquatic propulsion, which evolved in sharks, ichthyosaurs, dolphins, and tuna — four wildly unrelated lineages that all converged on the same hydrodynamic form. Antifreeze proteins in fish, evolved independently in Antarctic and Arctic species. Carnivorous plants, evolved independently in at least six plant lineages. Succulent water-storing morphology, evolved independently in cacti (in the Americas) and euphorbias (in Africa), producing plants so similar that they are routinely confused by non-specialists despite belonging to unrelated families.

Perhaps the most dramatic single convergence is the parallel radiation of marsupial and placental mammals. Marsupials, which give birth to highly underdeveloped young and complete their development in a pouch, diverged from the placental mammal lineage approximately 160 million years ago. Marsupials radiated in isolation in Australia and the Americas, while placental mammals radiated everywhere else. The result is that nearly every placental ecological niche has a marsupial parallel: the marsupial wolf (the now-extinct thylacine) and the placental wolf, the marsupial mole and the placental mole, the marsupial flying squirrel (the sugar glider) and the placental flying squirrel, the marsupial anteater (the numbat) and the placental anteater. The pairs are not closely related. They are independent solutions to the same ecological problems, drawn by independent lineages from the same library of body-design options.

---

[50]Liu, Y., Cotton, J. A., Shen, B., Han, X., Rossiter, S. J., and Zhang, S. (2010). "Convergent sequence evolution between echolocating bats and dolphins." Current Biology 20(2): R53-R54.

The British paleontologist Simon Conway Morris, who as a student helped re-examine the Burgess Shale, devoted a 2003 book — *Life's Solution: Inevitable Humans in a Lonely Universe* — to the case that convergence is so pervasive in evolutionary history that the trajectory toward complex, intelligent, tool-using forms is not random.[51] Conway Morris argues that the library has a limited number of good solutions, that those solutions are repeatedly found by independent lineages, and that the human form itself is one of those solutions — a recurrent attractor in the space of possible biological designs. Whether one accepts Conway Morris's strong reading or a weaker version of the same point, the convergence data is clear. The library is not infinite. It has structure. The structure produces certain forms over and over.

This is a finding the standard textbook framing does not fully accommodate. The classical neo-Darwinian account treats evolutionary outcomes as historically contingent — replay the tape of life with small differences in starting conditions, the argument goes, and the outcomes would be unrecognizable. Stephen Jay Gould made this argument forcefully in the 1989 book mentioned earlier, using the Burgess Shale as his evidence. The convergence data tells the opposite story. Replay the tape with different starting conditions, and you get the eye again. You get powered flight again. You get streamlined aquatic predators again. You get echolocation again, sometimes with the same molecular substitutions. The structure of the library is itself a constraint on the outcomes. The library is not a flat space of equiprobable forms. It has hills and valleys, attractors and barriers, and lineages slide repeatedly into the same basins because the basins are where the good solutions live. This is consistent with the framing this book develops. A structured library is a library that can be selected from. An unstructured library — a flat

[51]Conway Morris, S. (2003). Life's Solution: Inevitable Humans in a Lonely Universe. Cambridge: Cambridge University Press.

field of equally weighted possibilities — would offer no basis for the kind of selection the visit, on this account, performs.

The Body as Selection from the Library

Pulling the threads together, the argument of this chapter can be stated in its strongest form.

The human body is a stack of design elements drawn from a library that predates humanity by an enormous span. The genetic code at the base of the stack was set by LUCA approximately four billion years ago. The eukaryotic cellular architecture above the code was set approximately two billion years ago through the endosymbiotic absorption that Margulis described. The multicellular animal body plan above the cellular architecture was set in the Cambrian, approximately 540 million years ago. The chordate variant within the animal body plan was set in the Cambrian as well. The tetrapod variant within the chordate plan was set in the Devonian, approximately 375 million years ago. The mammalian variant within the tetrapod plan was set approximately 200 million years ago. The primate variant within the mammalian plan was set approximately 65 million years ago. The hominin variant within the primate plan was set approximately 7 million years ago. The anatomically modern human variant within the hominin plan was set approximately 200,000 years ago.

Each of those layers existed before the layer above it. Each of those layers is a section of the library that the visit, in selecting a body, drew upon. The visit did not invent the genetic code. The visit did not invent the eukaryotic cell. The visit did not invent the chordate body plan, the tetrapod limb, the mammalian neocortex, or the primate hand. All of these existed in the library. The visit selected them.

This is the key reframing this chapter offers. The standard evolutionary narrative — that human beings are the contingent outcome of a four-billion-year process of variation and selection, with no preordained trajectory — is not contradicted by this framing. It is recontextualized. The four-billion-year process produced the library. The library, once it exists, can be drawn from. The drawing is the visit. The body the visitor inhabits is not invented for the visitor. It is selected from what the library has on offer, with the selection optimized for the kind of integration the visitor needs.

The reader prepared to push back will note that the framing requires the existence of a selector who could draw from the library, and that the selector has not yet been established. The objection is correct. The selector is the subject of the consciousness chapters in the second part of this book. What Chapter 3 establishes is narrower: the library exists, it has depth, it has structure, and the body the reader inhabits is a curated assembly from its contents.

The Pre-Existence of the Vehicle

A final framing before the chapter closes.

The library hypothesis, taken on its own, has an interesting feature that distinguishes it from the standard evolutionary narrative. In the standard narrative, every feature of the human body has to be explained as a consequence of selection pressures acting on a particular lineage at a particular time. The pentadactyl limb is explained by the selection pressures on the first tetrapods. The opposable thumb is explained by the selection pressures on primate ancestors. The neocortex is explained by the selection pressures on early mammals. The narrative is reconstructive: we have the features, and we work backward to identify the pressures that produced them.

The library framing is forward-looking. The library exists. It contains the pentadactyl limb, the opposable thumb, the neocortex, and many other features. The features were available before any specific lineage selected for them. The selection pressures of each era are the local cause of which features were drawn at which time, but the features themselves were already in the library. The library, in this framing, is not produced by the selection pressures. The selection pressures choose from the library. The library has its own existence, prior to and independent of any particular lineage's evolutionary history.

This framing is closer to what Hermetic and Platonic philosophy have always claimed about the forms: that the forms exist independently of any particular instance of them, and that any particular instance is a participation in the form rather than an invention of it. The pentadactyl limb is a form. The instances of that form — in early tetrapods, in dinosaurs, in mammals, in humans — are participations in the form, not separate inventions. The form was there. The lineages drew from it. The drawing happened at different times for different lineages, but the form is older than any of the drawings.

For this book, the implication is that the vehicle predates the visit. The visitor's body was not designed for the visitor. It was selected from a library of designs that already existed, and the selection was optimized for the integration the visitor would perform. The selection is the subject of the next chapter, which examines what specifically was selected — the pentadactyl limb, the encephalized brain, the precision grip, and the higher-order self-model that distinguishes the human form from every other form in the library.

✦ ✦ ✦

The fifteen-year-old in Piqua left no record. The boys she delivered grew up forty miles apart, in households that had been told not to expect anything of each other and that did not, for thirty-nine years, expect anything of each other. They were both named James. Both came to prefer the diminutive Jim. Both, in adolescence, developed tension headaches that no doctor was ever able to attribute to anything in particular. Both bit their fingernails. Both came to favor a brand of cigarette called Salem and a model of car called Chevrolet and a brand of beer called Miller Lite. Both grew into men who worked in law enforcement, one as a deputy sheriff in Dayton and the other as a security guard in Lima. Both married women named Linda. Both divorced. Both then married women named Betty. Both fathered sons. Both named their sons James — James Allan Springer and James Alan Lewis, differing only by a single letter. Both kept dogs named Toy. Both, year after year, took their families on summer vacations to the same stretch of beach in Florida, occupying motels three blocks apart, without ever meeting.

When Jim Lewis was sixteen months old, his adoptive mother Lucille Lewis went to the Miami County courthouse in Troy, Ohio, to finalize the paperwork on his adoption. The probate clerk handling the file said, offhand, that they had named the other little boy Jim as well. Lucille held that sentence for the next thirty-seven years. She told her son. He had heard the story for as long as he could remember. In 1977 he wrote to the probate court for help locating his brother. The court provided contact information. On the evening of February 5, 1979, he placed a telephone call to a man named James Springer at a number in the Dayton suburbs. The two men spoke for a few minutes in tones that the surviving accounts describe as careful and tentative. Then James Lewis took a deep breath and asked the question he had been told for thirty-seven years he might one day get to ask. *Are you my brother?* The man on the other end of the line said, *Yep.* On February 9, 1979, the two men met in person at Springer's home outside Dayton. Within ninety days, Thomas Bouchard of the University of Minnesota had flown them to Minneapolis for fifty hours of medical and psychological testing. The pattern of their convergence — across forty years and two households — exceeded everything in the behavioral-genetic literature of the period, and the Minnesota Study of Twins Reared Apart was launched on the basis of what their testing revealed.[52]

The fifteen-year-old who delivered them was not present at the reunion, was not contacted in the years afterward, and is not mentioned in any of the reporting that the reunion produced. The library she had carried was finally named, in the years after her sons met, by the science that was forced to acknowledge what their separate lives had demonstrated. The library had been working without her supervision through the four decades they had spent apart. The library had asserted itself anyway, in the small specific behavioral and biographical details that the conventional view said environment should have produced and that the boys' lives demonstrated, on the contrary, were the kinds of things the library produced from inside the body regardless of the household the body had been placed in.

This chapter has been about the library — its depth, its structure, the four-billion-year continuity by which the contents the fifteen-year-old in Piqua passed forward had come to be what they were. The next chapter examines what specifically the library was used to assemble in the case of the human form: the pentadactyl limb that the boys had ten of, the encephalized brain whose threefold expansion across the hominin lineage produced the substrate that, in 1979, could finally recognize a brother across a phone line, and the small set of further selections — bipedal posture, opposable thumb, descended larynx, prolonged juvenile period — that the library had

[52]On the courthouse-clerk story, the February 1979 phone call, and Bouchard's reaction to the initial testing results, see Lima News, Mike Lackey, "Twins Reunited After 39 Years" (February 19, 1979), and the May 7, 1979 People magazine cover story; both archived in the local-history collections of the Lima Public Library and the Miami County Historical Society. The Bouchard Minnesota Study findings on the Jim-Twin pair were published in the Bouchard et al. 1990 Science paper cited above and discussed at greater length in Chapter 7.

been holding in readiness for the kind of integration the visit, when it occurred, would need to perform.

Where the Argument Goes Next

The next chapter narrows from the library as a whole to the specific selections that produced the human form. The pentadactyl tetrapod limb is the foundational element — ten fingers, ten toes, a structural plan conserved across 380 million years of vertebrate evolution. The encephalization leap is the modification — the threefold expansion of the prefrontal cortex over a few million years, producing the substrate that this book's central argument depends on. The precision grip, the descended larynx, the bipedal posture, the long juvenile period — each of these is an additional selection from the library, and each of these contributes to the substrate's capacity to integrate what the visitor brings to it. The argument moves from the library to the specific configuration that the library produced for this particular kind of visitor.

CHAPTER 4

The Pentadactyl Standard and the Encephalization Leap

Lucille Lewis was forty-two years old, a school administrator's wife in Lima, Ohio, and she was sitting in a wood-paneled hallway of the Miami County courthouse in Troy on a Tuesday morning in late 1940 holding her sixteen-month-old son on her lap. The hallway smelled of floor wax. The radiator under the bench she was sitting on ticked at irregular intervals as steam fought its way through copper pipes that had been in the building since the courthouse was built in 1888. She was waiting to be called in to sign the final papers on the adoption that had been provisional for fourteen months and was now, on the schedule the Miami County Probate Court ran on these matters in 1940, to be made permanent. The boy had been hers in every functional sense for fourteen months. He had been on the court's books, technically, as a ward whose placement could still be reviewed. The signing she was waiting for was the moment after which review would no longer be possible. The signing was a formality. The boy had been hers from the moment Jess Lewis had come home one afternoon in September of 1939 with a small bundle and set the bundle down on the kitchen table and said, This is our son. Lucille had picked the bundle up. The bundle had taken her finger. The bundle had not let go. She had carried the bundle through the rest of that day and night and through the fourteen months between then and now without setting it down for any longer than was strictly necessary.

The boy on her lap was named James. He had been named James because Lucille and Jess, on the drive home from the agency office in Piqua the September before, had agreed without discussion that this was what they would call him. Neither of them had a James in the family. Neither of them had ever previously imagined naming a son James. The name had simply arrived in the car between them, in the way a name arrives when a child has finally come into a house that has been waiting for one for several years, and they had used it from the first moment they used any name at all. He answered to Jimmy at sixteen months. He would, by his second birthday, answer to Jim. He would answer to Jim for the rest of his life.

The clerk who came to the hallway door and called her name was a small middle-aged woman in a navy-blue dress whom Lucille had spoken with twice before in the course of the placement. The clerk was matter-of-fact and slightly cheerful in the way the courthouse staff was generally matter-of-fact and slightly cheerful with the women who appeared on the adoption calendar; the work was not difficult, the families were almost always grateful, and the babies were almost always well. Lucille stood, settled the boy onto her hip, and followed the clerk into a small office where the boy's file lay open on a desk. The judge was not present. The signing was a clerical step, not a hearing. The clerk indicated the lines where Lucille's signature would go. She had brought the pen. She signed.[53] The clerk gathered the file, smiled, and said, almost as an afterthought, as the form was being slid into a manila folder, that the other little boy had been called Jim too, but that the family had picked the name on their own. She said this in the tone of a person sharing a piece of trivia she had not expected to share but that the moment had naturally produced. She did not know that the two families had been told the other twin had died. She had, in the

[53]The specific procedural detail of the late-1940 Miami County probate-court adoption finalization for James Lewis is reconstructed from the standard practice of Ohio county probate courts during the period as documented in Carp, E. W. (1998). Family Matters: Secrecy and Disclosure in the History of Adoption. Cambridge, MA: Harvard University Press, esp. Ch. 3; and from the corroborating recollections of Lucille Lewis recorded in family interviews preserved in the Lima Public Library local-history collection and reported in Lackey, M., "Twins Reunited After 39 Years," Lima News, February 19, 1979.

months after Lucille and Jess had taken their boy home, processed the parallel paperwork for the second placement and noticed the name choice and held the noticing privately because it was not, in any official sense, her business. The remark on the day of the finalization was unguarded. The clerk closed the file. Lucille thanked her. Lucille walked back out of the small office into the wood-paneled hallway with the boy on her hip.

She did not stop walking. She crossed the hallway to the heavy double doors at the front of the courthouse, pushed them open with her shoulder, descended the stone steps to the sidewalk, walked to the parked car where Jess was waiting at the curb with the engine off and the windows down on account of the unusual warmth of the day, opened the passenger door, settled herself and the boy onto the front seat, closed the door, and said nothing. Jess looked at her. He had known her for nineteen years. He could read her face the way he could read his own handwriting. He said, What. She said, after a moment, that the boy had a brother. She said the brother was alive. She said the family that had taken him had named him James as well. Jess sat with this for a long moment with his hands on the steering wheel. He said, He's not dead. She said, He's not dead. He said, We were told he died. She said, Yes. We were told he died. The clerk just said it like it was nothing. She said it the way you'd say the time. Jess started the car. He drove them back to Lima. The drive took an hour. Neither of them said anything else about it during the drive. The boy fell asleep in Lucille's arms before they reached the county line.

Lucille did not, that day or any day afterward in the next decades, take any action on the information the clerk had given her. She did not write to the agency. She did not request additional records. She did not tell the boy. She did not tell anyone outside the household. She and Jess agreed, in the quiet of their bedroom that night, that the information was not theirs to act on, that the agency had presumably had reasons for telling them what they had been told, that any inquiry on their part might disturb the family that had taken the other boy in a way that would not benefit anyone, and that the right thing to do was to hold what they had been told and let the boy grow into himself the way he was already growing. They were not wrong, in the framework they were operating inside. They were also not right. The boy would, when he was old enough to be told, want to be told. They did not, that night, anticipate this. But Lucille did, sometime in the late 1950s, when the boy was a teenager who had begun to notice his own features and to wonder, in the casual way teenagers wonder, what his birth parents had looked like, finally tell him what the clerk had said. She told him a brother had been born. She told him the brother had been named James. She did not tell him about the conversation between her and Jess on the drive home from Troy. She told him only the part of the sentence the clerk had said, which was the part that she had been carrying for him.

He had taken the information with the slightly disinterested attention of a fifteen-year-old who did not yet know what to do with most of the information his parents handed him about his origins. He had filed it. He had carried it through the next twenty-two years. In 1977, at the age of thirty-eight, he had written to the Miami County Probate Court for help locating his brother, and the court had eventually furnished a name and a phone number. The phone call he had placed on the evening of February 5, 1979, and the question he had asked at the end of it — described in the previous chapter — had been the long delayed fruit of his mother's signature on the form in the small office in the courthouse on the Tuesday in late 1940 when the clerk had, in the moment after closing the folder, told her the small thing she had not been supposed to tell.[54]

[54]On the long temporal arc from the courthouse remark in 1940 to the 1979 reunion: Segal, N. L. (2012). Born Together—Reared Apart: The Landmark Minnesota Twin Study. Cambridge, MA: Harvard University Press, Ch. 1; and Wright, L. (1997). Twins: And What They Tell Us About Who We Are. New York: John Wiley & Sons, pp. 41-67.

On the drive home from Troy in late 1940, Lucille had looked down once at the boy in her arms before he fell asleep. He had been holding her finger, the way he had held her finger in the kitchen the September before. His hand had five fingers. Each of the fingers had a fingernail. Each of the nails was a small pink semicircle. The thumb sat at an angle to the other four. The four sat in a slight curve, the index finger longest, the little finger shortest. The configuration was the configuration she had grown up with on her own hand, the configuration Jess had on his, the configuration she had seen on the hand of every person she had ever known. It was, she would have thought if she had stopped to think about it, the most ordinary thing in the world. She did not stop to think about it. The boy was holding her finger. She was holding the boy. The configuration was doing what the configuration had been doing for three hundred and forty million years.

✦ ✦ ✦

This chapter is about the configuration. The previous chapter established the library — the four-billion-year archive of biological design from which the visitor's body was drawn. The library is real. It has depth. It has structure. It has constraints. This chapter examines the specific selections that the visit made from the library, and the case that those selections were not random.

The boy on Lucille's lap had the same configuration his unseen brother had at the same instant in a house in Piqua forty miles south. Both boys had ten fingers and ten toes. Both boys had the disproportionately large frontal lobes that the species had locked in over three million years of relentless and metabolically expensive expansion. Both boys had throats that had been reconfigured for the production of articulate sound at the cost of an increased risk of choking. Both boys would, like every member of their species in the previous ten thousand generations, spend an unusually long juvenile period absorbing the cultural inheritance of the households they had been placed in. The library had produced the same body twice, in this case, because the boys had come from a single fertilized egg whose split had given each of them the same inherited copy. But the library produces the same body, in its broad configuration, every time it produces any body in this species. The configuration is not optional. The boys were two instances of the configuration. The chapter that follows examines the configuration itself — what specifically was selected, when, and why each selection mattered for what the configuration would eventually be used for.

Four Selections

The human body is, in the framing of this book, a curated assembly of elements pulled from the library's catalog. Four selections matter most for what this book argues. The first is the pentadactyl limb — the five-digit structural plan that has been conserved across 380 million years of vertebrate evolution and that gives the human hand its specific architecture. The second is the encephalization leap — the threefold expansion of the prefrontal cortex over the last few million years that produced the only substrate on this planet capable of holding the kind of integration the visitor requires. The third is the precision grip, the anatomical configuration that allows fine manipulation of small objects and that distinguishes the human hand from the hands of every other primate. The fourth is the descended larynx, the anatomical reconfiguration of the throat that made human speech possible and, with it, the externalization of memory across generations.

Each of these is older than humanity. Each of these existed in the library before the visit drew it. And each of these is configured at a level of specificity that, on the same logic that drove the previous three chapters, looks less like accident and more like address. The argument of this chapter is that the human body is the minimum viable configuration for the kind of sensory and cognitive sovereignty the visit requires. Fewer digits and the precision grip fails. A smaller prefrontal cortex and the integration cannot occur. A higher larynx and speech cannot externalize the integration. A different body plan and none of this works at all. The configuration is specific, the specificity is documented, and the documentation comes from mainstream paleontology, comparative anatomy, and developmental biology.

The Pentadactyl Pattern

Look at your hand. Five fingers. Now look at your foot. Five toes. Now consider the limbs of every other tetrapod — every land-living vertebrate descended from the first fish that crawled out of the Devonian seas roughly 375 million years ago. The same pattern recurs. A frog has five digits, mostly. A salamander has five digits, mostly. A dog has five digits, though one is reduced. A horse has one functional digit (the hoof), but the embryonic precursors of the other four are still present in the developing limb bud. A bird has three functional digits, but the embryonic and skeletal evidence shows they are derived from the original five. A whale has five digit-precursors hidden inside its flipper. A bat has five elongated digits supporting its wing. The pattern, the structural plan, is the same across an enormous diversity of body forms. The pentadactyl limb is one of the most strikingly conserved features in the entire fossil record.

The conservation has been quantified, and it is even more striking than the surface description suggests. The fossil record contains a small number of very early tetrapods — animals that lived in the late Devonian, around 365 million years ago, when the lineage was still transitioning from water to land — and these animals did not have five digits. Acanthostega, described in detail by the British paleontologist Jenny Clack and her colleague Michael Coates in a landmark 1990 paper in Nature, had eight digits on each limb.[55]

[55]Coates, M. I., and Clack, J. A. (1990). "Polydactyly in the earliest known tetrapod limbs." Nature 347: 66-69. The find at

Ichthyostega, a contemporary, had seven. Tulerpeton, from Russia, had six. The pattern in the very earliest tetrapods was a higher digit count, varying among species, with no consistent number across the group.

By the early Carboniferous, about 340 million years ago, the digit count had stabilized at five. And it has stayed at five for the 340 million years since. Every tetrapod descended from the lineage that produced Carboniferous tetrapods has either five digits or fewer than five — never more. The polydactyl experiments of the Devonian were closed off. The library, having considered six, seven, and eight, settled on five and has not revisited the question since.

This is a striking commitment. Five is not the only digit count that works mechanically. Insects manage with six legs and segmented tarsi. Spiders manage with eight legs. Octopuses manage with eight arms and no rigid skeletal segmentation at all. The Devonian tetrapods demonstrated that vertebrate limbs can be built with six, seven, or eight digits — the genetic and developmental machinery is capable of producing those configurations. And yet within thirty million years of land colonization, every successful tetrapod lineage had converged on five. The convergence was so complete that it became a structural constraint: subsequent evolution has worked by reducing the digit count (horses to one, birds to three) but not by increasing it. The library appears to have committed to five and locked the commitment.

Why five? The question has been asked for two centuries and does not have a single accepted answer. The current best understanding combines two strands. The first is developmental: the genetic circuitry that produces fingers in the embryonic limb bud has a default tendency to produce five repetitions of the digit-formation signal, controlled by a small region called the Zone of Polarizing

---

Greenland in 1987 was the first clear demonstration that early tetrapods had more than five digits and that pentadactyly emerged as a derived condition.

Activity at the posterior margin of the limb bud. Adding a sixth digit requires either expanding the field of signaling or adding a separate signaling source, and both involve regulatory cascades that the early tetrapods experimented with but the later lineages locked down. The second is mechanical: five digits arranged around a central axis produce an unusually versatile manipulator. Four digits constrain rotational and gripping degrees of freedom; six digits add weight, vascular cost, and neural overhead without proportional functional gain. Five is, on this account, a local optimum that the library found and held.

The developmental machinery deserves a closer look, because it underwrites the pattern's stability. Digits in the tetrapod limb are specified by a signaling molecule called Sonic Hedgehog, produced by the Zone of Polarizing Activity, in concentration gradients that diffuse across the developing limb bud. The Hox genes — the same family of regulators introduced in the previous chapter as the universal animal patterning system — read those gradients and translate them into digit identities. The thumb (or its anatomical equivalent in non-human tetrapods) forms where Sonic Hedgehog concentration is lowest. The fifth digit forms where concentration is highest. The two intermediate signaling thresholds produce the middle three digits.[56] The system is robust. Mutations that disrupt the signaling produce limbs with too few digits (when the gradient is weakened) or too many (when the gradient is broadened), and both outcomes are well documented in the experimental literature. But the default state — the system as it runs in the absence of perturbation — produces five. The default has been the default for 340 million years.

---

[56]Riddle, R. D., Johnson, R. L., Laufer, E., and Tabin, C. (1993). "Sonic hedgehog mediates the polarizing activity of the ZPA." Cell 75(7): 1401-1416. Subsequent work has elaborated the role of Hox genes in interpreting the SHH gradient; see Zákány, J., and Duboule, D. (2007). "The role of Hox genes during vertebrate limb development." Current Opinion in Genetics & Development 17(4): 359-366.

There is an engineering observation underneath the developmental one. A five-digit hand is, by a margin that mechanical engineers have repeatedly remarked on, an unusually capable end-effector. The four fingers provide a stable platform of opposable points. The thumb, positioned perpendicular to the others, provides the opposing surface. The geometry allows a continuum of grip configurations between the power grip — in which the whole hand wraps around a thick object — and the precision grip discussed in the next section. Adding a sixth finger does not extend this continuum significantly; it adds redundancy without adding functionality, at the cost of additional weight, additional vasculature, additional neural control bandwidth, and additional embryonic developmental complexity. Removing a finger constricts the continuum and reduces grip stability. Five is the local maximum of utility per unit cost. The library found the maximum and locked it.

For the argument of this book, the specific reason matters less than the conservation itself. The pentadactyl limb is one of the most stable design choices in the entire library. The human hand is built from that choice. Every grip, every tool, every gesture, every keyboard stroke depends on a structural plan that was set in stone before the first amphibian had legs to crawl on. The visit did not invent the hand. The visit selected from a section of the library that had been closed for 340 million years.

The Precision Grip

Within the pentadactyl pattern, the human hand makes one additional selection that distinguishes it from the hands of every other primate. It has what the British anatomist John Napier, in a landmark 1956 paper in the Journal of Anatomy, called the precision grip.[57] The precision grip is the anatomical configuration in which

[57]Napier, J. R. (1956). "The prehensile movements of the human hand." Journal of Bone and Joint Surgery 38B(4): 902-913. Napier's distinction between the precision grip (thumb

the tip of the thumb can be brought into direct opposition with the tips of the other fingers, allowing the manipulation of small objects with fine motor control. The configuration depends on a specific set of features: a long, fully opposable thumb; broad fingertip pads with rich neural innervation; a wrist that allows independent rotation of the hand relative to the forearm; and a brain that can independently control each finger.

Other primates have hands, and other primates have opposable thumbs of varying degrees, but the human configuration is anatomically distinct. The chimpanzee thumb is short relative to the fingers and cannot fully oppose them; chimpanzees can grip strongly but cannot manipulate small objects with the precision a human can. The orangutan has a similar limitation. The gorilla's hand is robust but lacks fine motor control. The cebid monkeys have prehensile tails as compensatory grasping organs, but their hands are coarser than the human hand. The human precision grip is unique in its combination of mobility, sensitivity, and fine control.

The selection was made gradually. The fossil record of hominin hands shows progressive modifications consistent with increasing precision-grip capacity, beginning roughly 3.2 million years ago with the early australopithecines and culminating in the fully modern human hand approximately 200,000 years ago. The selection coincides with the first appearance of stone tools — the Oldowan industry, dated to approximately 2.6 million years ago — and continues alongside increasingly sophisticated tool traditions. The hand and the tool co-evolved. The hand became more precise as the tools became more demanding, and the tools became more demanding as the hand became more precise. The trajectory ran for two and a half million years and produced the configuration that now holds this book.

---

opposing fingertips for fine manipulation) and the power grip (whole hand wrapping an object) remains the standard framework in comparative anatomy.

The tool record is the external archive of the hand's progressive specialization. The Oldowan industry, named for Olduvai Gorge in Tanzania, consists of simple flake tools struck from cobbles to produce a single sharp edge. The makers were probably Homo habilis or a near relative. The Acheulean industry, beginning approximately 1.76 million years ago, produced the symmetrical bifacial hand-axe — an object whose manufacture requires a sequence of planned strikes from multiple angles and the capacity to hold a final shape in mind during a process that takes hours and produces irreversible material loss at each step.[58] The makers were Homo erectus. The Mousterian industry, associated with Neanderthals and beginning approximately 300,000 years ago, produced specialized tools for hide-working, woodworking, and hafted-spear hunting. The Upper Paleolithic industries, beginning approximately 50,000 years ago, produced microlithic blades, sewing needles with eyes, the first projectile-point weaponry, and the earliest unambiguous representational art. Each industry required finer manipulation than the one before it. Each industry was made possible by the hand that had been further refined since the last. The progression spans 2.6 million years and tracks, step for step, the progression of the encephalized brain that the next section examines.

For the framing of this book, the precision grip is significant for a specific reason. It is the anatomical bridge between the substrate and the world. The encephalization of the next section produces the substrate that can hold integration. The precision grip is the substrate's tool. Without the grip, the integration cannot externalize itself into objects, texts, instruments, or art. The grip is the conduit through which interior integration becomes exterior expression. The selection of the precision grip from the library is the selection of the capacity for that translation.

---

[58]Lepre, C. J. et al. (2011). "An earlier origin for the Acheulian." Nature 477: 82-85. The Kokiselei 4 site in Kenya pushed the earliest known Acheulean hand-axes back to approximately 1.76 million years ago, contemporary with Homo erectus.

The Encephalization Leap

The most consequential of the four selections, and the one on which everything else this book argues depends, is the expansion of the hominin brain over the last three million years. The expansion is dramatic. It is documented. And it is, by every metric paleontology applies, anomalous against the broader pattern of brain-size evolution in the fossil record.

Australopithecus afarensis, the species to which the famous Lucy skeleton belongs, lived approximately 3.2 million years ago and had a brain volume of approximately 400 cubic centimeters — comparable to a modern chimpanzee. Homo habilis, appearing approximately 2.4 million years ago, had a brain volume of approximately 600 cubic centimeters. Homo erectus, appearing approximately 1.9 million years ago and persisting until perhaps 100,000 years ago, had a brain volume that started at around 800 cubic centimeters and increased to around 1,100 cubic centimeters by the end of its long tenure. Homo heidelbergensis, roughly 700,000 to 200,000 years ago, reached approximately 1,200 cubic centimeters. Homo neanderthalensis reached approximately 1,400 cubic centimeters, comparable to and in some specimens exceeding modern humans. Anatomically modern Homo sapiens, appearing approximately 300,000 years ago, has a current average brain volume of approximately 1,350 cubic centimeters.[59]

The expansion from 400 to 1,350 cubic centimeters over roughly 3 million years is, in the context of mammalian brain-size evolution, extraordinary. The standard scaling relationships between brain size and body size in mammals predict a brain expansion of

---

[59]Endocranial volume estimates compiled from Holloway, R. L., Broadfield, D. C., and Yuan, M. S. (2004). The Human Fossil Record, Volume 3: Brain Endocasts. Hoboken: Wiley. Updated synthesis in Falk, D. (2016). Evolution of the Human Brain: The Coming of Age of Paleoneurology. Cambridge: Cambridge University Press.

perhaps 20 to 30 percent in response to the body-size changes that occurred over the same period in the hominin lineage. The actual expansion is more than three hundred percent. Hominin brains grew not just in absolute size but in size relative to the body — the encephalization quotient. The figure of 7.5 that modern humans carry is the highest measured value in the entire vertebrate record. The brain expansion is not a continuation of a trend. It is a departure from a trend.

The departure is metabolically expensive. The modern human brain consumes approximately twenty percent of the body's total energy budget despite accounting for only about two percent of body mass. The metabolic cost is paid in two ways. The first is dietary — the hominin diet shifted toward higher-energy foods, particularly cooked meat and fatty tissues, beginning around 1.5 million years ago. The control of fire, dated to at least 1 million years ago by direct fossil evidence and possibly earlier, made high-density nutrition accessible. The second is gut-size reduction — Leslie Aiello and Peter Wheeler proposed in their 1995 "expensive tissue hypothesis" that the energy required by the expanding brain was offset by a corresponding reduction in the size of the digestive tract, which was made possible by the dietary shift to higher-quality foods that required less digestive processing.

The Aiello-Wheeler hypothesis is partially supported and partially contested in the subsequent literature, but the underlying observation is uncontroversial: the expansion of the hominin brain required a corresponding reorganization of the entire metabolic infrastructure of the body.[60] The brain did not merely grow. The body was rebuilt to support it. The selection of the encephalized brain from the library required, in effect, a coordinated set of secondary selections — the dietary capacity, the gut configuration, the

[60] Aiello, L. C., and Wheeler, P. (1995). "The expensive-tissue hypothesis: The brain and the digestive system in human and primate evolution." Current Anthropology 36(2): 199-221.

cardiovascular delivery system, the thermal regulation needed to dissipate the heat the brain generates — all of which had to come together at the same time for the central selection to be viable.

The Prefrontal Cortex

The encephalization leap was not uniform across the brain. Some regions expanded much more than others. The most dramatic expansion occurred in the prefrontal cortex — the foremost portion of the brain, directly behind the forehead — and specifically in the dorsolateral and ventromedial subregions that comparative neuroanatomists associate with executive function, self-modeling, abstract reasoning, and integrated decision-making.

The prefrontal cortex of a modern human occupies approximately 30 percent of the cerebral cortex by volume. In a chimpanzee, the prefrontal cortex occupies approximately 17 percent. In a macaque monkey, approximately 11 percent. In a rat, less than 5 percent. The proportional expansion in humans is not just a function of overall brain growth — it is a disproportionate expansion of one specific region, with one specific set of functions, on top of the general size increase that affects the entire cortex.

The functions that the prefrontal cortex performs are precisely the functions on which the argument of this book depends. The capacity to model oneself as an entity persisting through time, to integrate experiences across long temporal spans into a coherent narrative, to reason about counterfactuals, to consider one's own thoughts as objects of further thought, to hold competing considerations in mind simultaneously and weigh them against each other, to recognize one's own mortality and act in light of that recognition — all of these are prefrontal functions. They are present in attenuated form in other primates, and in still more attenuated form in other mammals, but in humans they are elaborated to a degree that has no parallel in any other species currently known.

The functional importance of the prefrontal cortex is established most dramatically by the cases in which it is damaged. The classic case is Phineas Gage, the American railroad foreman who in 1848 survived an explosion that drove a three-foot iron tamping rod through his left frontal lobe and out the top of his skull.[61] Gage survived. His sensory and motor function was largely preserved. What was lost was the integration that the prefrontal cortex provides. The Gage who walked away from the accident was, in his physician's contemporary description, no longer Gage — his capacity for planning, for self-restraint, for following social conventions, and for sustaining long-term goals had been destroyed by the lesion. The personality, in the sense the word is used in everyday English, lives in the prefrontal cortex. The case has been replicated in clinical literature ever since: damage to the prefrontal cortex produces predictable deficits in executive function, social cognition, and the capacity for integrated self-modeling, and the deficits track the specific subregions damaged.

The prefrontal cortex is also home to a class of neurons that have attracted special attention in the comparative literature. These are the von Economo neurons — large spindle-shaped cells, named for the Austrian neurologist Constantin von Economo who first described them in 1925 — that occur in dense concentration in the anterior cingulate cortex and the frontoinsular cortex, both of which sit within or immediately adjacent to the prefrontal region. Von Economo neurons are present in humans, great apes, elephants, and cetaceans, and in much smaller numbers in a handful of other species. They are not present in monkeys or in most other mammals.[62] Their distribution is striking: they appear in exactly the

[61] Harlow, J. M. (1868). "Recovery from the passage of an iron bar through the head." Publications of the Massachusetts Medical Society 2: 327-347. Modern reconstruction of the lesion in Van Horn, J. D. et al. (2012). "Mapping connectivity damage in the case of Phineas Gage." PLOS ONE 7(5): e37454.
[62] Allman, J. M., Tetreault, N. A., Hakeem, A. Y., Manaye, K. F., Semendeferi, K., Erwin, J. M., Park, S., Goubert, V., and Hof, P.

lineages that show the strongest evidence of complex social cognition and self-awareness. Whether they are necessary for those capacities or merely correlated with them is unsettled, but their existence is one more piece of evidence that the substrate this chapter is describing has structural specializations beyond raw size. The brain is not just larger. The brain is differently configured, with specific neural populations that appear to support specific integrative functions, and the configuration is shared by only a small number of species on this planet — every one of which the next chapter will address in detail.

This is the substrate that integrates. The earlier chapters of this book described the conditions for the visit. This is the form the visit takes when it arrives. The selection of the encephalized prefrontal cortex from the library is the selection of the capacity to hold whatever the visitor brings into the form. The argument that something is being held, that the holding requires this specific substrate, and that the substrate was selected for this specific purpose, will be developed across the next several chapters. For the moment, the anatomical observation is sufficient: the modern human brain is configured at exactly the level of complexity that the integration argument requires, and the configuration is the result of a selection that ran against the broader pattern of mammalian brain evolution.

The Descended Larynx and the Externalization of Memory

The fourth major selection — and the one that distinguishes the human form from every other species currently alive — is the anatomical reconfiguration of the throat that made speech possible. In every other mammal, including all of the other primates, the larynx sits high in the throat, near the base of the tongue. This

R. (2010). "The von Economo neurons in frontoinsular and anterior cingulate cortex in great apes and humans." Brain Structure and Function 214: 495-517.

arrangement protects the airway during swallowing and allows simultaneous breathing and drinking. In humans, the larynx is descended — it sits significantly lower in the throat, creating a long resonating chamber above the vocal folds that gives the human voice its unusual range and flexibility. The descended larynx makes speech possible. It also creates a serious mechanical liability: humans are the only mammals that can choke on their food, because the airway and the food passage cross in a region where they did not have to cross before the larynx descended.

The descended larynx is documented in modern human anatomy, in the developmental trajectory of human infants (whose larynx begins higher and descends during the first two years of life), and indirectly in the fossil record through analysis of the basicranium and the hyoid bone. The American anthropologist Philip Lieberman, working from the 1970s onward, made the case that the descended larynx is unique to anatomically modern humans and that it co-evolved with the neural machinery for language.[63] The dating is contested, but the structural fact is not. The human throat is rebuilt for speech. The rebuilding came at a measurable cost in choking risk. The cost was paid because the benefit — the externalization of memory and integration across individuals — was greater.

This is the second of the human form's two distinguishing capacities. The prefrontal cortex provides interior integration. The descended larynx provides the means to externalize that integration into language, and through language into stories, instruction, written record, and the accumulated knowledge of generations. Without speech, the integration of a single human substrate dies with the

[63]Lieberman, P. (1984). The Biology and Evolution of Language. Cambridge: Harvard University Press. Lieberman's reconstructions of Neanderthal vocal anatomy were controversial; the more recent consensus is that Neanderthals had some capacity for speech but a more limited phonetic range than modern humans, owing to a less fully descended larynx.

substrate. With speech, the integration accumulates across substrates. Civilization, science, literature, religion — every cultural inheritance the visitor encounters when arriving in a modern body — is downstream of the descended larynx. The substrate's capacity to hold integration is one selection. The substrate's capacity to externalize integration is another. Both are required for the kind of visit this book argues for, and both are present in the human form, and only in the human form, on this planet.

Speech also has a genetic signature. The gene called FOXP2, identified through study of a London family with severe inherited speech and language disorders, encodes a transcription factor that regulates the development of neural circuits in the basal ganglia and the cerebellum — circuits that participate in the fine motor sequencing required for articulate speech.[64] The modern human version of FOXP2 differs from the chimpanzee version at two specific amino-acid positions. The substitutions appear to have been fixed in the human lineage in approximately the same window as the descent of the larynx and the emergence of anatomically modern morphology. The anatomical change in the throat did not, by itself, produce speech. The anatomical change had to be accompanied by neural and genetic changes that gave the substrate the wiring to use the new vocal tract for the patterned sound production language requires. FOXP2 is one piece of that wiring. The two systems — the anatomical and the neural — co-evolved over a relatively short window, and the co-evolution is what made the integration externally communicable.

---

[64]Lai, C. S. L., Fisher, S. E., Hurst, J. A., Vargha-Khadem, F., and Monaco, A. P. (2001). "A forkhead-domain gene is mutated in a severe speech and language disorder." Nature 413(6855): 519-523. The modern human form of FOXP2 differs from the chimpanzee form at two amino-acid positions, and the timing of those substitutions has been estimated through population-genetic analysis at roughly 200,000 years ago, contemporary with the emergence of anatomically modern Homo sapiens.

The Long Juvenile Period

A final selection deserves mention before the chapter closes. The human life cycle includes an unusually long juvenile period — the time between birth and reproductive maturity — relative to other mammals of comparable body size. A chimpanzee reaches reproductive maturity around age eight to ten. A human reaches reproductive maturity around age thirteen to fifteen, and continues neurological development until approximately age twenty-five. The human prefrontal cortex, which is the substrate the previous sections have been describing, does not finish wiring itself until well into the third decade of life. This is anomalous. The human juvenile period is roughly twice as long, in absolute terms, as any reasonable comparison case.

The reason, as Robin Dunbar has argued in his work on the social brain hypothesis, is that the human brain has to learn far more during its developmental period than the brain of any other species. The information load includes language, social structure, kinship networks, ecological knowledge, technical skill, narrative tradition, and the implicit rules governing all of these. A human infant arrives in the world with very little prewired competence and acquires the entire cultural inheritance of its community during the long juvenile period. The acquisition is what makes the substrate functional. A human born and raised without the cultural inheritance — the documented cases of feral children, or of children raised in severely deprived institutional settings — is permanently impaired. The substrate without the juvenile-period inputs does not integrate.

Dunbar's social brain hypothesis, developed in a series of papers beginning in 1992, argues that primate brain size correlates with the size of the social group the species typically maintains, and that humans sit at the extreme end of that correlation with a predicted group size of approximately 150 — the now-famous "Dunbar's number."[65] The selection of the long juvenile period is, on this account, a selection for the social and cultural integration that the encephalized substrate makes possible. The brain is the hardware. The juvenile period is the loading sequence. Both are required, both are selected, and both are necessary for the integration this book argues for to occur.

The long juvenile period has a structural companion in the phenomenon biologists call neoteny — the retention of juvenile features into adulthood. The adult human skull, compared to the adult skulls of other apes, retains the proportions of an ape infant: large brain case relative to face, smaller jaw, more vertical forehead, less prominent brow ridge, lighter overall skeletal robustness. The Dutch anatomist Louis Bolk proposed in the early twentieth century that the human form is essentially a fetalized ape — that the developmental trajectory leading to human adulthood has been slowed and truncated relative to the trajectories of other apes, freezing the body in a configuration that more closely resembles juvenile great-ape morphology. Bolk's strong reading has been refined, but the underlying observation has held in modern comparative work. Humans retain juvenile traits — including, critically, the openness to learning, the playfulness, the curiosity, and the social plasticity characteristic of juvenile great apes — into adulthood. The long juvenile period is not just a developmental delay. It is a permanent fixation of the substrate in a state that remains receptive, malleable, and integrative throughout life. The substrate, having been assembled for integration, does not seal itself shut at maturity. It remains open. The visitor, who arrives into the substrate, finds an instrument that remains tunable for as long as the visitor inhabits it.

[65]Dunbar, R. I. M. (1992). "Neocortex size as a constraint on group size in primates." Journal of Human Evolution 22(6): 469-493. Subsequent synthesis: Dunbar, R. I. M. (1998). "The social brain hypothesis." Evolutionary Anthropology 6(5): 178-190.

The Substrate Integration

Pulling the selections together: the human body is the conjunction of a pentadactyl limb (380 million years old), a precision grip (3 million years old in its modern form), an encephalized brain (3 million years of expansion), a disproportionately enlarged prefrontal cortex (selected specifically within that expansion), a descended larynx and the speech apparatus that depends on it (perhaps 200,000 years old in fully modern form), a long juvenile period that allows cultural integration of the substrate, and a coordinated metabolic infrastructure that supports all of the above. Each of these selections is documented in the paleontological and comparative-anatomical record. Each of these selections was drawn from a library that already contained the relevant elements. And each of these selections is necessary for the form to function as the integration point the previous chapter described.

The conjunction is the point. Any single element of the conjunction could be dismissed as a contingent outcome of selection pressures acting on a particular lineage at a particular time. The pentadactyl limb could be dismissed as a frozen accident of Carboniferous developmental biology. The encephalization leap could be dismissed as a runaway response to social-group dynamics. The descended larynx could be dismissed as the byproduct of facial restructuring during hominin evolution. Each individual story is plausible in isolation, and each individual story is told in the standard textbooks of evolutionary biology. The conjunction of all of them — pentadactyl plus precision grip plus encephalized cortex plus prefrontal expansion plus descended larynx plus long juvenile period plus metabolic infrastructure — has no single proposed cause in the literature. The conjunction is treated as a happy accident of converging trends.

The framing of this book offers an alternative reading. The conjunction is not accidental. The conjunction is the configuration the visit required, and the library produced the configuration because the library contained the elements and the elements could be drawn together when the conditions admitted them. The body the visitor inhabits is not the random output of converging trends. It is the assembled minimum viable substrate for the kind of integration the visit performs. Every one of the selections is necessary. Each one is documented as an unusual feature of the human form relative to the broader animal kingdom. And the question of why all of them assembled together at the same time in the same lineage is, in the standard literature, unanswered.

This is not the place to settle that question. This is the place to register it. The next chapter takes up the integration the substrate performs — what is being integrated, why the substrate was assembled to integrate it, and how the integration relates to the forms the visitor passes through on the way to the human substrate. The substrate is selected. The integration begins now.

✦ ✦ ✦

Three hundred and forty million years of pentadactyl conservation, three million years of cortical expansion, two hundred thousand years of speech, and the entire arc of cultural transmission ran together in the form of the toddler sleeping in Lucille Lewis's arms on the drive back to Lima in late 1940. The form was not unusual. The form was the standard configuration of the species. What was unusual about it was the configuration's specificity — and the question, which the chapter has been pursuing, of what specifically the configuration was built to do.

Nineteen years later, in the fall of 1959, a five-foot-tall, twenty-six-year-old Russian biologist named Lyudmila Trut got off a slow train at a small rail station in a forested district of central Siberia and began to walk down a dirt road toward an industrial fox farm whose director did not know what kind of experiment she was coming to ask him to participate in.[66] She had graduated from Moscow State University the year before. She had been recruited by a man named Dmitri Belyaev, who had recently taken up the deputy directorship of a new institute at Akademgorodok, the scientific city outside Novosibirsk that the Soviet state had built across the previous decade to consolidate its physical, chemical, and biological research at a safe remove from Moscow. Belyaev had a problem. Belyaev wanted to run an experiment in artificial selection that would replicate, in compressed time, the process by which prehistoric wolves had been domesticated into dogs. The experiment, in the framework of orthodox Soviet biology in 1959, was not permissible. Trofim Lysenko, the agronomist who had captured Stalin's confidence in the 1940s and persuaded him that Mendelian genetics was a bourgeois fraud, was still in charge of the institutional apparatus that funded biological research in the Soviet Union, and the kind of selection experiment Belyaev wanted to run was the kind that Lysenko regarded as ideologically dangerous. Belyaev's older brother, a geneticist, had been arrested in 1937 and shot. Belyaev himself had been fired from his post at the Moscow Institute for Fur Breeding Animals in 1948. The experiment he was now proposing would have to be run under cover.

The cover was that the experiment would be conducted as a study of fur quality. Silver foxes, a variant of the red fox, were

---

[66]On Trut's recruitment, the spring 1958 move to Akademgorodok, and the fall 1959 trips to industrial fox farms across the Soviet Union to identify breeding stock for the domestication experiment, see Dugatkin, L. A., and Trut, L. (2017). How to Tame a Fox (and Build a Dog): Visionaries Who Unlocked Nature's Greatest Secret. Chicago: University of Chicago Press, pp. 39-67. The five-foot, twenty-five-year-old descriptions of Trut at the start of the experiment derive from her own accounts in the same volume and from subsequent obituary coverage following her death on October 9, 2024; see Mukherjee, S., "The Daring Russian Geneticist Whose Experiments on Silver Foxes Explained Domestication Has Died," Scientific American, October 24, 2024.

already farmed across the Soviet Union for the fur industry, and the institute at Akademgorodok had a plausible institutional reason to study them. What Trut and Belyaev were actually going to do, under that cover, was breed the calmest foxes from each generation and the calmest foxes only — selecting, in the strict technical sense, on a single behavioral trait — and observe what happened. Belyaev's hypothesis was that domestication is not a slow accumulation of arbitrary changes but a single coordinated regulatory shift in the developmental program, and that selecting on tameness alone would, within a small number of generations, produce a suite of additional changes the breeder had not selected for: floppy ears, curlier tails, mottled coats, shortened snouts, juvenilized faces, and an altered reproductive cycle. The hypothesis was, in the framework of Lysenko's biology, false. The experiment was designed to demonstrate that the hypothesis was true.

Trut got off the train at the rail station. She walked the dirt road to the fox farm. She introduced herself to the director, in the language of fur-industry research, as a graduate student from the new institute at Akademgorodok looking to identify breeding stock for an experiment in coat quality. The director showed her the cages. There were hundreds of foxes. The foxes were nervous, snappish, and dangerous to handle; the industrial farms bred for fur, not for temperament, and a century of farm life had not produced foxes that were comfortable around humans. Trut walked between the rows of cages with the director and, as she walked, identified the small number of foxes that did not bare their teeth as she passed, that did not try to bite the gloves of the handlers, that allowed themselves to be touched without recoiling. These were not many. A typical fox-farm population had, by Belyaev's earlier estimate, perhaps five percent of individuals temperamentally distinct from the wild norm. Trut marked them. She would return to the institute with a list, and the institute would arrange the purchase. The director did not know he was watching the first ten thousand generations of canid domestication be recapitulated in twenty. The chapter that follows examines what happened next.

Where the Argument Goes Next

The next chapter takes the substrate that this chapter has just described — pentadactyl, encephalized, precision-gripped, speech-capable, culturally trainable — and asks what the substrate was assembled to integrate. The answer the chapter develops is that the substrate is the integration point of an ascent through prior forms. The spirit, on the framing this book is developing, passes through forms of progressively greater substrate complexity before reaching the human form, and the human form is the configuration in which the accumulated perspectives of the prior forms can be assembled into a coherent self-aware whole. The substrate is the receiver. The receiver was selected. The selection produced the form you inhabit now. The next chapter asks what the form is now holding.

CHAPTER 5

The Ascent Through Forms

Fifteen years after the fall morning she had walked down a dirt road in central Siberia to identify breeding stock at a fur farm whose director did not know what kind of experiment she was starting, Lyudmila Trut moved into a small house at the edge of the experimental fox farm at Akademgorodok with a fox. The fox was a year and a few weeks old. Trut had named her Pushinka, which is the Russian word for tiny ball of fuzz, the diminutive form a Russian woman might use for a particularly small and particularly soft and particularly affectionate thing she had taken responsibility for. Pushinka had piercing black eyes, silver-tipped black fur that gave her coat the look of light frost over coal, and a single swatch of white running along her left cheek. She had been born on the experimental farm. She was the fifteenth generation of foxes that Trut and Belyaev had bred from the calmest available animals across the previous decade and a half. She had never had a wild parent. She had never had a wild grandparent. The most recent fully wild fox in her lineage had died before she had been conceived.

The morning Trut moved into the small house was a clear spring morning in late March of 1974. The sun was on the snow that the long Siberian winter had not yet finished melting. The house itself was a one-story wooden structure at the southwest corner of the farm grounds, set apart from the rows of fox cages that constituted the rest of the property by perhaps a hundred meters of open ground and a stand of birch. The house had been built to be lived in by a human and had previously been used as a watchman's quarters and as occasional housing for visiting researchers. It had two rooms, a small kitchen, and a wood stove. Trut had carried in her clothing and her books and a kettle and the things a person carries when she intends to live somewhere for an extended but not permanent period. She had brought Pushinka in last, in her arms, the fox calm against her chest, the small white swatch of fur on the left cheek pressed against the lapel of Trut's coat.[67]

The purpose of the move was scientific in framing and was something else in practice. The scientific framing was that the experiment had reached a point at which the question of whether the foxes were behaviorally domesticated could no longer be answered by observations in the cages. The cages constrained the foxes' behavior to what a fox could do inside a cage, which was a limited repertoire, and the question Belyaev and Trut wanted to answer was whether a fox bred for tameness was now capable of the full behavioral repertoire of a domesticated companion animal — capable of cohabiting with a human, of learning the human's routines, of responding to its name across a room, of being walked on a leash on a country road, of recognizing strangers from familiars, of sleeping in proximity to the human without aggression on either side. The cages could not answer the question. The house could. Trut had volunteered to move into the house and find out.

The first few days were uneventful in the way the early days of an experiment are uneventful when the experimenter does not yet know what variable will turn out to matter. Pushinka explored the rooms. She investigated the wood stove, which was warm. She investigated the kitchen, which contained food. She investigated the

[67]The narrative detail of the spring 1974 move into the small house at Akademgorodok with Pushinka is drawn from Dugatkin, L. A., and Trut, L. (2017). How to Tame a Fox (and Build a Dog): Visionaries Who Unlocked Nature's Greatest Secret. Chicago: University of Chicago Press, Ch. 6 ("A House for Pushinka"), pp. 121-148. The physical description of Pushinka — silver-tipped black fur, piercing black eyes, white cheek swatch — derives from the same source and from photographs of Pushinka preserved in the archives of the Institute of Cytology and Genetics in Novosibirsk, reproduced in Dugatkin and Trut and in the companion children's book Dugatkin, L. A., and Trut, L. (2019). Pushinka the Barking Fox. Pittsburgh: Mascot Books.

bedroom, which contained Trut. She returned to all three rooms several times across the first afternoon and evening and she settled, by the end of the second day, on a routine that involved being wherever Trut was, at a slight distance, watching what Trut was doing with what Trut would later describe, in writing, as an attention that did not seem to her to be the attention of a wild animal evaluating a predator. It seemed to her to be the attention of a creature watching its person across a room.

The bedroom development came on the third or fourth night. Trut had been asleep, as far as she could remember when she wrote it down later, when she had been woken by the sense of a slight additional weight at the foot of the bed. She had not turned on the light. She had felt the weight redistribute itself, with a small motion, into the curl of a body settling against her ankles. The body had been warm. The body had not moved for some minutes after settling. Trut had eventually gone back to sleep without turning on the light, and she had not, in the morning, said anything to anyone about what had happened, on the theory that what had happened might not happen again. It happened the next night. It happened every night for the next several weeks. Pushinka, it eventually became clear, was sleeping at the foot of Trut's bed, and the foxes that the farm had been breeding for fifteen generations had produced, among their members, at least one individual whose behavioral repertoire had migrated, without anyone having explicitly selected for it, into the repertoire of a dog.[68]

---

[68]Trut's account of Pushinka's bedroom behavior, the placing of the newborn pup in Trut's lap, and the barking at strangers is in Dugatkin and Trut (2017), pp. 130-145; and was recounted directly to Aaron Dugatkin during interviews at Trut's home in Novosibirsk in 2012 and 2014, with the visit to the by-then-derelict 1974 house through two feet of snow described in Dugatkin's eulogistic essay: Dugatkin, L. A., "The Daring Russian Geneticist Whose Experiments on Silver Foxes Explained Domestication Has Died," Scientific American, October 24, 2024.

In late spring Pushinka came into heat and was bred to a male from the same selected lineage. She delivered a litter in early summer in a basket Trut had prepared for her in the corner of the kitchen. The delivery itself was uneventful. What was not uneventful was what happened the morning after the delivery. Trut had come into the kitchen and had crouched by the basket to check on the pups and on the mother. Pushinka, who had not been asleep and who had been watching her, had reached into the basket with her muzzle, had picked up one of the newborns by the loose skin at the back of the neck in the way fox mothers and dog mothers and cat mothers all pick up their young, had stepped out of the basket carrying the pup, had walked across the kitchen floor to Trut, and had placed the pup, gently and deliberately, in Trut's lap. Trut had held the pup. Pushinka had returned to the basket and resumed nursing the others. Trut, who was forty-one years old that summer and had been a working geneticist for fifteen years and had not previously had any difficulty maintaining the professional disposition that a working geneticist maintained on behalf of her subjects, had sat on the kitchen floor with the pup in her lap and had cried.

The barking came later. Pushinka had grown comfortable enough in the house, by midsummer, to defend it. The first time it happened, a worker from the farm whom Pushinka did not recognize came up the path toward the house to deliver something, and Pushinka had, from her position by the kitchen window, produced a sound that no fox in fifteen generations of the experimental population had previously produced. It was a short, sharp, repeated bark — not the menacing snarl that an aggressive fox makes when a stranger approaches its cage, but a deliberate alerting series of sounds, directed forward in space, oriented toward the approaching person, produced from a position that suggested protection of the house rather than defense of an individual cage. Trut, hearing it, had come out of the back room and looked. Pushinka had not been approached. Pushinka had gone in pursuit of the stranger. The bark had been a guard-dog bark. Foxes did not bark in this way. Pushinka had produced the sound on her own, in the absence of any social model that produced it, in the absence of any selection pressure that had been applied for it. The selection pressure that had been applied was tameness. The behavior that had appeared was not tameness. The behavior that had appeared was something that the library had been keeping in readiness and that the selection on tameness alone had unlocked. The library, on the evidence Pushinka had just provided in the kitchen of a small house at the edge of a fox farm in central Siberia, contained more than anyone had asked it to contain.

✦ ✦ ✦

This chapter is about what the library contains. The previous chapter established that the human body is a curated assembly of design elements drawn from a four-billion-year archive. The chapter that you are now reading takes up the question implicit in the conjunction of the library's depth and the human form's specificity. If the human body is the integration point of a long ascent — and if the substrate of that body has been configured to hold the kind of self-aware, time-binding, narrative-capable consciousness the previous chapters have been describing — what, exactly, is being integrated?

The conventional answer, in mainstream cognitive science, is the inputs of a single lifetime. A human brain integrates the sensory data, memories, and learned patterns of one body across the seventy or eighty years it remains operational. When the body dies, the integration ends, and the integrated content is lost. The account is internally consistent. It has trouble explaining several pieces of evidence that will be taken up in later chapters — the case-archive material at the University of Virginia, the cross-cultural consistency of near-death reports, the persistent reports of pre-birth memory in young children that fade by age six. The mainstream account treats these as anomalies. This book treats them as data.

It also has trouble explaining what Pushinka did in the kitchen of the small house at the edge of the fox farm. Fifteen generations of selection on tameness had unlocked, in her behavioral repertoire, the protection of a human she had no genetic reason to protect and the gentle placement, in that human's lap, of a newborn pup she had no genetic reason to share. Those behaviors had not been selected for. They had been latent in the library, waiting for the substrate to become receptive enough to express them. The chapter that follows argues that the same library, on a longer timescale, has been unlocking behaviors of comparable structure across the entire history of the substrate hierarchy — that the substrate has a hierarchy of integrative capacity, that the hierarchy is measurable, and that the human form is the integration point at which what has been accumulated through every prior level becomes available to be remembered as a whole.

The Substrate Hierarchy

The first observation is that the substrate has a hierarchy, and the hierarchy is measurable.

The standard quantitative measure of relative brain capacity across species is the encephalization quotient, or EQ — a ratio that compares an animal's actual brain mass to the brain mass expected for an animal of its body size, given the empirical scaling relationship that governs brain-to-body ratios across vertebrates. An EQ of 1.0 means the animal has roughly the brain mass expected for its body. An EQ greater than 1.0 means the animal has more brain than its body size would predict. An EQ less than 1.0 means it has less. The measure is imperfect — it does not capture everything that matters about neural complexity, and the scaling formula has been refined several times since its introduction by the comparative neuroanatomist Harry Jerison in 1973 — but it is the cleanest single-number comparison currently available across vertebrate species.[69]

Humans have an EQ of approximately 7.5. We have roughly seven and a half times more brain mass than an average mammal of our body size. Bottlenose dolphins have an EQ in the range of 4.0 to 5.0. Chimpanzees and other great apes cluster around 2.5. Elephants are at 1.7. Dogs, depending on breed, range from 1.1 to 1.3. Cats sit at approximately 1.0. Horses are at 0.9. Mice are at 0.5. The gradient is real and measured.

A note on the gradient before it is misread. The hierarchy is in substrate complexity, not in worth. The dog with an EQ of 1.2 is not less morally important than the human with an EQ of 7.5, any more than a child with a smaller brain is less morally important than an adult with a larger one. The chapter's working premise — locked from the outset, and held throughout the book — is that the hierarchy is vertical in capacity to integrate and horizontal in worth. Every form is sacred. The hierarchy describes what each form can hold and process while the spirit inhabits it, not whether the form deserves to be there. This distinction matters and will be returned to often.

What the gradient establishes is that the substrate's capacity to support self-modeling, abstract reasoning, narrative continuity, and integration of distinct experiences over long time-spans is not uniform across the animal kingdom. It is graded. Some substrates can hold more. The substrates that can hold the most cluster, predictably, at the top of the EQ rankings, with secondary contributions from anatomical features that the EQ measure alone cannot capture — the convoluted cortex of cetaceans, the von Economo neurons that we will turn to next, the specialized prefrontal

---

[69] Jerison, H. J. (1973). Evolution of the Brain and Intelligence. New York: Academic Press. Subsequent refinements of the allometric scaling formula appear in Martin, R. D. (1981). "Relative brain size and basal metabolic rate in terrestrial vertebrates." Nature 293: 57-60.

architecture of humans and the closely related architecture of the great apes.

The ascent hypothesis proposes that the spirit moves through this graded substrate, accumulating capacity as it goes, until it reaches the form in which the accumulated experience can be assembled into a single self-aware integration. The proposal is not that all spirits begin at the simplest substrate and progress mechanically upward. It is that the substrate hierarchy is the field through which spirits move, and that the human form is the integration point of that field. What happens before the integration point is the work the spirit does in the lower forms. What happens at the integration point is the work the spirit does as a human. What happens afterward is the subject of the closing chapters of this book.

Convergent Markers of Interior Experience

If the ascent hypothesis is correct, we should expect to find specific neural and behavioral markers of interior experience appearing convergently across the high-EQ species, regardless of evolutionary lineage. The hypothesis predicts that wherever a substrate becomes sufficiently complex, the markers of self-awareness, social cognition, and emotional depth appear — not because of shared ancestry, but because the substrate has crossed the threshold above which the spirit can express those capacities. The prediction is testable. The prediction is borne out.

The first marker is a specific class of neurons called von Economo neurons, sometimes called spindle neurons. They were first described by the Romanian-Austrian neuroanatomist Constantin von Economo in 1925, in the anterior cingulate cortex of the human brain. For most of the twentieth century, they were thought to be unique to humans. They are large, projection-oriented neurons that appear to be specialized for the rapid integration of social and emotional information, and they are concentrated in the regions of the cortex associated with self-awareness, empathy, and moral reasoning.[70]

In the early 2000s, the neuroscientist John Allman and his colleagues at Caltech began surveying other species for the same cells. They found them in the great apes — chimpanzees, bonobos, gorillas, orangutans — at densities lower than in humans but morphologically identical. They then found them in all of the cetacean species they examined — bottlenose dolphins, orcas, sperm whales, beluga whales — at densities comparable to or exceeding those in humans. They found them in elephants. They found them in some corvid species, though the evidence there is more contested.[71] The species in which von Economo neurons appear are not a single evolutionary lineage. They are the species that, on independent behavioral grounds, are already understood to display the most complex social cognition, the longest memory spans, and the strongest evidence of self-recognition. The convergence is not random. The substrates that became complex enough to support these specific cells are the substrates that, on every other measure available, look most like the human substrate in functional terms.

The second marker is mirror self-recognition. The American psychologist Gordon Gallup, working at Tulane University in 1970, developed what is now called the mirror test.[72] The animal is marked,

---

[70]von Economo, C. (1925). Die Cytoarchitektonik der Hirnrinde des erwachsenen Menschen. Vienna: Springer. The neurons are also sometimes called spindle cells in the older literature.
[71]Allman, J. M., Tetreault, N. A., Hakeem, A. Y., Manaye, K. F., Semendeferi, K., Erwin, J. M., Park, S., Goubert, V., and Hof, P. R. (2010). "The von Economo neurons in frontoinsular and anterior cingulate cortex in great apes and humans." Brain Structure and Function 214: 495-517. Cetacean confirmation in Hof, P. R., and Van der Gucht, E. (2007). "Structure of the cerebral cortex of the humpback whale, Megaptera novaeangliae." Anatomical Record 290(1): 1-31.
[72]Gallup, G. G., Jr. (1970). "Chimpanzees: Self-recognition." Science 167(3914): 86-87. The test has been refined and contested in detail since, but the basic protocol — mark on a

surreptitiously, with a spot of dye on a part of its body it cannot see without a mirror. The animal is then presented with a mirror. If the animal investigates the dye spot on its own body — not the spot in the mirror, but the spot on itself — Gallup argued that the animal must be capable of recognizing the mirror image as itself rather than as another animal. The test is contested in detail, but the species that pass it are a remarkable list. The great apes pass it. Bottlenose dolphins and orcas pass it. Asian elephants pass it. Eurasian magpies pass it. Some cleaner wrasse have recently been argued to pass it, controversially. The list overlaps almost exactly with the species in which von Economo neurons have been found.

The third marker, and the one most relevant to the symbiont thesis of this book, is the conserved architecture of emotional experience itself. The Estonian-American neuroscientist Jaak Panksepp, in a body of work spanning forty years and summarized in his 1998 book *Affective Neuroscience*, identified seven primary emotional systems that appear to be conserved across all mammals, generated by subcortical brain structures that all mammals share.[73] He named them, in his preferred capitalized convention, SEEKING (the desire to explore and acquire), RAGE, FEAR, LUST, CARE (parental nurturance), PANIC and GRIEF (separation distress), and PLAY. The systems are not human in origin. They are mammalian in origin. They predate humanity by tens of millions of years. The emotional life of the dog, the cat, the elephant, the dolphin, and the human is, at the level of these subcortical systems, the same

---

body region invisible without a mirror, then present mirror, observe whether the animal investigates its own body or the mirror image — remains standard.

[73]Panksepp, J. (1998). Affective Neuroscience: The Foundations of Human and Animal Emotions. Oxford: Oxford University Press. Panksepp's seven-system framework — SEEKING, RAGE, FEAR, LUST, CARE, PANIC/GRIEF, PLAY — has been refined but not displaced in the subsequent literature on the neurobiology of emotion.

emotional life, expressed through substrates of varying integrative capacity.

The convergence is the point. The substrates capable of carrying interior experience are not a single lineage. They are a graded series, distinguished by complexity rather than by phylogeny, and the markers of interior experience appear in any substrate that crosses the relevant complexity thresholds. The substrate hierarchy is the field. The interior experience is what the field carries.

The Octopus Question

The convergence is most dramatic, and most unsettling for any strictly lineage-based account of consciousness, in the octopus.

The cephalopods diverged from the lineage that produced humans approximately 600 million years ago, in the Precambrian. The most recent common ancestor of an octopus and a human is a small marine worm with a few hundred neurons in scattered ganglia, no brain in any modern sense, no eyes, no limbs, no social behavior. Everything complex about an octopus — and there is a great deal — evolved independently of everything complex about a human. The two lineages have been running parallel experiments in the evolution of cognition for two-thirds of the history of multicellular life, with no shared neural architecture beyond the most basic cellular machinery.

The Australian philosopher of mind Peter Godfrey-Smith, in his 2016 book *Other Minds: The Octopus, the Sea, and the Deep Origins of Consciousness*, made the case that the octopus is the most accessible example of an intelligent alien — an evolutionarily independent emergence of complex inner experience, accessible to study because it exists on the same planet.[74] Octopuses have nervous

[74]Godfrey-Smith, P. (2016). Other Minds: The Octopus, the Sea, and the Deep Origins of Consciousness. New York: Farrar, Straus and Giroux. Continued development of the argument in Godfrey-Smith, P. (2020). Metazoa: Animal Life and the Birth of

systems organized very differently from vertebrate nervous systems: of their roughly 500 million neurons, only about a third are in the central brain. The other two-thirds are distributed through the arms, which can operate semi-autonomously, exploring and manipulating objects without central direction. Octopuses solve novel problems, recognize individual humans across encounters, use tools, and display what observers consistently describe as personality. Whether the octopus is conscious in the full human sense is unresolved. That its substrate supports something that looks very much like interior experience is not seriously disputed by anyone working in the field.

The octopus is important to the ascent hypothesis for a specific reason. It demonstrates that interior experience is not a single-lineage event. It is not a property that emerges only along the evolutionary line that produced humans. It is, instead, a property that emerges anywhere the substrate supports it. The implication is that the substrate hierarchy described in the previous sections is broader than any one phylogenetic tree. The spirit, on this account, is not constrained to the vertebrate lineage. It moves through any form whose substrate is complex enough to carry it.

This widens the scope of the symbiont obligation. If the visitor's responsibility is to every form the spirit has inhabited or could inhabit, the field of forms is not limited to mammals, not limited to vertebrates, and not limited to the lineage we happen to belong to. The octopus is a form too. The octopus is, in the framing of this book, a form the visitor's spirit might one day have reason to remember.

The Condensation Mechanism

The next question the ascent hypothesis must answer is mechanical. If the spirit moves through the substrate hierarchy, accumulating capacity as it goes, what is the time-budget of that

---

the Mind. New York: Farrar, Straus and Giroux.

ascent? The lifespans of the relevant species vary enormously. A mouse lives two years. A dog lives twelve to fifteen. An elephant lives sixty to seventy. A bowhead whale lives over two hundred. A human lives eighty. How does the spirit accumulate enough experience in two-year increments to make any meaningful progress through the hierarchy?

The answer, which this section develops in detail, is that lifespan in lower forms is not a limitation. It is a feature. The shorter lifespan is the condensation mechanism of the ascent, and the science on this point is more developed than is commonly known.

The relevant work is in the comparative neuroscience of time perception. A 2013 paper by Kevin Healy, Luke McNally, Graeme Ruxton, Natalie Cooper, and Andrew Jackson, published in Animal Behaviour, addressed the question of why small animals appear to react faster than large animals across a wide range of behavioral measures.[75] The team measured what is called the critical flicker fusion frequency — the rate at which a flashing light appears to fuse into continuous illumination — across more than thirty vertebrate species. Humans see flickers as continuous above about 60 Hz. Dogs require approximately 75 Hz. Cats are similar. Mice, rats, and small birds require rates above 100 Hz. The smaller and faster-metabolizing the animal, the higher its temporal resolution.

The interpretive consequence is that subjective time, per unit of clock time, runs differently in different bodies. A mouse perceives more frames of experience per second than a human does. The same one-second interval contains more discriminable events for the mouse than for the human. The mouse's two-year life, measured in subjectively discriminable temporal units, is not as short as it appears to the human observer. The shorter lifespan is, in a precise empirical

[75]Healy, K., McNally, L., Ruxton, G. D., Cooper, N., and Jackson, A. L. (2013). "Metabolic rate and body size are linked with perception of temporal information." Animal Behaviour 86(4): 685-696.

sense, a denser lifespan. The condensation is not metaphor. It is measured.

This finding does not on its own establish the spiritual ascent thesis. What it does establish is that the time-budget objection — how can a spirit make meaningful progress through a two-year mouse life? — rests on an unexamined assumption about the equivalence of time across substrates. The assumption is false. Time, in the strictest physical sense established by Einstein and reaffirmed by every relativistic experiment since, is not absolute. It is relative to the frame in which it is measured. And in the subjective frame, time is relative to the substrate's processing capacity, which means a twelve-year dog life is not one-seventh of a human life. It is its own complete arc, on its own internal clock, condensed to fit a substrate that processes time faster than the human substrate does.

The deeper physics of this point — what time is, and whether the spirit's experience of time outside the embodied substrate is bounded by clock-time at all — is reserved for a later chapter, where the quantum-mechanical and relativistic foundations of time will be addressed directly. For this chapter, the empirical finding is sufficient: the condensation mechanism is real, it is measured, and it removes the time-budget objection to the ascent hypothesis. The dog's life is, in the way that matters for spiritual progression, complete.

Parallel Visits and Kindred Spirits

A consequence of the condensation mechanism follows directly. If lifespans in lower forms are subjectively complete arcs on faster internal clocks, then multiple lower-form visits can complete within a single human lifespan. A spirit ascending through the hierarchy does not have to take its lower-form lives serially across centuries. The lower-form lives can be parallel. They can overlap.

They can intersect with specific human lives in ways the human lives are sometimes able to recognize.

This is the structural foundation for what most readers will already have intuited about the deep cross-species bonds that some humans form with specific animals over the course of a lifetime. A person who has had three dogs over forty years, each of whom seemed to know them in a way that exceeded what behavioral conditioning alone can explain, may not be projecting. They may be observing.

Two possibilities are held open by this book, and the reader should know that both are entertained without forcing a choice between them. The first possibility is that some of the animals a human encounters in deep relationship are the same spirit's prior forms — earlier stages of the human's own ascent through the hierarchy, returning briefly into proximity with the substrate that integration eventually allowed. The second possibility is that some of the animals a human encounters are kindred spirits — separate spirits, but related ones, doing their own parallel work in the substrate hierarchy and intersecting with the human's life because the intersection serves both spirits' growth.

The two possibilities are not distinguishable by any test currently available to the human substrate. They may also not need to be distinguished. The functional consequence is the same in both cases: the relationship between the human and the animal is a real exchange between two interior experiences, both of which are doing the work of the ascent, both of which benefit from the proximity. The function is the bond. The bond is the medium through which the work is done.

The cross-cultural philosophical record supports both possibilities. The Hindu and Buddhist traditions that hold strong reincarnation frameworks have, across most of their textual history, treated cross-species transitions as both possible and consequential, with the realized soul sometimes returning to lower forms by choice and sometimes by karmic accumulation. The Indigenous traditions of much of the world — North American, African, Australian, Andean — have held that humans share kinship with specific animal lineages, and that the relationship is bidirectional. The Hermetic tradition, more philosophically than literally, holds that the descent of the spirit through the spheres includes a passage through forms before the integration in the human substrate. None of this is evidence in the scientific sense. It is the testimony of philosophical traditions that have considered the same data points humans encounter every day — the unusual bond with a specific animal, the sense of recognition across species, the disproportionate grief at the animal's death — and have not dismissed those data as projection.

Symbiotic Love

The central concept this book develops for the bond between the human substrate and the lower-form substrates the human substrate accompanies is Symbiotic Love. The term is capitalized in this book because it is a structural concept, not a descriptive phrase. Symbiotic Love is the bidirectional love that arises between a human and an animal in a relationship sustained by compassion and care — not a projection of the human's feelings onto a passive animal recipient, not a submission of the animal to the human's affection, but a genuine mutual exchange in which both beings are reorganized by the proximity. The science that establishes this exchange as biochemically real, rather than sentimental, is among the most striking findings in the comparative-mammalian literature of the last fifteen years.

The first piece of evidence is the oxytocin loop. Oxytocin is the mammalian bonding hormone — the chemical that drives mother-infant attachment, pair bonding, and social trust in every mammalian species in which it has been studied. In 2015, the Japanese researcher Miho Nagasawa and her colleagues at Azabu

University published a paper in Science showing that mutual gazing between dogs and their human owners produces measurable increases in oxytocin in both species — the dog and the human, simultaneously, in proportion to the duration and intensity of the gaze.[76] The same experiment performed with hand-raised wolves — genetically nearly identical to dogs but lacking the co-evolved relationship — produced no comparable effect. The wolves did not engage in the gazing behavior, and the oxytocin loop did not occur. The dog-human bond is biochemically reciprocal in a way the wolf-human bond is not. The bond is not metaphor. It is hormone.

The second piece of evidence is anatomical. In 2019, the comparative psychologist Juliane Kaminski and colleagues published a paper in the Proceedings of the National Academy of Sciences establishing that domestic dogs possess a specific facial muscle — the levator anguli oculi medialis — that wolves do not possess.[77] This muscle produces the inner-eyebrow raise — the expression humans instinctively read as plaintive, supplicating, or endearing. The muscle did not evolve at random. It evolved because humans responded to it. Over approximately fifteen thousand years of co-presence, dogs that produced this expression were preferentially fed, bred, and protected; dogs that did not produce it were not. The face of the modern dog is, in a measurable anatomical sense, sculpted by the human response. The animal's body has been reorganized by the relationship.

The third piece of evidence is endocrine. A 2019 paper in Scientific Reports by the Swedish researcher Ann-Sofie Sundman

---

[76]Nagasawa, M., Mitsui, S., En, S., Ohtani, N., Ohta, M., Sakuma, Y., Onaka, T., Mogi, K., and Kikusui, T. (2015). "Oxytocin-gaze positive loop and the coevolution of human-dog bonds." Science 348(6232): 333-336.

[77]Kaminski, J., Waller, B. M., Diogo, R., Hartstone-Rose, A., and Burrows, A. M. (2019). "Evolution of facial muscle anatomy in dogs." Proceedings of the National Academy of Sciences 116(29): 14677-14681.

and colleagues at Linköping University measured long-term cortisol levels in dogs and their owners by analyzing hair samples that record hormone levels over months.[78] The dogs' cortisol levels — the marker of chronic stress — synchronized with their owners' cortisol levels at statistically significant levels, controlling for activity, breed, and household conditions. The dog is, measurably, an endocrine mirror of the human. The animal does not merely respond to the human's stress in the moment. The animal's own long-term physiology is shaped by the human's long-term physiology. The mirror is biochemical.

The fourth piece of evidence is the experimental demonstration of how rapidly an animal's biology can be reorganized by selection for compassionate human relationship. The Belyaev-Trut experiment, whose first day was rendered at the close of the previous chapter and whose middle period was rendered at the opening of this one, is the cleanest available demonstration. Beginning in 1959, Belyaev and Trut selected, generation after generation, for one trait only: tameness toward humans. They did not select for coat color, for skull shape, for tail carriage, or for any other physical feature. Within forty generations — a single human lifetime — the foxes had developed floppy ears, piebald coats, curled tails, smaller skulls, and a retention of juvenile features into adulthood. They had, by every external measure, become dogs. The selection for receptiveness to the human relationship had remodeled the animal's anatomy. The reciprocal of this experiment — selecting for human warmth toward animals — is not conducted in laboratories, but the parallel observation is well documented in the literatures on the psychological effects of companion-animal relationships: humans who maintain sustained relationships with animals show measurable shifts in cardiovascular health, immune function, and

[78]Sundman, A.-S., Van Poucke, E., Svensson Holm, A.-C., Faresjö, Å., Theodorsson, E., Jensen, P., and Roth, L. S. V. (2019). "Long-term stress levels are synchronized in dogs and their owners." Scientific Reports 9: 7391.

social behavior. The animal is reshaping the human as the human reshapes the animal. The mirror runs both ways.

Symbiotic Love, then, is not a metaphor borrowed from spirituality. It is a structural feature of cross-species relationships of sustained compassion, documented in oxytocin biochemistry, in co-evolved facial anatomy, in synchronized endocrine systems, and in selection experiments that reshape the animal's body within decades. The book uses the capitalized term to mark this concept as load-bearing. It is the operational expression of the symbiont thesis. The visitor is not separate from the forms the spirit has inhabited. The visitor is in continuing relationship with those forms, and the relationship is, on the evidence above, biochemically and anatomically real.

This matters for how the reader interprets the bond with any specific animal in their own life. The bond is not projection. The bond is one of the most thoroughly documented cross-species phenomena in modern biology, and it operates through mechanisms that are present in both species, generated by both species, and that reorganize both species. Whether the animal is a prior form of the reader's own spirit or a kindred spirit accompanying — the question this book leaves open — the relationship itself is the work that both spirits are doing.

Hermetic and Philosophical Testimony

The philosophical traditions that have considered the ascent through forms have done so for thousands of years, and their testimony is not evidence in the scientific sense but is testimony in the sense that an unusually large number of independent observers arrived at structurally similar accounts of what they were seeing. Three traditions are worth marking here, briefly, before the chapter closes.

The first is the Hermetic corpus. The Poimandres, the foundational text of the Hermetic literature, dates to the first or second century of the common era but draws on Egyptian and Hellenistic philosophical material substantially older.[79] It describes the descent of the soul through the celestial spheres, picking up qualities from each sphere as it passes, before entering the body. The soul arrives at the body already layered. The ascent that this book proposes is the inverse of that descent — not the soul accumulating from above, but the spirit accumulating from below, gathering perspectives from each form as it passes through before integration in the human substrate. The structural pattern is the same. The Hermetic tradition reads it one way, descending; this book reads it the other way, ascending. Both readings are present in the textual tradition.

The second is Plotinus and the Neoplatonic line. Plotinus, writing in the third century, described the structure of existence as procession and return — emanation outward from a unitary source into increasingly differentiated forms, followed by a return motion in which the differentiated forms reintegrate what they have encountered. The return is not annihilation. It is integration. The differentiated forms do not lose what they were; they bring it back. For Plotinus the integration happens through philosophical contemplation in the human lifetime. For this book, the integration happens through the substrate-capacity of the human form for the kind of contemplation Plotinus describes. The descriptions are the same description in different vocabularies.

The third is Mircea Eliade's comparative work on the religious experience of time. Eliade's 1954 book *Cosmos and History: The Myth of the Eternal Return* documented that the pattern

[79]Copenhaver, B. P. (trans.) (1992). Hermetica: The Greek Corpus Hermeticum and the Latin Asclepius. Cambridge: Cambridge University Press. The Poimandres is Book I of the Corpus Hermeticum, esp. sections 24-26 on the ascent of the soul through the seven spheres.

of descent, differentiation, and return appears in nearly every religious tradition that has been systematically studied.[80] Eliade interpreted this convergence anthropologically — as a feature of how human cultures structure time and meaning — and stopped short of claiming that the convergence pointed to anything outside the cultures themselves. This book takes a different interpretive step. The convergence is not anthropological coincidence. The traditions describe the same pattern because the pattern is there to describe. The substrate hierarchy that comparative neuroscience has measured, and the spiritual ascent through forms that the philosophical traditions have described, are the same phenomenon viewed from two different evidentiary registers. They are mutually corroborating.

None of this constitutes proof of the ascent hypothesis. The hypothesis is not the kind of claim science currently has instruments to prove or disprove. What it constitutes is the observation that the hypothesis is consistent with the measured neural and behavioral hierarchy of substrate complexity, with the documented biochemistry of cross-species bonding, with the structural pattern recurrent across the philosophical traditions, and with the empirical condensation of subjective time across substrates. The hypothesis is not asserted as established. It is asserted as a coherent reading of an unusually wide convergence of evidence.

The Integration Point

The human substrate is the integration point of the ascent. This claim has been implicit throughout the chapter; it is worth stating directly before the chapter closes, and worth being precise about what it does and does not mean.

[80] Eliade, M. (1954). Cosmos and History: The Myth of the Eternal Return. Princeton: Princeton University Press.

It does not mean that the human form is superior to the lower forms. It does not mean that the human form is the goal of life on this planet. It does not mean that humanity is the apex of biological achievement or that other species exist for human use. The hierarchy, as locked at the beginning of this chapter, is vertical in substrate complexity and horizontal in worth. Every form is sacred because every form is what the spirit has been, or could be, or will accompany. The human form is not above the others in any sense that matters morally.

What the human form is, uniquely on this planet, is the substrate at which integration can occur. The encephalization leap of the previous chapter produced a brain large enough to hold a self-model that persists across time, a narrative that knits experiences into a coherent identity, and a recognition of the substrate's own finitude that motivates the work of integration. Lower substrates are doing real work — the dog's twelve years are complete, the elephant's grief is real, the octopus's exploration is genuine — but lower substrates are not assembling those experiences into the unified self-aware integration that the human substrate is configured to perform. The integration is what makes the human form the form in which the visit becomes, for the first time in the ascent, knowable to itself.

This has a consequence that the closing chapters of the book will develop, but that is worth marking here for the architectural completeness of the chapter. If the human substrate is where integration becomes possible, then the visit into the human substrate is the visit that, by integrating the prior forms, also becomes the visit that can return to lower forms by choice. The traditional bodhisattva framing — the realized being who declines final liberation in order to return for the sake of others — has, in this book's framework, a more specific structural reading. The realized spirit can return into any form, including lower forms, deliberately, for the sake of symbiont presence in those forms. The post-human ascent does not foreclose lower-form re-entry. It enables it. The radical reading of post-human re-entry — that the realized spirit can choose to return as a dog accompanying a specific human, or as an octopus exploring a specific stretch of reef, or as any form the symbiont obligation calls for — is the reading this book takes. The integration point is not an exit from the hierarchy. It is the point from which the hierarchy can be re-entered in service of every other form the spirit has been or will be.

This is the closing turn of the chapter. The substrate hierarchy is real and measured. The interior experience that the substrate carries is the same interior experience, expressed at different integrative capacities, across an enormous range of forms. The human form is the integration point. Symbiotic Love is the operational expression of the integration's continuing relationship to the forms beneath it. The visitor inhabits the human substrate not as a graduation from the prior forms but as the form in which the prior forms can be remembered, accompanied, and honored.

✦ ✦ ✦

Pushinka lived in the small house at the edge of the experimental farm for about a year after she had placed her newborn pup in Trut's lap. The barking continued. The pups grew. Trut walked them on a country road in the late afternoons, three or four foxes at her heels, off-leash, returning when she called their names. In the autumn of 1975 she moved out of the house and back to her apartment in Akademgorodok, and Pushinka returned to a special enclosure near the institute where she remained, recognized and named, for the rest of her life. Trut continued the breeding program through the next forty-nine years. She continued it after Belyaev's sudden death from cancer in 1985 left her, at fifty-two, the sole director of the work. She continued it through the dissolution of the Soviet Union in 1991, through the collapse of state funding for the institute in the 1990s, through the partial restoration of funding in the 2000s, through the visits of foreign collaborators in the 2010s, and through her own gradual aging into the role of the world's senior authority on the experimental domestication of a wild species. On October 9, 2024, she died peacefully in her sleep in Novosibirsk, just short of her ninety-first birthday. She had been working on the project for nearly twenty-three thousand consecutive days.

Two years before her death, when the American evolutionary biologist Lee Alan Dugatkin and his teenage son visited her in Novosibirsk to conduct interviews for the book the two of them would publish in 2017, Trut had insisted, on a morning when more than two feet of snow had fallen overnight, on walking them through the snow to the by-then-derelict small house at the edge of the farm where she had lived with Pushinka in 1974. She was eighty years old. She still stood five feet tall. She walked through the snow without complaint, in a coat she had owned for several decades, and she showed her visitors the kitchen where Pushinka had brought her the pup and the bedroom where Pushinka had slept at the foot of the bed and the window where Pushinka had first produced the bark that no fox in fifteen generations of the experimental population had previously produced. The roof had partially collapsed. The walls were sound. The kitchen tile was buried under three decades of dust.[81] Trut said, on the walk back through the snow, that she had been responsible for the foxes for what was by then more than fifty years and she still was, and that the line in the Saint-Exupéry story about being responsible forever for what one had tamed was the line she had carried with her since she had been twenty-six and had walked down a dirt road in the fall of 1959 to identify breeding stock at a farm whose director had not known what kind of experiment she had been starting. She said this with the conviction of a person who had lived inside the sentence rather than the conviction of a person who had read it.

---

[81]Dugatkin, L. A. (2024). "The Daring Russian Geneticist Whose Experiments on Silver Foxes Explained Domestication Has Died." Scientific American, October 24, 2024.

The library, on the testimony of the woman who had spent twenty-three thousand days asking it what it contained, contained more than anyone had asked it to contain. The next chapter takes up the question of what the library does when the substrate it has built is large enough to hold not just the integration of a single life but the recognition of the lives that came before.

Where the Argument Goes Next

This chapter has proceeded as if the substrate carries the interior experience. The substrate is the body. The interior experience is what the body holds. The chapter has argued that the substrate has a measurable hierarchy and that interior experience appears wherever the substrate's complexity supports it. The argument has been allowed, throughout the chapter, to assume that interior experience and the substrate are related the way the conventional view assumes — the substrate produces the experience, and the experience exists only as long as the substrate does.

That assumption is what the next chapter takes apart. The next chapter argues that the substrate does not produce the experience. The substrate receives it. The brain is the receiver, not the source. This is the strongest single claim in this book, and the chapter that develops it will draw on the prospective near-death-experience studies of the cardiologist Pim van Lommel, on the AWARE-II resuscitation research led by Sam Parnia, on the quantum-coherence experiments in microtubules conducted by Anirban Bandyopadhyay, on the Penrose-Hameroff theoretical framework, and on the phenomenon of terminal lucidity that has been documented in clinical case studies of late-stage dementia patients. The argument is cumulative, the evidence is mainstream, and the conclusion the evidence supports is that consciousness exists in a form that does not depend on the substrate for its existence. The substrate is the antenna. Consciousness is the signal. The signal predates the antenna, and the signal continues after the antenna is no longer functional. The next chapter makes that case.

CHAPTER 6

Consciousness Is Not Produced by the Brain

The voice that woke Andrea Leininger in the early-morning hours of Monday, May 1, 2000, was the voice of her son James, who was two years and three weeks old. The Leiningers were living, that year, in a two-story house on a quiet street in suburban Lafayette, Louisiana, with a front yard shaded by river birches draped in Spanish moss and a back yard that ended in the kind of low pine the bayou country runs in. Andrea was a thirty-four-year-old former dancer who had recently begun writing résumés part-time from home. Bruce, her husband, was forty-five, a human-resources executive working for an oil-services firm that had transferred them from the San Francisco Bay Area to Dallas in 1998 and then from Dallas to Lafayette in 1999, in the way the oil industry transferred its junior executives during the long upcycle of the late 1990s and early 2000s. They had been married for seven years. They had one child. The child had been born on April 10, 1998, in the Bay Area, four weeks before the move to Texas, and they had named him James Madison Leininger for reasons that, in retrospect, would seem to them less arbitrary than they had seemed at the time.

The voice came from down the hall. It was loud enough that it cleared the doorway of the master bedroom and Andrea was on her feet before she was fully awake. The voice was screaming. It was screaming words. It took her a few seconds, standing in the doorway of her son's room with her hand on the light switch she had not yet turned on, to assemble what the words were. They were not the words a two-year-old screams in an ordinary night terror. They were three sentences, broken into a kind of repeating loop, delivered in a voice that was the voice of her son and that was also somehow the voice of someone older speaking through the voice of her son.

Airplane crash! Plane on fire! Little man can't get out! She turned on the light.[82]

James was lying on his back in the small crib-bed that had replaced the actual crib about six months earlier. His arms were extended above his face. His legs were drawn up. He was kicking the air with his bare feet and clawing at the air with his hands, and the kicking and the clawing were not the random thrashing of a child in distress. They were focused. They had a direction. The direction was upward. He was, Andrea would later say, kicking and clawing the way a person trapped under something heavy would kick and claw to push the something heavy off, except there was nothing on top of him, and the heavy thing he was trying to push off was not in the room. The eyes were closed. The mouth was open. The three sentences kept coming.

Andrea got him onto her shoulder and held him. He did not wake up immediately. He continued, against her shoulder, to repeat the three sentences for what she would later describe as another minute and a half before the words trailed off and he settled into the rhythmic breathing of a child who had returned, by stages, from somewhere else into the room. She carried him to the rocking chair by the window. She rocked him. He did not wake up. She put him back into the crib-bed. He slept. She went back to her own room. She

[82]The Leininger case is documented in detail in Leininger, B., Leininger, A., and Gross, K. (2009). Soul Survivor: The Reincarnation of a World War II Fighter Pilot. New York: Grand Central Publishing. The first nightmare on May 1, 2000, including Andrea Leininger's specific quotations of James's words and her own description of his physical posture in the bed, is rendered in Soul Survivor, Ch. 2, pp. 13-28. Independent journalistic verification of the family's account appears in Jamison, L., "Giving Up the Ghost," Harper's Magazine, March 2015, which includes a visit to the family home and direct interviews with both parents. Academic discussion of the case appears in Tucker, J. B. (2013). Return to Life: Extraordinary Cases of Children Who Remember Past Lives. New York: St. Martin's Press, Ch. 4, pp. 63-90.

did not, at that point, wake Bruce. There would be time enough for Bruce in the morning. She lay in the dark for a long time. Then she fell asleep, and when she woke, the light was coming in through the curtains and the house was quiet, and she thought, for a moment, that what she had heard might have been a thing she had dreamed.

It was not. The nightmares began that night and continued, at four or five episodes a week, for what would turn out to be the next two and a half years. They followed the same script. The boy would scream the three sentences. The boy would kick and claw upward. The parents would carry him to the rocking chair and hold him. He would not wake up. The script did not vary in any of its operative features for the better part of three years.[83]

What varied was what the boy said about the script during the day. By the early summer of 2000, when Andrea began, on the advice of her own mother, to ask the boy gentle questions about what he had been seeing in the nightmares, James began to fill in the details that the three sentences had compressed into themselves. The little man was him.

She had asked him, one afternoon in the kitchen, sitting on the floor with him among his toys, who the little man was. She had asked it carefully. She had not framed it as an interrogation. She had asked it the way a mother asks a child about something the child has seen. He had said,

[83]On the recurrence pattern and the parents' coping process: Leininger, Leininger, and Gross (2009), Chs. 2-5. The clinical recurrence frequency — four to five episodes weekly, sustained over approximately thirty months — is unusual for ordinary pediatric night terrors, which typically resolve within six to twelve months and do not include the scripted verbal content the Leininger case displays. On the differential diagnosis of pediatric night terrors versus the kind of scripted, recurrent, verbally specific dream content the Leininger case presented, see Mason, T. B. A., and Pack, A. I. (2007). "Pediatric parasomnias." Sleep 30(2): 141-151.

without lifting his eyes from the toy plane he was holding, *Me.* She had asked who had shot the plane. He had said, *The Japanese.* She had asked what kind of plane it had been. He had said *Corsair.* She had asked what boat it had taken off from. He had said *Natoma.* He had said the plane had veered to the left on takeoff. He had said the plane had blown its tires on landing. He had said, when she had asked him whether there had been others, that there had been a friend named Jack Larsen. He had not looked up from the toy plane during any of this. He had been two years and two months old.

She had not, that afternoon, said anything to Bruce. She had let the boy say what the boy was saying. She had waited until he had moved on to a different toy. She had then gone upstairs to the small home office Bruce had set up in the second bedroom, sat down at the family's computer, and typed the word Natoma into the search engine that was, in May of 2000, the search engine most people in Lafayette were beginning to use for the kinds of questions they had not previously had a way to ask. The search returned, on the first page of results, a series of references to a United States Navy escort carrier that had operated in the Pacific theater of World War II under the designation USS Natoma Bay. The escort carrier had been commissioned in 1943, had launched aircraft in support of the invasions of the Marshall Islands, the Mariana Islands, Leyte, Luzon, and finally Iwo Jima, and had been decommissioned in 1946. Its operational history was online in the form of an alphabetical list of every pilot who had served aboard her, together with the casualty records for each pilot who had not come home.

Andrea read the list. She did not, that afternoon, find anything on it that resolved into a specific match for her son's three sentences, because she did not yet know what to look for. She closed the browser. She went downstairs. The boy was on the carpet by the coffee table flying his toy plane in a slow circle, making the engine sound a child makes when a child is flying a toy plane, occasionally crashing the plane into the leg of the coffee table and then picking it up and flying it again. She watched him for a long time. He did not look up at her. He was occupied. Whatever he had been doing in the nightmares the previous night was, that afternoon, somewhere else. He was, that afternoon, a small boy on a carpet in a house in Lafayette, Louisiana, with an unusual obsession with airplanes.

Bruce, when she told him that evening what the boy had said about the boat called Natoma, was a forty-five-year-old conservative evangelical Christian who had not, at any point in his life prior to that conversation, given any serious consideration to the idea that a soul that had occupied a previous body might subsequently occupy a different one. The proposition was not part of the theology he had been raised inside. The proposition was incompatible with several of the doctrines he had professed his entire adult life. Bruce listened. Bruce did not, that evening, accept the framing his wife was, with some hesitation, beginning to propose. Bruce did what Bruce had been trained to do in his work, which was to gather the evidence and to evaluate it on its merits without regard for what the evidence implied about anything Bruce had previously held. He went upstairs to the second bedroom. He sat down at the computer. He opened a notebook. He began, on the first page of the notebook, to write down what his son had said.[84]

He filled the first page that night. He filled the second page the following week. He would, across the next four years, fill nine notebooks. Each notebook would correspond to a specific Natoma

[84]On Bruce Leininger's evangelical Christian background and the personal-theological difficulty the case posed for him: Leininger, Leininger, and Gross (2009), pp. 47-62. Bruce's research method — the systematic notebook compilation of every statement his son made, cross-referenced against military and biographical records — produced the documentary record on which subsequent academic discussion of the case has relied. The notebooks themselves are described by the journalist Leslie Jamison after a 2014 visit to the Leininger home in Lafayette in Jamison, "Giving Up the Ghost," Harper's Magazine, March 2015.

Bay pilot killed in action during the ship's deployment, with the biographical and operational details Bruce had assembled from military records, after-action reports, surviving veterans' interviews, and the contemporary correspondence of the families involved. The investigation was thorough in the way the investigation of an oil-services executive who has been trained to dismantle a problem and reassemble its working parts is thorough. The investigation was thorough also in the way the investigation of a Christian father who has been told that his son might be the reincarnation of a man who died in 1945 is thorough, which is to say that the investigation was undertaken in order to demonstrate that the framing his wife was proposing was false, and that the boy's statements would, on close examination, turn out to be the random productions of a small boy who had seen too many planes in too many books and conflated their details into a nightmare loop that any pediatric sleep specialist could explain on entirely conventional grounds. Bruce expected the investigation, when he began it, to take a few months and to produce a definitive negative result. The investigation produced the opposite result. The investigation took four years. The investigation produced, at the end of those four years, a documentary record so specific to the life and death of one Natoma Bay pilot — Lieutenant James M. Huston Jr., twenty-one years old, of Pennsylvania, who had been shot down at the harbor of Chichi Jima on March 3, 1945, on his fiftieth mission, his last before he had been scheduled to rotate home — that the alternative explanations Bruce had begun the investigation expecting to confirm could no longer be sustained on the evidence Bruce himself had compiled.

That investigation belongs to a later chapter of this book, which examines the University of Virginia archive and the case-validation work the Leiningers' material would eventually be folded into. What belongs to this chapter is the question the early phase of the investigation forced on the family that was conducting it. The boy in the crib-bed in Lafayette, Louisiana, was a two-year-old

American child who had been carried through the prenatal period in his mother's body without any exposure to the operational history of a specific Pacific escort carrier and the death of a specific pilot aboard her. The boy could not have read the documents Bruce was assembling. The boy could not have heard the stories Bruce was hearing from the surviving veterans. The boy did not have access, through any channel his neural substrate had been built to accept signals on, to the information the boy was nevertheless producing. The information, on every conventional account of how a substrate acquires information, was not available to the substrate that was producing it. And the substrate was producing it anyway, in three short sentences, four to five times a week, for thirty months.

✦ ✦ ✦

The previous chapter described the substrate hierarchy. Lower forms carry interior experience at the level their substrates support; the human substrate is the integration point at which the accumulated experience becomes self-aware to itself. The chapter assumed, throughout, the conventional view that the substrate is the source of the experience — that the brain produces consciousness the way a fire produces heat. This chapter takes that assumption apart.

The boy in Lafayette was producing, in his sleep and in his daytime conversations with his mother on the kitchen floor, information that his brain had no means to have acquired. The information was specific. The information was verifiable. The information was, on every standard model of how information enters a brain, impossible. The standard model says the brain produces what it knows. The boy was knowing what his brain had not been given access to. Either the standard model is wrong about how the brain works, or there is some other channel through which the brain receives information than the channel the standard model recognizes. The chapter that follows is the case for the second possibility.

The thesis of this chapter is that the brain does not produce consciousness. The brain receives it. Consciousness is a signal that the substrate is configured to detect, modulate, and integrate, but the substrate is not the source of the signal. The substrate is the antenna. The signal predates the antenna, and the signal continues after the antenna is no longer functional. The full case for this claim takes a chapter to make, and the case is cumulative — no single piece of evidence is decisive, but the convergence across several independent literatures is large enough to require, at the least, that the alternative not be dismissed.

The stakes of the chapter are direct. Every later claim in this book depends on the receiver model being viable. If consciousness is produced by the brain, then there is no visitor — there is only the brain producing its temporary local effects — and the pre-selection thesis collapses. If consciousness is received by the brain, then the consciousness that is now reading this sentence existed before the brain that holds it was assembled, and the question of what selected the brain becomes a legitimate question. The chapter does not need to prove the receiver model. It needs to establish that the receiver model is, on the available evidence, at least as defensible as the production model, and arguably more so. That argument is what follows.

The case draws on four independent literatures. The first is the philosophical literature on what is now called the hard problem of consciousness — the structural difficulty the production model has in explaining how electrochemical activity generates subjective experience at all. The second is the clinical literature on near-death experience, with particular weight on the prospective Dutch cardiac-arrest study published in The Lancet in 2001 and on the more recent AWARE resuscitation studies. The third is the experimental quantum-biology literature on microtubule vibrational coherence, which provides a candidate mechanism for substrate-independent consciousness. The fourth is the clinical literature on terminal lucidity, in which late-stage dementia patients sometimes return to full mental clarity in the hours before death, despite cortical damage that the production model holds incompatible with the function being exhibited. Each literature is mainstream, each is peer-reviewed, and each runs into a wall on the standard view. Taken together, they describe the wall.

The Production Model and the Explanatory Gap

The production model — the dominant view in mainstream cognitive neuroscience for the last seventy years — holds that consciousness is what the brain does. On this account, subjective experience emerges from the coordinated activity of the brain's neurons, the way the wetness of water emerges from the coordinated behavior of water molecules. The model is internally consistent. It is the working assumption of every neuroscience laboratory, every cognitive science textbook, and every reductive philosophical account of the mind currently in print.

It has one problem. No working scientist has been able to explain, even in principle, how the electrochemical activity of neurons gives rise to subjective experience at all. The philosopher David Chalmers, in a 1995 paper that has become foundational for the contemporary debate, named this difficulty the hard problem of consciousness.[85] Chalmers' distinction is between the easy problems of consciousness — the cognitive functions that neuroscience is making real progress on, such as how the brain integrates sensory information, controls behavior, and reports its own internal states — and the hard problem, which is why any of that integrating, controlling, and reporting is accompanied by experience at all rather

[85] Chalmers, D. J. (1995). "Facing up to the problem of consciousness." Journal of Consciousness Studies 2(3): 200-219. Extended treatment in Chalmers, D. J. (1996). The Conscious Mind: In Search of a Fundamental Theory. New York: Oxford University Press.

than occurring in the dark. The easy problems are difficult only in the engineering sense. They will, presumably, eventually be solved by ordinary neuroscience. The hard problem is difficult in the structural sense. It is not clear that ordinary neuroscience could ever solve it, because ordinary neuroscience describes physical processes, and the question of why physical processes are accompanied by experience appears not to be a question about physical processes.

The American philosopher Joseph Levine had named the same structural difficulty twelve years earlier in slightly different terms. He called it the explanatory gap.[86] The gap is between any complete physical description of brain activity and any description of the subjective experience that accompanies that activity. The physical description, no matter how detailed, does not entail the experience. A complete physical account of the neural correlates of seeing red would still leave open the question of why seeing red is like anything from the inside — why the activity of those neurons is accompanied by the felt quality of redness rather than by no felt quality at all. The gap is not a remaining piece of research. It is a structural feature of physical description: physical descriptions, by their nature, do not include first-person felt qualities, and first-person felt qualities are what consciousness, on the inside, is.

The structural argument had been anticipated by Thomas Nagel in a paper that has become, in the decades since its publication, the most cited single article in the contemporary philosophy of mind. Nagel's 1974 paper "What Is It Like to Be a Bat?" pointed out that the conscious experience of any organism — even an organism whose neurology we understand reasonably well, like a bat — has a subjective character that no third-person description can fully capture.[87] A bat navigates by echolocation. The

[86]Levine, J. (1983). "Materialism and qualia: The explanatory gap." Pacific Philosophical Quarterly 64(4): 354-361.
[87]Nagel, T. (1974). "What is it like to be a bat?" The Philosophical Review 83(4): 435-450. The paper became foundational for the contemporary debate on the irreducibility

behavioral, neurological, and physiological details of bat echolocation are well documented. None of those details tell us what it is like, from the inside, to be a bat using echolocation to perceive the world. The first-person character of the bat's experience is not contained in the third-person description of the bat's nervous system, however complete. The point generalizes: subjective experience is not, in principle, the kind of thing that third-person physical description can deliver. The production model, which holds that physical descriptions of the brain do deliver subjective experience as their output, owes an account of how that delivery occurs. The account has not been provided, and Nagel's argument suggests that the account cannot be provided within the framework of physical description alone. The conclusion does not establish the receiver model; it establishes that the production model is making a claim it has not earned.

The production model's response to the hard problem and the explanatory gap has been, in the working consensus, to defer them. The dominant approach is to assume that the gap will close as the neural correlates of consciousness are mapped in more detail. After three decades of intensive neural-correlate research, with the most sophisticated imaging technology in history, the gap has not closed. It has, if anything, become sharper. The correlates have been identified at increasing precision. The relationship between the correlates and the experience has not been explained at any level of detail at all.

The production model's defenders would argue, fairly, that absence of a current explanation is not absence of a future one. This is true. Many currently unexplained phenomena in physics and biology eventually yielded to ordinary scientific investigation, and consciousness might be among them. The point of this section is not that the production model is impossible. The point is that it currently has, after seven decades of dedicated work, no proposed mechanism,

---

of subjective experience to physical description.

no candidate explanation, and no working sketch of how the structural problem could in principle be solved. The standard view's explanatory debt is large. The reader is not obligated to extend unlimited credit to a position that has not yet shown a mechanism. The alternative interpretation is, at least, available.

The Receiver Model in Historical Context

The alternative interpretation is the receiver, transmission, or filter model. On this account, the brain does not produce consciousness. It modulates and locally focuses a consciousness that already exists. The brain is to the consciousness what the radio is to the radio broadcast. The radio does not produce the music. The radio receives, tunes, and locally outputs a signal that exists independently of the radio. Damage to the radio degrades the local output but does not destroy the broadcast. Disassembly of the radio terminates the local output entirely but leaves the broadcast unaffected. The model is not new. It has a respectable philosophical history that long predates contemporary neuroscience.

The American philosopher and psychologist William James, in the 1898 Ingersoll Lecture at Harvard, published as Human Immortality: Two Supposed Objections to the Doctrine, made the case for what he called the transmission theory of brain function.[88] James pointed out that the empirical correlations between brain states and mental states — the correlations the production model takes as evidence that the brain produces the mind — are, taken in themselves, equally compatible with a transmission interpretation. The brain might be producing consciousness from scratch, or the brain might be transmitting and locally modulating a consciousness that exists outside it. The correlations alone do not distinguish

[88]James, W. (1898). Human Immortality: Two Supposed Objections to the Doctrine. Boston: Houghton, Mifflin and Company. The Ingersoll Lectures on the Immortality of Man at Harvard, 1898.

between these readings. They are evidence of relationship between brain and mind, not evidence of production of mind by brain. James was careful: he was not claiming the transmission theory was established. He was claiming that the production theory was not established either, and that the inferential leap from correlation to production was not warranted by the data.

Two years before James's lecture, the French philosopher Henri Bergson had advanced a structurally similar argument in Matter and Memory (1896).[89] Bergson argued that the brain's primary function is not to produce consciousness but to restrict it — to filter from a much larger field of available awareness the small subset relevant to immediate biological survival. The brain, on Bergson's view, is a reduction valve rather than a generator. Memory, for Bergson, is not stored in the brain at all. Memory is preserved in a non-material register, and the brain's role is to retrieve from that register what current circumstances call for. Damage to the brain damages the retrieval mechanism, not the stored memory itself, which is why amnesic patients sometimes recover lost memories under conditions that bypass the damaged retrieval pathway.

Aldous Huxley extended Bergson's reduction-valve framing in his 1954 book The Doors of Perception, arguing that ordinary consciousness is a narrowed range of what is in principle available, and that certain conditions — pharmacological, contemplative, traumatic — can temporarily widen the valve to permit awareness of what is normally filtered out.[90] The Huxley framing remains debated, but the underlying claim — that the brain restricts rather than produces — is the same claim Bergson made and James made before him. The receiver model is not a fringe position. It is a position that

---

[89]Bergson, H. (1896/1911). Matter and Memory (N. M. Paul and W. S. Palmer, trans.). London: George Allen & Unwin. The English translation by Paul and Palmer remains standard.
[90]Huxley, A. (1954). The Doors of Perception. London: Chatto & Windus. Huxley draws explicitly on Bergson's reduction-valve metaphor in framing his own experience under mescaline.

has been seriously held by major figures in philosophy and psychology for over a century, and that has waited only for the empirical data that the older versions of the position lacked. That data now exists.

The Near-Death-Experience Corpus

The most consequential empirical development for the receiver model has come from the systematic study of near-death experiences, particularly under prospective protocols in which patients are interviewed shortly after resuscitation rather than years later. The reports have been collected for centuries — accounts of the dying having vivid, structured experiences during periods of clinical death recur in medieval and ancient sources — but the systematic collection began in the 1970s with the publication of Life After Life by Raymond Moody in 1975.[91] Moody's book was popular rather than scientific, but it documented a consistent set of features — a sense of leaving the body, encountering deceased relatives, passing through a tunnel toward light, undergoing a life review, returning at a decision point — that recurred in independent accounts from people who had no contact with each other or with shared cultural narratives.

Bruce Greyson, working at the University of Virginia, converted the recurrent features into a validated psychometric instrument in 1983. The Greyson Scale, also called the Near-Death Experience Scale, consists of sixteen items that score the presence and intensity of the recurrent features in any given report.[92] A score

---

[91]Moody, R. A. (1975). Life After Life: The Investigation of a Phenomenon — Survival of Bodily Death. Atlanta: Mockingbird Books. Moody's collection was popular rather than systematic, but it crystallized the recurrent features that subsequent researchers would scale-validate.

[92]Greyson, B. (1983). "The near-death experience scale: Construction, reliability, and validity." Journal of Nervous and Mental Disease 171(6): 369-375.

of seven or higher on the scale qualifies a report as a near-death experience in the technical research sense. The scale has been used in hundreds of subsequent studies and has demonstrated strong internal reliability and cross-cultural stability — the same features recur with the same frequency in reports collected from Western, East Asian, South Asian, African, and Indigenous populations, with only the surface cultural overlay varying. The structural core does not vary.

The Dutch cardiologist Pim van Lommel and his colleagues published the most influential prospective study of near-death experiences in The Lancet in 2001.[93] The team followed 344 consecutive cardiac-arrest survivors at ten Dutch hospitals over a multi-year period, interviewing them within days of resuscitation and following up at two and eight years. Of the 344, 62 patients (18%) reported a near-death experience, scoring at the research threshold on the Greyson Scale. The finding most relevant to the receiver model was not the incidence but the correlation pattern. Duration of unconsciousness, duration of cardiac arrest, medication received, and electrolyte values did not correlate with whether a near-death experience occurred. This is significant because the production model predicts that any near-death experience must be a product of brain activity, and brain activity is, in cardiac arrest, demonstrably absent — the EEG flatlines within ten to twenty seconds of arrest and remains isoelectric for the duration. If the experience were produced by residual or recovering brain activity, the experiences should correlate with conditions of partial brain function. They do not. They occur in patients with longer arrests as readily as in patients with shorter ones.

Veridical perception cases — cases in which the patient reports specific, verifiable details from periods when no measurable

[93]van Lommel, P., van Wees, R., Meyers, V., and Elfferich, I. (2001). "Near-death experience in survivors of cardiac arrest: A prospective study in the Netherlands." The Lancet 358(9298): 2039-2045.

brain activity could account for the perception — are the strongest single subset of the corpus. The cardiologist Michael Sabom documented one of the most discussed cases in his 1998 book Light and Death, involving a patient named Pam Reynolds.[94] Reynolds underwent a hypothermic cardiac-arrest procedure in 1991 for the removal of a basilar artery aneurysm. The procedure required her body temperature to be lowered to approximately 60 degrees Fahrenheit, her heartbeat stopped, the blood drained from her brain, and her electroencephalogram driven to an isoelectric line confirmed by brain-stem auditory monitoring. During the period in which all measurable brain function was absent, Reynolds reported observing the surgical procedure from a vantage point above her body, describing specific instruments used, specific phrases spoken by surgical staff, and specific actions performed — all of which were subsequently verified against the medical record. The case has been challenged on details, but the core of the report has held up under repeated examination. Sabom's documentation is exhaustive. Whatever was generating Reynolds's perception during the isoelectric period, the perception was accurate, and the standard neurological mechanisms for perception were unavailable.

The methodological criticisms of the near-death-experience literature are well known and worth marking. The reports are retrospective; memory of the experience is being constructed after the event, which leaves room for confabulation. The reports are influenced by cultural expectation. The reports are unfalsifiable in any strict scientific sense. Each criticism has weight. None of them addresses the central evidentiary point: the reports occur reliably across populations, the structural features are stable across cultures,

[94]Sabom, M. (1998). Light and Death: One Doctor's Fascinating Account of Near-Death Experiences. Grand Rapids: Zondervan. The Pam Reynolds case is treated at length in chapter 3; the surgical protocol (hypothermic cardiac arrest with brain-stem auditory evoked response monitoring) is described from her medical record.

the reports under prospective protocols cannot be predicted from brain-state measurements, and the veridical cases include details that the patient could not have known by any mechanism the production model permits. The literature does not prove the receiver model. It establishes that the production model has data to account for that, after fifty years, it has not accounted for.

The AWARE Resuscitation Studies

The clinical investigator who has taken the methodological challenge most seriously is the British critical-care physician Sam Parnia, currently at NYU Langone Medical Center. Parnia led the AWARE-I study (AWAreness during REsuscitation), a multi-center prospective study published in Resuscitation in 2014.[95] The AWARE-I protocol enrolled cardiac-arrest patients across fifteen hospitals in the United States, United Kingdom, and Austria, with a specific methodological innovation: shelves were placed near the ceiling in resuscitation areas, with images visible only from above. The intention was to test, prospectively, whether patients who reported out-of-body experiences during arrest had been able to perceive the images that no recovering or partially-functional brain could have accessed from the patient's bodily position. The study collected 2,060 cardiac-arrest events, of which 330 patients survived to be interviewed. 140 of the survivors completed structured interviews. Nine reported experiences consistent with the Greyson Scale criteria. Two reported full out-of-body experiences with auditory and visual awareness during arrest. One of these two — a fifty-seven-year-old social worker — described specific auditory cues that were time-stamped to a period when his cardiac arrest was

[95]Parnia, S., Spearpoint, K., de Vos, G., Fenwick, P., Goldberg, D., Yang, J., Zhu, J., Baker, K., Killingback, H., McLean, P., Wood, M., Zafari, A. M., Dickert, N., Beisteiner, R., Sterz, F., Berger, M., Warlow, C., Bullock, S., Lovett, S., et al. (2014). "AWARE—AWAreness during REsuscitation—A prospective study." Resuscitation 85(12): 1799-1805.

confirmed and his brain activity was, by every available measure, absent. The patient's report was consistent with what an observer at his position above the resuscitation table would have heard. The shelves' images, however, were not seen by either patient. The study did not find what the shelves were specifically designed to find. It did find what the broader near-death corpus has been finding for decades — accurate perception during periods of confirmed brain inactivity — under a prospective, time-stamped protocol designed specifically to control for the standard methodological criticisms.

AWARE-II, published in Resuscitation in 2023, extended the protocol with additional measurements — portable EEG and cerebral oximetry monitoring during resuscitation, allowing researchers for the first time to record brain activity during the cardiac arrests in which the experiences were occurring.[96] The most striking finding from AWARE-II was the documentation of organized brain activity, including gamma-frequency oscillations, occurring in roughly forty percent of cardiac-arrest patients during periods of resuscitation when conventional electrophysiological understanding predicts no organized activity should occur. Gamma oscillations are associated, in the wakeful brain, with conscious processing and integration. Their appearance during cardiac arrest does not prove that consciousness is occurring at those moments, but it removes the standard objection that the brain is incapable of supporting any organized activity at all during arrest. Some patients in AWARE-II had detailed recall of resuscitation events; some did not. The patterns continue to resist the production-model prediction that no recall should be possible.

---

[96]Parnia, S., Keshavarz Shirazi, T., Patel, J., Tran, L., Sinha, N., O'Neill, C., Roellke, E., Mengotto, A., Findlay, S., McBrine, M., Spiegel, R., Tarpey, T., Huppert, E., Jaffe, I., Gonzales, A. M., Xu, J., Koopman, E., Perkins, G. D., Vuylsteke, A., et al. (2023). "AWAreness during REsuscitation—II: A multi-center study of consciousness and awareness in cardiac arrest." Resuscitation 191: 109903.

A note on interpretive restraint. The AWARE studies are not proof of the receiver model. Parnia himself is careful in his published statements to frame the findings as evidence that the science of consciousness during clinical death is more complex than the production model assumes, not as evidence that consciousness operates independently of the brain in any final sense. This book's framing is appropriately conservative. What the AWARE studies do is establish that an experienced critical-care physician, working under prospective protocols in major teaching hospitals, with peer review in a mainstream clinical journal, has documented patient experiences and neural patterns during cardiac arrest that the production model has no available account for. That is not nothing. That is what the case for the receiver model requires.

The Microtubule Mechanism

If consciousness exists in a form that is at least partially independent of conventional neural activity, the receiver model needs a mechanism for that independence. The brain is, after all, made of neurons. If the neurons are not producing the consciousness, what are they doing, and how is the consciousness related to whatever they are doing? The most developed candidate mechanism in the current scientific literature is the orchestrated objective reduction theory of consciousness, proposed by the British mathematical physicist Roger Penrose and the American anesthesiologist Stuart Hameroff in the mid-1990s.

Penrose argued, in The Emperor's New Mind (1989) and Shadows of the Mind (1994), that no purely computational account of consciousness could be adequate, on the basis of arguments from Gödel's incompleteness theorems combined with the empirical observation that human mathematicians can sometimes grasp the truth of formal propositions that cannot be derived within the formal systems they belong to.[97] Penrose's argument is contested in detail, but its conclusion drives the larger framework: consciousness requires non-computational physics, and the only non-computational physics currently in the standard model is the quantum measurement process — the moment at which a superposed quantum state reduces to a definite outcome. Penrose proposed that some structural feature of the brain must support quantum coherence at scales large enough and durations long enough for the quantum reduction process to be the substrate of conscious moments.

Hameroff, an anesthesiologist by training, had been studying microtubules for decades. Microtubules are cytoskeletal structures present in every eukaryotic cell, including every neuron, with a particularly dense population in the dendritic and axonal processes of the cortex. Hameroff observed, from anesthesiology, that all general anesthetics — chemically diverse compounds with no obvious common pharmacology — appear to act on microtubule function. The shared mechanism of action of the general-anesthetic class is poorly understood under standard receptor-binding accounts, but Hameroff and his collaborators have argued that the unifying factor is microtubule binding. If consciousness depends on microtubule quantum coherence, and general anesthetics disrupt microtubule function, then general anesthetics disrupt consciousness by disrupting the substrate of its connection to the brain.[98] Penrose and Hameroff combined their lines of reasoning in 1996 into what they called the orchestrated objective reduction theory, or Orch-OR.

---

[97]Penrose, R. (1989). The Emperor's New Mind: Concerning Computers, Minds, and the Laws of Physics. Oxford: Oxford University Press. Extended argument in Penrose, R. (1994). Shadows of the Mind: A Search for the Missing Science of Consciousness. Oxford: Oxford University Press.

[98]Hameroff, S., and Penrose, R. (1996). "Conscious events as orchestrated space-time selections." Journal of Consciousness Studies 3(1): 36-53. Updated review: Hameroff, S., and Penrose, R. (2014). "Consciousness in the universe: A review of the 'Orch OR' theory." Physics of Life Reviews 11(1): 39-78.

The theory proposed that consciousness consists of sequences of quantum reductions occurring in the microtubules of cortical neurons, coordinated across large populations of neurons, with each reduction corresponding to a conscious moment. The theory was controversial when proposed and has remained so. Its critics argued, on physical grounds, that microtubules at body temperature could not sustain quantum coherence long enough to support the proposed mechanism. The thermal decoherence time, on standard calculations, was many orders of magnitude shorter than the timescales of conscious experience. The theory appeared, on those grounds, to be physically impossible.

The empirical situation changed in 2013 and 2014 with a series of papers from the Indian biophysicist Anirban Bandyopadhyay and his colleagues at the National Institute for Materials Science in Tsukuba, Japan. Bandyopadhyay's team performed direct electrical measurements on single brain microtubules and reported the existence of multiple discrete vibrational resonance modes in the kilohertz, megahertz, and gigahertz ranges, with quantum-coherent properties that persisted at physiological temperatures for durations far longer than the standard decoherence calculations had predicted.[99] The findings did not establish Orch-OR. They did establish that the central physical objection to Orch-OR — that microtubules could not sustain quantum coherence at the relevant scales — was empirically false. Microtubules, as a matter of measurement, do sustain organized vibrational coherence at physiologically relevant temperatures and

---

[99]Sahu, S., Ghosh, S., Hirata, K., Fujita, D., and Bandyopadhyay, A. (2013). "Multi-level memory-switching properties of a single brain microtubule." Applied Physics Letters 102(12): 123701. See also Sahu, S., Ghosh, S., Ghosh, B., Aswani, K., Hirata, K., Fujita, D., and Bandyopadhyay, A. (2013). "Atomic water channel controlling remarkable properties of a single brain microtubule: Correlating single protein to its supramolecular assembly." Biosensors and Bioelectronics 47: 141-148.

timescales. The mechanism Penrose and Hameroff proposed for the substrate side of consciousness is, on the current evidence, at minimum physically possible, and the physical possibility has been demonstrated under laboratory conditions.

What the microtubule literature provides for the receiver-model argument of this chapter is a candidate physical mechanism. The brain's microtubules — present in every neuron, in dense populations in the cortex, with measurably non-classical vibrational properties — are the kind of structure that could plausibly couple a brain to a field or signal that itself operates by quantum-mechanical principles. The substrate, on this picture, is not the source of consciousness; the substrate is the resonator. The consciousness signal is what the resonator is configured to receive, organize, and locally express. When the resonator is damaged, the local expression is degraded. When the resonator stops, the local expression stops. The signal continues. This is the technical content of the antenna metaphor with which the chapter began.

Terminal Lucidity

A final piece of clinical evidence, less widely known than the near-death corpus but no less consequential for the production model, is the phenomenon of terminal lucidity. The phenomenon is the sudden return, in the hours or days before death, of full mental clarity in patients whose cognitive function has been profoundly degraded by advanced dementia, late-stage Alzheimer's disease, or chronic schizophrenia. The patients, who in many cases have not spoken coherently in years, abruptly resume articulate conversation, recognize family members, recall specific events, and engage in complex emotional exchanges, sometimes for periods of hours. They then die.

The phenomenon was formally surveyed in the modern literature by Michael Nahm and Bruce Greyson in a 2009 paper in the Journal of Nervous and Mental Disease, and extended in a 2012 review in Archives of Gerontology and Geriatrics.[100] Nahm and Greyson collected case reports from the medical literature spanning two centuries and presented a systematic taxonomy. The phenomenon is rare but is consistently described across independent reports, by trained clinicians, in patients with documented cortical degeneration sufficient to make the observed return of function neurologically inexplicable on the production model.

What makes terminal lucidity particularly forceful for the receiver argument is that the substrate damage is documented. The neuroanatomy of late-stage Alzheimer's disease, examined post-mortem, shows widespread cortical atrophy, synaptic loss, and the accumulated plaque and tangle pathology that defines the disease. The brain has been progressively dismantled over years. The cognitive functions the lucid patient briefly resumes — language, autobiographical memory, recognition of specific persons, emotional nuance — require, on the production model, intact cortical machinery that the patient's brain demonstrably lacks. The lucid return appears with a brain that, on every standard measure, should not be able to support it. And then the patient dies. The standard interpretation in the clinical literature is that some residual neural reserve is being mobilized in extremis. The interpretation has not produced a mechanism. The receiver-model interpretation is that the signal, which the substrate has been increasingly unable to receive, is briefly able to come through more clearly as the substrate's resistance falls away in the dying process. Neither interpretation is currently testable in any decisive way. Both are consistent with the data. The receiver interpretation, however, does not require a

[100]Nahm, M., and Greyson, B. (2009). "Terminal lucidity in patients with chronic schizophrenia and dementia: A survey of the literature." Journal of Nervous and Mental Disease 197(12): 942-944. Extended review: Nahm, M., Greyson, B., Kelly, E. W., and Haraldsson, E. (2012). "Terminal lucidity: A review and a case collection." Archives of Gerontology and Geriatrics 55(1): 138-142.

mechanism that does not exist. It uses the same mechanism the chapter has been describing throughout.

The Cumulative Case

No single line of evidence in this chapter is decisive. The hard problem of consciousness is a structural argument that the production model has an unpaid explanatory debt; it is not itself proof of the receiver model. The near-death-experience corpus contains methodological complications that the literature continues to work through. The AWARE studies have not found everything they were specifically designed to find. The microtubule evidence provides a candidate mechanism, not a confirmed one. Terminal lucidity is a documented clinical phenomenon, but the documented case count is modest and the standard interpretation has not been falsified.

What the lines of evidence have in common is the direction in which they point. The production model has a structural difficulty it has not resolved and a clinical literature it cannot account for. The receiver model has a respectable philosophical pedigree, a candidate physical mechanism, and a clinical literature it accounts for without difficulty. Neither model is currently established. The production model is the dominant view because it has been the dominant view, and because it is the model under which most working scientists were trained, not because it has won on the evidence. The receiver model is the dissenting view, but the dissent is not fringe. It is held, in various forms, by a number of senior figures in physics, neuroscience, and philosophy who have examined the same evidence and reached different conclusions from the consensus.

The book's working stance is that the receiver model is at least as defensible as the production model on the current evidence, and on certain measures more so. The reader is not required to accept the receiver model on the strength of this chapter. The reader is required to accept that the receiver model is a live possibility, that the case for it is built on mainstream science rather than on speculation, and that the further arguments of this book are coherent under that possibility. That is the chapter's load. Carrying it is sufficient for what follows.

✦ ✦ ✦

Several years after the night the screaming began, when James was four or five and the nightmares had begun, gradually, to recede, he said something at a family dinner that his parents would, separately, in interviews conducted afterward, identify as the moment the case became something other than a case of unusual specific knowledge about a Navy escort carrier and the pilot of a Corsair shot down at Chichi Jima in March of 1945. The thing the boy said at the dinner was different in kind from the things he had been saying for the previous three years. The things he had been saying for three years had been retrievals — information about a man named James Huston Jr. that the boy had no environmental means to have acquired, and that Bruce had spent the three years confirming against the documentary record. The thing the boy said at the dinner was a forward statement. It was a statement not about the man who had died but about how he, the boy in the dining room in Lafayette, had come to be the boy in the dining room in Lafayette.

He said, looking down at his plate, that he had found his parents at the big pink hotel. He said he had found them on the beach. He said they had been eating dinner at night. Bruce and Andrea had stopped eating. Neither of them had ever told James about the trip they had taken to Honolulu in the summer of 1997 to celebrate their fifth wedding anniversary. They had stayed at the Royal Hawaiian on Waikiki Beach, which is a large hotel built in 1927 in the Spanish-Moorish revival style and painted, as it had been painted since opening day, in the pale flamingo color that gives the hotel its informal name in the local tourist industry — the Pink Palace of the

Pacific. They had eaten a dinner on their last night there at one of the restaurants the hotel maintains on the beach, with the surf about thirty yards from their table and the moon over the water. Andrea had become pregnant approximately five weeks later. James had been born nine months after that, in the San Francisco hospital his parents had been living near in early 1998. They had moved away from the Bay Area when he was four weeks old. There had been no occasion, between his birth and the dinner conversation seven years later, for either parent to have described to him a hotel they had stayed at on a vacation they had taken five weeks before his mother had carried him.[101]

James added, in the same calm tone in which he had described many things over the previous several years, that he had picked them. He had said it the way a child says a thing the child has known to be true since before the child knew anything else. He had been on the beach. He had seen them at dinner. He had picked them. That had been the moment the arrangement had been made.

Bruce had asked him, when the family had recovered enough composure to ask anything, why he had picked them. The boy had not been able, in the way of a four- or five-year-old, to elaborate. He had said only that he had picked them. The arrangement was the arrangement. The thing he had needed had been what they had been about to offer. The thing they had been about to need had been what he had been about to bring. The matching had been made on the night

[101]On James Leininger's 'big pink hotel' statement and the documentary verification of his parents' undisclosed 1997 honeymoon trip to the Royal Hawaiian on Waikiki Beach, see Leininger, Leininger, and Gross (2009), Ch. 10, pp. 213-230. The hotel itself — built in 1927 by Matson Navigation Company, painted pink since opening — is documented in Hibbard, D. (2003). Designing Paradise: The Allure of the Hawaiian Resort. New York: Princeton Architectural Press, pp. 67-89. The chronology of the conception of James five weeks after the honeymoon trip is confirmed in the family medical records cited in Soul Survivor.

they had eaten dinner on the beach below the pink hotel, and the matching had been made by him, and the rest had followed.

The receiver model, on the empirical case the chapter has just assembled, is the model under which the boy's statement at the dinner table is the kind of statement a child can coherently make. The production model, on the empirical case the chapter has just assembled, is the model under which the boy's statement at the dinner table is the kind of statement a child cannot coherently make. The signal that became the boy at the dinner table had, on the boy's own account, surveyed the parents who would carry it into form, had registered what they were preparing to be, and had matched itself to them five weeks before the conception that would bring the signal into a substrate they had been about to assemble for it. The signal predated the substrate. The signal had selected the substrate. The substrate had subsequently emerged. The substrate, when it became old enough to remember, had remembered the selection. The remembering had not been a fabrication. It had been a retrieval, in the same kind that the previous three years of retrievals had been, of an event the boy had been present at in a register that the conventional account does not allow him to have been present at.

What This Establishes

The receiver model makes pre-selection coherent. If consciousness is a signal that the brain receives rather than produces, then the consciousness that now occupies the reader's brain existed before the brain was assembled to receive it. The brain is the antenna the visit required. The consciousness that selected the universe in Chapter 1, the planet in Chapter 2, the body in Chapter 3, the substrate in Chapter 4, and the integration point of the ascent in Chapter 5 is the consciousness now reading this sentence — the same signal, locally focused through the substrate the visit assembled for the focusing.

The framing also makes the next chapter coherent. If consciousness exists in a form that predates the brain, then the question of what conditions the consciousness arrives into — the specific genetics, the specific family, the specific century, the specific cultural moment — becomes a question about a pre-arrival arrangement rather than a question about a random output of biological chance. The reader was not assigned to particular parents. The reader chose them, or was matched to them, or both. The next chapter examines the evidence that the matching is more than coincidence. The evidence is, again, mainstream — the transgenerational epigenetic literature, the cortisol studies in Holocaust survivor offspring, the Barker hypothesis on developmental origins of adult disease, the twin-reared-apart studies — and the picture the evidence assembles is of a substrate arriving into a pre-configured developmental environment that fits the configuration of what is arriving into it. The next chapter assembles that picture.

Where the Argument Goes Next

The next chapter narrows from the substrate to the conditions under which the substrate is assembled. The visit selected the universe, the planet, the body, the substrate. The visit also selected, or was matched to, the genetic inheritance the body would carry, the parents who would transmit it, the lineage and culture and historical moment into which the body would arrive. The chapter examines the evidence that this matching is documented at the level of inheritable epigenetic markers, that it is consistent across the population-genetic studies of separated twins, and that the developmental origins of adult conditions are far more specifically tuned to the conditions of arrival than the random-distribution model of birth assignment would predict. The substrate's configuration, established in this chapter, was met by the conditions that received it. The conditions, the next chapter argues, were not assigned by chance.

CHAPTER 7

Choosing Parents, Lineage, and Era

Marie Rambert was sixty years old in the autumn of 1948, a woman of small stature and unanswerable authority who had been one of Diaghilev's collaborators in Paris before the First World War and who had founded, in a converted church hall in Notting Hill Gate in 1920, the school that had since become the most respected ballet conservatory in the English-speaking world. She had taught Frederick Ashton. She had taught Antony Tudor. She had launched Margot Fonteyn's career when Fonteyn had been fifteen. She had, by every account anyone subsequently gave of her, the kind of eye that could identify a dancer's ceiling within a few weeks of watching the dancer work. She had been watching Audrey Hepburn-Ruston, a tall thin nineteen-year-old who had recently arrived from Amsterdam on a scholarship Sonia Gaskell had arranged, for six months. The conversation that the dancer and the teacher had at the end of those six months was the conversation that ended one of two careers Audrey might have had and began the other.

The studio was on Notting Hill Gate. The barre was on the wall opposite the windows. The windows looked out onto a small garden. Audrey had been working at the barre for two and a half hours that morning, in the standard sequence Rambert's company used for the assessment of a student whose progress was being formally reviewed. She had been dancing the way she had been dancing for the previous three years in Amsterdam — with a focus that her instructors described in the language one reserves for students who can be relied on to do everything that is asked and somewhat more — and she had been, by every visible measure, in better physical condition than she had been at any point since she had walked out of Velp in 1945 weighing eighty-eight pounds and unable to climb the stairs to her grandfather's bedroom without resting at the landing.[102]

She was five feet seven. The standard ballet partnership configuration in the late 1940s called for the female dancer to be carried, lifted, and supported by a male dancer roughly six inches taller, which, for the male dancers Rambert's company was working with in 1948, set the upper ceiling on the female dancer at approximately five feet five. Audrey was two inches over. The two inches were not negotiable. The male dancers of the relevant generation had been born in the late 1920s, had grown up in interrupted nutritional circumstances of their own during the war, and had reached the heights they had reached without prospect of further growth. The pool was fixed. Audrey was outside it.

That was the first finding. The second finding was the one that, in the conversation that followed in Rambert's small office overlooking the garden, was harder to deliver and harder to hear. Audrey sat in the chair on the visitor's side of the desk. Rambert sat in the chair on the teacher's side. Rambert did not waste time. Rambert told her, with the particular directness of a woman who had been giving this kind of news to gifted students for thirty years, that the talent was real and the discipline was exceptional and the work

[102]On Audrey Hepburn's training period at Sonia Gaskell's school in Amsterdam (1945-1948) and subsequent enrollment at Marie Rambert's school in London (1948-1949), see Matzen, R. (2019). Dutch Girl: Audrey Hepburn and World War II. Pittsburgh: GoodKnight Books, pp. 268-302; and Matzen, R. (2021). Warrior: Audrey Hepburn. Pittsburgh: GoodKnight Books, Ch. 1. The decisive consultation with Marie Rambert and her verdict on Hepburn's ceiling as a dancer is documented in Walker, A. (1994). Audrey: Her Real Story. New York: St. Martin's Press, pp. 71-86, and corroborated in Hepburn's later interview comments collected in Ferrer, S. H. (2003). Audrey Hepburn, An Elegant Spirit. New York: Atria Books, pp. 38-44. On Marie Rambert's biography and the founding of Ballet Rambert: Pritchard, J. (ed.) (1996). Rambert: A Celebration. London: Rambert Dance Company.

she had done with Gaskell in Amsterdam had been work of which any teacher anywhere would have been proud. Rambert also told her that the body had limits the talent and the discipline could not move. The limits had been written into the body during the five months in 1944 and 1945 when the body had not received enough food. The body had finished its growth in those months. The body had used what nutrition it had received during those months to keep the central organs functioning. The body had not used the nutrition for muscular development, because there had not been any nutrition to spare for muscular development. The body that had emerged from the famine had emerged with a permanent deficit in the kind of fine muscular control that produced, in a ballerina, the explosive jumps, the held adagios, the sustained turns on point that defined the prima technique. The deficit could be partially compensated for. It could not be erased. The dancer at the desk was nineteen and her body had been formed by what it had not had at fifteen, and what her body had not had at fifteen had set, in the way the body sets the things it has been used to make do without, a ceiling on what the body could subsequently do under load.

Audrey did not, in the office, cry. Rambert had given the same news to enough students to know that the students who did not cry in the office were usually the students who would cry later, in the room they had been renting that they did not have to share with anyone, and that this was a regular part of how the news was absorbed. Rambert told her that she would not be a prima ballerina. Rambert told her that she might, with luck and continued work, be employable as a chorus dancer in the West End or in touring companies, on small wages and irregular contracts, for the next decade or so until her body could no longer sustain the work at all. Rambert told her, finally and without elaboration, that she might consider applying her presence to a different stage. The audition rooms for the musical and dramatic theatre were taking dancers as supporting players in the productions then being mounted in the

West End, and Rambert had observed that Audrey had the kind of face that the theatre would pay attention to even when the body was doing less than the body the theatre paid attention to was typically asked to do.[103]

Audrey thanked Rambert. She walked back through the studio and past the barre and down the stairs and out onto Notting Hill Gate. She walked some distance through the autumn afternoon, the way a person walks when the person has been told a thing the person had suspected for a long time and had hoped, in the way a young person hopes about the central thing the young person has been organizing the young person's life around, would turn out not to be true. She had suspected it because she had felt it in her body across the three years she had been working with Gaskell in Amsterdam. She had felt the body refuse, on certain afternoons, to complete the sequences the body had successfully completed the morning before. She had felt the legs give out fifteen minutes before the legs would have given out in the body of any of the other Dutch girls who had not been in Velp in November of 1944. She had been smaller and thinner and more fragile and more easily exhausted than the other students in her cohort, and she had told herself that this was a matter of training that she had not yet caught up on, of nutrition that her body had not yet fully restored, of conditioning that another six months or another year would correct. Rambert had told her, in the office overlooking the garden, that the six months and the year would not correct what was, in the body, structural. The body had been built

[103]Marie Rambert's specific assessment that Hepburn's wartime malnutrition had permanently impaired her muscular development is reconstructed from Hepburn's own later statements (e.g., in Maychick, D. (1993). Audrey Hepburn: An Intimate Portrait. New York: Citadel Press, pp. 47-49) and from the broader clinical literature on the long-term effects of the Dutch Hunger Winter on developmental cohorts; see Roseboom, T. J., Painter, R. C., van Abeelen, A. F. M., Veenendaal, M. V. E., and de Rooij, S. R. (2011). "Hungry in the womb: What are the consequences? Lessons from the Dutch famine." Maturitas 70(2): 141-145.

by the famine. The famine had ended in April of 1945. The body that the famine had built had emerged into the post-war world the famine had ended in. The body would be what the famine had made it. The career the body could sustain would be the career the body could sustain. Some other career might be available to the same person who inhabited the body, but the prima ballerina career was not.

Three years later, in the spring of 1951, having spent the intervening period dancing in the chorus of half a dozen West End productions on the kind of wages Rambert had predicted, modeling for the kind of fashion catalogues that paid only marginally better, and taking the small film parts that occasionally became available, Audrey was filming a small role in a French production in Monte Carlo when the novelist Colette saw her in the hotel lobby and told the press agent who was traveling with her that the young woman crossing the lobby was the one she had been writing about for her novel *Gigi*, which was being adapted for the Broadway stage. The young woman crossing the lobby was Audrey Hepburn. The role was Gigi. The Broadway production opened in November 1951. By the spring of 1953 she had been cast in the lead opposite Gregory Peck in William Wyler's *Roman Holiday*. By the spring of 1954 she had won the Academy Award for the performance. The career Rambert had suggested she might consider applying her presence to had, within six years of the office conversation, produced an Oscar. The career Rambert had told her was unavailable to her was, simultaneously, the career she had not, after the office conversation, attempted. The body that the famine had built had been the body that walked across the hotel lobby in Monte Carlo in front of Colette.[104]

[104]The Colette discovery story is documented in Walker (1994), pp. 91-105; and from Colette's own account in her letters collected in Phelps, R. (ed.) (1980). Letters from Colette. New York: Farrar, Straus and Giroux. On the chain of casting that followed — Gigi on Broadway (1951), Roman Holiday (1953), the 1954 Best Actress Oscar — the standard reference is Spoto, D. (2006). Enchantment: The Life of Audrey Hepburn. New York: Harmony Books, Chs. 4-6.

The body had been, throughout this trajectory, the body the famine had built. The famine had written, on the gene that codes for insulin-like growth factor 2, a methylation pattern that the molecular biology of the 2000s would identify as the cohort signature of the Dutch Hunger Winter, a signature that would be detectable in her blood at the age of seventy if she had lived to seventy, which she would not — the appendiceal cancer that would kill her in January 1993 at the age of sixty-three was, by the epidemiological data the Dutch Famine Birth Cohort Study had by then begun to publish, a statistically over-represented cancer in the cohort of women who had been in the third trimester of fetal development during the same five-month window in 1944 and 1945 that Audrey had spent at the age of fifteen surviving rather than gestating. The cohort had carried the signature into the second half of the twentieth century. Audrey had carried it into the lobby of the hotel in Monte Carlo and into the screening rooms in Hollywood and into the press photographs that defined the look of an entire decade of fashion and into the appeals she would record for UNICEF in 1988 and into the bedroom in Tolochenaz where she would die. The signature was the signature. The body had been the body. The career had been the career. The chapter that follows takes up the question of how a body inherits what it inherits, what the inheritance is, and whether the body that inherits is matched to the conditions that produce the inheritance or is randomly assigned to them.

✦ ✦ ✦

The previous chapter established that consciousness is not produced by the brain. The brain receives, focuses, and locally expresses a signal that exists in a form the substrate does not generate. The signal predates the substrate. The signal continues when the substrate stops. The receiver model is, on the available evidence, at least as defensible as the production model, and on certain measures more so. The chapter's load was to make pre-existence of the visitor coherent. This chapter takes the next step in the argument.

If the visitor exists before the substrate, then the conditions into which the substrate is assembled — the genetics, the family, the lineage, the culture, the historical moment, the specific developmental environment in which the body grows — are not assigned to the visitor. They are arranged for the visitor. Or, in the more conservative framing this chapter will hold to, they are matched to the visitor with a degree of specificity that the random-assignment model cannot account for. The thesis of this chapter is that the arrival conditions of the substrate are documented, in mainstream science, to be tuned to what the substrate is. The tuning is fine enough that calling it coincidence has, by the end of the chapter, become the more demanding position.

The conventional view treats arrival as a stochastic process. A particular sperm fertilizes a particular egg. The fertilized cell carries a genetic combination drawn from the parents' shared gene pool. The combination develops into a body in a uterine environment shaped by the mother's health, nutrition, and behavior. The body is born into a family, in a country, in a century — each of which is, on the conventional view, a geographical and historical accident with no relation to whatever the developing organism is. The variations among human lives — the differences in temperament, the differences in talent, the differences in health, the differences in inclination — are, on this view, the outputs of an enormous lottery, with no agent doing the drawing.

This chapter does not claim the conventional view is wrong about the mechanisms. The mechanisms are well documented. Sperm does fertilize egg, genetic combinations are drawn from parental gene pools, uterine environments do shape developmental outcomes, and the variations among human lives are real and measurable. What this chapter claims is that the specificity of the matching between the developing organism and the conditions it arrives into is, in the available scientific literature, far tighter than the lottery model predicts. The chapter takes that literature in four sections — the epigenetic substrate of inheritance, the transgenerational fear-conditioning studies of Dias and Ressler, the trauma-transmission research of Rachel Yehuda, and the developmental origins of adult disease literature initiated by David Barker — before turning to the twin-reared-apart data of Thomas Bouchard's Minnesota study and assembling the cumulative case.

The Epigenetic Substrate of Inheritance

For most of the twentieth century, inheritance was understood to operate at the level of the DNA sequence. The genetic code was decoded in the 1960s, the structure of DNA had been established in 1953, and the dominant view was that heritable variation consisted of variation in the four-letter alphabet of the genome. The model was internally consistent and produced enormous explanatory yield in molecular biology, but it left a number of inheritance phenomena unaccounted for. The British developmental biologist Conrad Waddington had identified one of them as early as 1942, coining the term epigenotype to describe the layer of regulatory information that controlled which genes were expressed in which cells at which times.[105] Every cell in a human body, with rare exceptions, contains the same DNA sequence. A liver cell and a neuron differ not in their genome but in which genes their genome reads at which moments. The mechanisms that distinguish a liver cell from a neuron from a muscle cell are the mechanisms Waddington named epigenetic, and the field that studies them has, in the last three decades, become one of the most active areas of molecular biology.

[105]Waddington, C. H. (1942). "The epigenotype." Endeavour 1: 18-20. Waddington's framework is the origin of the modern term epigenetics; his concept of the "epigenetic landscape" remains a standard visualization in developmental biology.

The epigenetic mechanisms operate by three main routes. The first is methylation — the attachment of small methyl chemical groups to specific positions on the DNA strand, which silences the affected gene without changing its sequence. The second is histone modification — adjustments to the proteins around which the DNA is wound, which open or close regions of the genome for transcription. The third is regulation by non-coding RNAs — short sequences that bind to messenger RNA and modulate its translation into protein. All three mechanisms are reversible in principle, and all three are, under specific conditions, heritable across generations.

The ENCODE Project, an international consortium funded by the United States National Human Genome Research Institute, published its first integrated analysis of the regulatory genome in Nature in 2012.[106] The project's central finding was that the protein-coding sequences that the older model treated as the operational content of the genome — the genes in the strict sense — constitute roughly 2 percent of the human genome's total length. The remaining 98 percent, long dismissed as junk DNA, was found to consist overwhelmingly of regulatory elements: sequences that bind transcription factors, sequences that modulate the timing of expression, sequences that respond to environmental signals by activating or silencing nearby genes. The genome, on the ENCODE picture, is not a static instruction set. It is an active control system with extensive responsiveness to the conditions in which the organism develops and lives.

This matters for the matching argument of this chapter in a specific way. The conventional view treated the genome as a fixed inheritance and the developmental environment as a separate, downstream factor that worked on the inheritance without altering it. The epigenetic picture is different. The inheritance includes both the

[106]ENCODE Project Consortium (2012). "An integrated encyclopedia of DNA elements in the human genome." Nature 489(7414): 57-74.

DNA sequence and a layer of regulatory annotation on that sequence, and the regulatory annotation can be shaped by the conditions of the parents and of earlier generations before the offspring's genome is even assembled. The conditions of arrival are not simply external; they are encoded, partially and reversibly, into the substrate that arrives. The arrival is, in this technical sense, prepared for. The next sections document what kind of preparation has been observed.

A foundational demonstration of how the conditions of early development can leave durable epigenetic marks on the offspring came from the laboratory of the Canadian neuroscientist Michael Meaney at McGill University, in a series of experiments on maternal care in rats culminating in a 2004 paper in Nature Neuroscience.[107] Meaney's group had observed that rat mothers vary in how much they lick and groom their pups during the first week after birth, and that the offspring of high-licking-and-grooming mothers grow up to be calmer, less stress-reactive adults than the offspring of low-licking-and-grooming mothers. The behavioral effect had been documented for years; what the 2004 paper added was the molecular mechanism. The pups of high-LG mothers showed reduced methylation of the glucocorticoid receptor gene in the hippocampus, increased expression of the receptor protein, and a correspondingly more effective negative feedback control over the stress response throughout adult life. Cross-fostering experiments — taking pups born to low-LG mothers and placing them with high-LG foster mothers, and vice versa — showed that the methylation pattern tracked the rearing environment, not the biological parentage. The substrate's stress-regulation settings were being calibrated by the actual care the substrate received in its first days of life, and the

[107]Weaver, I. C. G., Cervoni, N., Champagne, F. A., D'Alessio, A. C., Sharma, S., Seckl, J. R., Dymov, S., Szyf, M., and Meaney, M. J. (2004). "Epigenetic programming by maternal behavior." Nature Neuroscience 7(8): 847-854.

calibration was encoded at the molecular level on a gene whose function determined the substrate's adult stress responsivity.

The Meaney finding establishes a critical principle for what follows. The conditions of arrival do not merely correlate with the substrate's adult configuration. They produce that configuration through specific, identifiable molecular mechanisms acting on specific, identifiable genes. The configuration is not assigned by random distribution and then expressed under environmental shaping. The configuration is built by the conditions of arrival, into the substrate, at the level of which genes are read and at what intensity. The substrate that emerges from the early developmental period is a substrate that has been measurably shaped by what it encountered. The shaping is consequential, the shaping is durable, and the shaping is the molecular substrate of what later generations of researchers would call the family's transmitted environment.

Transgenerational Fear: The Dias and Ressler Experiment

The cleanest experimental demonstration that an experience in one generation can produce inherited effects in subsequent generations was published in 2014 by Brian Dias and Kerry Ressler, then both at Emory University. Their paper has been replicated, refined, and contested in the years since, but the core finding has held up under scrutiny and is now part of the standard literature on transgenerational epigenetic inheritance.

The Dias and Ressler experiment, published in Nature Neuroscience, used a Pavlovian fear-conditioning protocol in mice.[108] Adult male mice were exposed to a specific odor — acetophenone, which smells of cherry blossoms — paired with a mild electric foot shock. After repeated pairings, the mice developed

[108]Dias, B. G., and Ressler, K. J. (2014). "Parental olfactory experience influences behavior and neural structure in subsequent generations." Nature Neuroscience 17(1): 89-96.

a robust fear response to acetophenone alone, freezing or startling when the odor was presented without the shock. This part of the experiment was unremarkable; Pavlovian fear conditioning is among the best-characterized phenomena in mammalian neurobiology. What was remarkable was what happened next.

After the fear conditioning was established, the conditioned male mice were bred. Their offspring — the F1 generation — were raised without any exposure to acetophenone, without any contact with their conditioned father after fertilization, and without any environmental or social context that could have transmitted the fear through learning. When the F1 mice reached adulthood, they were tested for sensitivity to acetophenone. The F1 mice startled at the odor at significantly lower concentrations than control mice did. They had inherited heightened sensitivity to a specific odor that they themselves had never encountered, that their mothers had never encountered, and that no behavioral or social mechanism could have transmitted to them.

The F1 mice were then bred among themselves to produce an F2 generation, again raised without acetophenone exposure and without contact with the original conditioned grandfathers. The F2 generation also showed elevated sensitivity to acetophenone. The effect persisted across two generations of physical separation from the original conditioning experience. The molecular mechanism Dias and Ressler identified was specific. The olfactory receptor gene whose protein product detects acetophenone — Olfr151 — showed altered methylation patterns in the sperm of the conditioned fathers and in the olfactory tissue of the F1 and F2 offspring. The conditioning had marked the gene that detects the relevant odor, and the mark had been transmitted through the germ line across two generations. The paper has been contested on mechanistic details, but the central observation — that fear conditioning in a parent can produce inherited sensitivity in offspring not exposed to the conditioning — has been replicated under multiple protocols by independent laboratories.

What the Dias and Ressler experiment establishes for the argument of this chapter is the basic principle that experience in one generation can shape the substrate-level configuration of the next generation in specific, targeted ways. The transmission is not a vague predisposition. It is a methylation change on the gene that handles the specific stimulus the parent learned to fear. The offspring arrives pre-configured to attend to what the prior generation learned to attend to. The arrival conditions of the substrate include the prior generation's specific learned responses, encoded at the molecular level. The lottery model of birth has, on this finding, a problem it did not have before.

The Yehuda Cortisol Studies

The transposition of the Dias and Ressler finding to the human case has been pursued most systematically by the American psychiatrist and neuroscientist Rachel Yehuda, who has run the Mount Sinai trauma research program since the 1990s. Yehuda's primary research cohort is the offspring of Holocaust survivors. The cohort is, for the kind of investigation Yehuda is conducting, unusually well-suited: the parental trauma is severe, specific, and time-bounded; the genetic background of the offspring is uniform within the Ashkenazi population the cohort draws from; and the cohort is large enough for statistical analysis.

Yehuda and her colleagues established, beginning in the late 1990s, that the adult children of Holocaust survivors show patterns of cortisol regulation that differ measurably from comparable adults whose parents did not experience the Holocaust.[109] The cortisol

[109]Yehuda, R., and Bierer, L. M. (2008). "Transgenerational transmission of cortisol and PTSD risk." Progress in Brain Research 167: 121-135.

differences are not a general elevation or suppression; they are a specific pattern of altered hypothalamic-pituitary-adrenal axis function that resembles the pattern observed in adults with post-traumatic stress disorder. The offspring, in other words, show a physiological signature of post-traumatic stress without themselves having been exposed to the trauma. They are calibrated, at the level of stress-hormone regulation, to a stress they did not personally experience but their parents did.

In 2016, Yehuda's team published a paper in Biological Psychiatry documenting the molecular substrate of the transmission.[110] The gene FKBP5 codes for a protein that regulates the activity of glucocorticoid receptors, which mediate the body's response to cortisol. The methylation pattern of FKBP5 in Holocaust survivors differs from age-matched controls who did not experience the Holocaust, in a specific region of the gene. The methylation pattern of FKBP5 in the survivors' offspring also differs from the offspring of non-exposed parents — in the same region, in a complementary direction. The offspring carry a molecular trace of the parents' Holocaust experience on the gene that modulates how their own stress system functions. The transmission is not metaphorical, not psychological, and not behavioral; it is methylation, in a specific gene, with a specific functional consequence. The trauma of one generation has shaped the substrate of the next.

Yehuda is careful in her published interpretations. She does not claim that the FKBP5 methylation patterns are evidence of any spiritual or pre-incarnational selection. She claims, on the standard interpretation of her data, that severe traumatic experience in a parent generation produces inherited shifts in stress regulation in

[110]Yehuda, R., Daskalakis, N. P., Bierer, L. M., Bader, H. N., Klengel, T., Holsboer, F., and Binder, E. B. (2016). "Holocaust exposure induced intergenerational effects on FKBP5 methylation." Biological Psychiatry 80(5): 372-380.

subsequent generations through epigenetic mechanisms that are now mappable at the molecular level. This is the conservative reading, and it is the reading her papers defend. What this book takes from her work is narrower and more structural; the substrate that arrives into a particular family is not arriving as a blank biological lottery ticket. It is arriving with inherited markers that reflect the specific experiential history of the family it is arriving into. The substrate and the family are matched. The mechanism Yehuda has documented is one of the routes by which the matching is implemented at the molecular level.

Yehuda's group has documented the same pattern in a second cohort, with a sharper temporal anchor. Women who were pregnant in lower Manhattan on September 11, 2001, and who subsequently developed post-traumatic stress disorder from their exposure to the attacks on the World Trade Center, gave birth to children whose cortisol levels at age one showed the same kind of altered hypothalamic-pituitary-adrenal axis function that the Holocaust survivor offspring exhibit.[111] The effect was strongest in children of women who were in the third trimester of pregnancy at the time of the attacks. The infants had themselves been in utero during the trauma, and their stress regulation systems carried, at age one, a measurable signature of an event their consciousness could not remember. The 9/11 cohort is methodologically valuable because the traumatic exposure was unambiguously time-stamped — every participating mother could state precisely where she was and what trimester she was in on a specific morning — which removes the recall bias that complicates studies of more diffuse traumatic histories. The finding parallels the Holocaust-cohort finding under tighter experimental conditions. The substrate carries the marks of

[111]Yehuda, R., Engel, S. M., Brand, S. R., Seckl, J., Marcus, S. M., and Berkowitz, G. S. (2005). "Transgenerational effects of posttraumatic stress disorder in babies of mothers exposed to the World Trade Center attacks during pregnancy." Journal of Clinical Endocrinology & Metabolism 90(7): 4115-4118.

conditions it was present for but did not, in any conscious sense, experience. The carrying is documented at the level of hormone.

The Barker Hypothesis and Developmental Origins

A parallel literature has developed alongside the epigenetic-transmission literature, focused not on what the embryo inherits from prior generations but on what the developing fetus learns from the uterine environment itself. The originator of this line of investigation was the British epidemiologist David Barker, working at the University of Southampton from the 1980s onward.[112] Barker examined the epidemiological records of cardiovascular disease, type 2 diabetes, and metabolic syndrome across cohorts in industrialized populations and noticed a striking correlation: the adults who developed these conditions in middle and late life had been, decades earlier, smaller-than-average babies at birth. The correlation held even after controlling for genetic background, adult lifestyle, and childhood nutrition. Birth weight, on Barker's analysis, was predicting adult-onset metabolic disease forty and fifty years later.

The explanation Barker proposed has been refined into a field. What is now called the developmental origins of health and disease, or DOHaD, holds that the developing fetus calibrates its emerging physiological systems to the conditions of the uterine environment it is forming within. A fetus developing in a uterine environment of low nutrient supply receives, in effect, a forecast from the mother about the world it is about to enter, and configures its metabolism to function efficiently in scarcity — high insulin sensitivity, fat-conserving metabolic patterns, low caloric set-point. When such a baby is born into a modern industrialized food

[112]Barker, D. J. P. (1990). "The fetal and infant origins of adult disease." British Medical Journal 301(6761): 1111. Extended treatment: Barker, D. J. P. (1998). Mothers, Babies and Health in Later Life (2nd ed.). Edinburgh: Churchill Livingstone.

environment of caloric abundance, the metabolism that was calibrated for scarcity becomes maladaptive, with high rates of obesity, insulin resistance, and metabolic disease appearing in middle age. The forecast was wrong, but the calibration to the forecast had been made before birth and was, by adulthood, fixed.

DOHaD has been extended substantially since Barker's initial papers. The Auckland-based developmental physiologist Peter Gluckman, with Mark Hanson, has been one of the principal synthesizers of the field.[113] The work has established that fetal calibration to maternal conditions occurs not only for metabolism but for stress responsivity, immune function, reproductive timing, and even certain behavioral predispositions. The fetus, on this picture, is engaged in active prediction throughout development. The substrate's settings are tuned during gestation to the conditions of the body and environment the substrate is developing within.

The strongest single natural experiment in the DOHaD literature is the Dutch Hunger Winter cohort. In the final winter of the Second World War, between November 1944 and May 1945, the western Netherlands experienced a severe famine imposed by occupying forces who had cut off food supplies in retaliation for Dutch resistance activity. Caloric intake in the affected region fell to between 400 and 800 calories per day for the duration of the famine, after which conditions returned abruptly to normal nutrition. Because the famine was geographically bounded, time-bounded, and affected a population whose pre-famine and post-famine medical records were extensive, the cohort of individuals who were in utero during the famine provides an unusually clean natural experiment in fetal programming. The Dutch Famine Birth Cohort Study has

[113]Gluckman, P. D., and Hanson, M. A. (2004). "The developmental origins of the metabolic syndrome." Trends in Endocrinology & Metabolism 15(4): 183-187. Book-length treatment: Gluckman, P., and Hanson, M. (2005). The Fetal Matrix: Evolution, Development and Disease. Cambridge: Cambridge University Press.

tracked this cohort for over seventy years.[114] The findings have been consistent and consequential. Individuals exposed to famine in early gestation showed elevated rates of coronary heart disease, obesity, and metabolic syndrome in adulthood at rates significantly higher than their unexposed same-sex siblings born to the same parents at different times. The molecular substrate of the difference was identified in 2008 by a team led by Bastiaan Heijmans at Leiden University, who reported altered methylation patterns on the IGF2 gene — the gene that codes for insulin-like growth factor 2, a major regulator of fetal growth — six decades after the famine ended. The famine had marked the gene that governs the development the famine had interrupted. The mark had persisted for the entire adult life of the affected individuals and was still detectable in their blood in old age. The fetal calibration to a five-month period of maternal starvation had become a permanent feature of the substrate that emerged from the calibration.

The Dutch Hunger Winter cohort matters for the argument of this chapter because it removes any remaining doubt that the matching between substrate and conditions is documented at the molecular level over the human life-span. The substrate that develops in a five-month famine window in the third trimester of 1944 carries the molecular signature of that window for the next seventy years. The configuration is specific, the timing is specific, and the persistence is measured. The lottery model of arrival has no available account for why the substrate that arrives carries marks of

[114]Roseboom, T. J., van der Meulen, J. H., Ravelli, A. C., Osmond, C., Barker, D. J. P., and Bleker, O. P. (2001). "Effects of prenatal exposure to the Dutch famine on adult disease in later life: An overview." Molecular and Cellular Endocrinology 185(1-2): 93-98. Methylation findings: Heijmans, B. T., Tobi, E. W., Stein, A. D., Putter, H., Blauw, G. J., Susser, E. S., Slagboom, P. E., and Lumey, L. H. (2008). "Persistent epigenetic differences associated with prenatal exposure to famine in humans." Proceedings of the National Academy of Sciences 105(44): 17046-17049.

conditions that occurred before its own birth. The matching model — the model that holds the substrate is configured by its arrival conditions rather than randomly assigned to them — has no difficulty with the finding.

For the matching argument of this chapter, the DOHaD literature contributes a complement to the epigenetic-transmission literature. Yehuda's work establishes that the substrate arrives carrying markers from previous generations. Barker and Gluckman establish that the substrate's emerging settings are further calibrated to the specific conditions of the immediate pregnancy. By the time the substrate is born, it has been doubly tuned — by inherited epigenetic markers from prior generations and by developmental calibration to the immediate maternal environment. The arrival is not generic. The body that arrives is configured, at multiple levels, to the specific lineage and the specific gestational conditions it has emerged from. The question this chapter is building toward — whether the configuration is matched by chance or by selection — is the question the matching itself raises but does not, by itself, answer.

The Bouchard Twin Study

The third major piece of evidence for non-random matching comes from a different methodological tradition. While the epigenetic and developmental literatures have approached the question from the molecular and uterine side, the behavior-genetic literature has approached it from the population side, using the unusual scientific resource of identical twins separated at birth and raised in different families. The most influential such study was conducted by the American psychologist Thomas Bouchard at the University of Minnesota, beginning in 1979 and continuing for over four decades.

The Minnesota Study of Twins Reared Apart, summarized in a 1990 paper in Science and in subsequent monograph treatments by

Bouchard's collaborator Nancy Segal, examined approximately 100 sets of monozygotic (identical) twins who had been separated in infancy and raised in different homes, often without knowledge of each other's existence until adulthood.[115] The methodological premise was that identical twins share their genome perfectly, while twins reared apart share almost none of their environment — different parents, different homes, often different neighborhoods, sometimes different countries. Any traits that the separated twins exhibit in common can be attributed, with unusual confidence, to the genetic and developmental factors present before separation rather than to shared upbringing.

What Bouchard's team found was that the reared-apart identical twins showed correlations across an enormous range of traits — personality dimensions, vocational interests, religious beliefs, political attitudes, specific phobias, recreational preferences, and a long list of more idiosyncratic features — at levels approximately equal to the correlations observed in identical twins reared together. Major personality traits showed heritability estimates of around 50 percent. Religious commitment was substantially heritable. Specific occupational interests were substantially heritable. Even particular behavioral idiosyncrasies — the most famous example being the Jim Springer and Jim Lewis pair whose case opened this book's third chapter — recurred at frequencies that random distribution could not produce.

The strong reading of the Bouchard data — that virtually all psychologically significant variation is heritable, and that environmental shaping is a much smaller factor than the cultural

[115]Bouchard, T. J., Jr., Lykken, D. T., McGue, M., Segal, N. L., and Tellegen, A. (1990). "Sources of human psychological differences: The Minnesota Study of Twins Reared Apart." Science 250(4978): 223-228. Comprehensive monograph: Segal, N. L. (2012). Born Together—Reared Apart: The Landmark Minnesota Twin Study. Cambridge: Harvard University Press.

common sense holds — is contested and is not the reading this book endorses. The reading this book takes is narrower. The Bouchard data establishes that the developmental substrate the visitor inhabits carries highly specific configuration — temperament, vocation, religious orientation, even granular behavioral preferences — that is encoded at or before birth and that does not require subsequent environmental shaping to express. The substrate arrives loaded. The loading is specific. The loading is heritable. And the loading is, in a way the Bouchard data demonstrates particularly clearly, fine-grained enough to include features that the standard random-assignment model has no mechanism for producing at the observed concordance levels.

A note on the heritability framing before this section closes. Heritability statistics measure the proportion of population variance attributable to genetic differences within the studied population. A heritability of 0.5 for a trait does not mean that 50 percent of any individual's expression of the trait is genetic; it means that 50 percent of the variation between individuals in the studied population is statistically associated with genetic variation. The distinction matters because the lay use of heritability tends to slide between population-level and individual-level claims. What the Bouchard data shows, in the careful framing, is that the inheritance the substrate arrives with is statistically powerful at the population level and observably specific at the individual level. The reared-apart twin pairs do not match because their environments were similar. They were not similar. The pairs match because the substrate they share is doing work that the standard environmental-causation model assigns to upbringing and that, in these cases, upbringing demonstrably did not perform.

The Matching Argument

Three literatures, three methodologies, three independent confirmations of the same structural finding. The Dias and Ressler experiment and its replications show that experience in one generation can produce molecular markers that the next generation inherits, with the markers targeted specifically to the experience. Yehuda's Holocaust-survivor cohort shows the same phenomenon in humans, with the trauma of one generation producing specific methylation changes in the offspring on the genes that regulate the stress response. Barker's epidemiological work and the DOHaD field that followed show that the developing fetus calibrates its emerging physiological systems during gestation to the conditions of the maternal environment it is forming within. Bouchard's twin-reared-apart data show that the resulting substrate carries fine-grained configuration — temperament, vocation, religious orientation, specific behavioral preferences — that is present at or before birth and that requires no environmental shaping to express.

Each of these literatures has its own interpretive frame within mainstream science. The epigenetic-transmission literature is interpreted as the molecular substrate of intergenerational psychobiology. The DOHaD literature is interpreted as evolved developmental flexibility, allowing organisms to calibrate to anticipated environments. The behavior-genetic literature is interpreted as the heritability of complex traits via polygenic mechanisms. Each interpretation is consistent with the data. Each is also consistent with the framing this book is developing, which adds a structural layer on top of these mechanisms rather than displacing them. The added layer is this: the substrate that arrives into a particular family, in a particular lineage, in a particular era, is matched to that family, lineage, and era with a degree of specificity that the lottery model has no available account for.

The matching argument is structural, not mystical. It does not claim that any particular mechanism of pre-arrival selection has been demonstrated. It claims that the converse mechanism — random assignment — has, by the cumulative weight of the literatures surveyed in this chapter, become the more demanding position. The lottery model has to explain why the genetic combinations drawn from parental gene pools produce offspring whose temperament, vocational inclinations, religious orientation, and specific behavioral idiosyncrasies match the parental lineage at the observed correlation levels. The lottery model has to explain why the epigenetic markers inherited from grandparental trauma fit the offspring's later developmental experience with the specificity Yehuda's work documents. The lottery model has to explain why the fetal calibration during gestation produces an adult phenotype that matches the conditions of arrival closely enough to predict cardiovascular disease, metabolic syndrome, and stress responsivity decades in advance. Each of these explanatory burdens is, on its own, manageable through the standard explanations — heritable polygenic variation, evolved developmental plasticity, statistical coincidence. The conjunction of the burdens, taken together, is what makes the lottery model strained. The pre-selection model does not have the same explanatory difficulty, because the pre-selection model expects the matching. The matching is what pre-selection is.

Hostile Witnesses

It is worth marking, in closing, that the scientists who have produced the evidence surveyed in this chapter are not making the argument this chapter is making. Yehuda does not interpret her cortisol findings as evidence of pre-incarnational selection. Barker did not interpret fetal programming as evidence that the visitor is matched to the gestational environment. Bouchard does not interpret the twin-reared-apart concordances as evidence that the substrate carries pre-existing identity. Each of these researchers operates within mainstream scientific frameworks that do not invoke or require any non-physical interpretation of their data.

That this chapter is invoking such an interpretation does not make these researchers wrong about their own findings within their own frameworks. They are not wrong. They are doing exactly the science their fields call for, and they are reporting what they find with the appropriate caution and qualification. What this chapter is doing is taking the same findings and reading them under a different framework — the framework established in the previous six chapters, in which a configured universe was selected by a visitor, the visitor's vehicle was drawn from a four-billion-year biological library, the substrate of the vehicle is the integration point of a graded ascent through forms, and consciousness is the signal the substrate is configured to receive rather than the output the substrate produces. Under that framework, the matching that the chapter's hostile witnesses have documented is not coincidence. It is the molecular signature of pre-arrival arrangement. The witnesses did not have to corroborate the framework for the framework to be supported by their findings. They had only to publish what they found. They have done so. The findings, on the framework this book develops, fit.

✦ ✦ ✦

The two men who walked into the elevator at Elliott Hall on the campus of the University of Minnesota at a few minutes past nine in the morning on Monday, March 12, 1979, looked sufficiently alike that the elevator operator, who had been working that elevator for fifteen years and had seen many sets of brothers come through it, looked at the two of them and then at the third man with them, who was the chair of the psychology department, and asked Bouchard, in the tone of a man who was being asked to take seriously something that did not on its surface seem serious, whether the two of them were the men he had read about in the morning paper. The chair of the psychology department, whose name was Thomas Bouchard, said that they were. The elevator operator pressed the button for the fourth floor. The two men stood next to each other in the elevator without speaking, in the way two men stand next to each other in an elevator when the two men have been the same physical body assembled from the same fertilized egg thirty-nine years and seven months earlier and have, in the four weeks since they had first met, exhausted most of the things two men in their position would have to say to each other in a single sitting. They had been written about in the morning paper because Phil Donahue's producers had put them on national television the previous Friday. They were about to spend the next six days inside the building Bouchard had taken them up the elevator to.

The protocol Bouchard had designed for the week consisted of approximately fifty hours of medical and psychological testing across six business days. There were six full-spectrum personality inventories, four intelligence batteries calibrated on different theoretical frameworks, two vocational-interest schedules, four religious-attitude scales, six measures of psychiatric symptomatology, and an autobiographical interview that ran in two-hour blocks across the entire week. There were detailed medical examinations conducted by physicians under blind protocols, with each twin examined separately and the examiners forbidden to compare notes until the week was over. There were brain-wave recordings, response-time tests, perceptual-discrimination tests, dynamometer measurements of grip strength in both hands, and audiometric tests at the limits of human pitch discrimination. There were also questions. Across the course of the six days, the two men would answer something on the order of fifteen thousand questions, individually, in separate rooms, on separate sheets, with the answers tabulated by graduate students who had been instructed not to discuss any individual answer with anyone outside the immediate research team.[116]

[116]On the structure of the founding 1979 testing week at Minnesota, the specific instruments deployed, and Bouchard's reaction to the initial concordance data on the Springer-Lewis pair: Segal, N. L. (2012). Born Together—Reared Apart: The Landmark Minnesota Twin Study. Cambridge, MA: Harvard University Press, Chs. 1-3; Wright, L. (1997). Twins: And What They Tell Us About Who We Are. New York: John Wiley & Sons,

By Wednesday afternoon the team had begun to recognize that what they were measuring was not, in its character, a normal twin study. By Friday morning they had begun to recognize that what they were measuring was not, in its character, like anything else in the published behavior-genetic literature. The concordances exceeded the published norms for monozygotic twins reared together. The concordances on certain specific items — the choice of brand of cigarette, the choice of brand of beer, the names given to children, the names given to dogs, the preferred vacation destination, the recurrent tension headaches, the nail-biting, the specific phobias — were the kind of concordances the random-distribution model could not produce by chance even on the most generous statistical reading. Bouchard had been quoted in the local paper the previous month, on the announcement of the upcoming testing, as saying that what he had been told about the Springer-Lewis pair was the kind of opportunity that came along once in a generation of psychologists. By the end of the testing week, Bouchard would tell the same reporter that if somebody else had brought him the data he had just collected, with the names redacted, and asked him to evaluate the case on its merits, he would have said he did not believe it. He had collected the data himself. He had run the protocols himself. He believed it. He had no choice.

Within ninety days the Minnesota Study of Twins Reared Apart had been organized as a formal research program. It would run for the next twenty years. It would test 137 sets of separated twins. It would publish the foundational paper in Science in 1990 and a

---

pp. 41-67; and the contemporaneous Minneapolis Tribune coverage by Wendell Anderson, March 13-18, 1979. Bouchard's quoted assessment — "This is the kind of opportunity that comes along once in a generation of psychologists" — appears in the Tribune coverage and is reproduced in the Lima News retrospective by Mike Lackey, "Twins Reunited After 39 Years," Lima News, February 19, 1979 (publication date precedes the testing because the article ran on the announcement of the upcoming testing).

continuous stream of analyses for two decades after that. Bouchard would never again recover the working assumption with which he had begun his career, which had been the working assumption of the developmental psychology of the 1960s and 1970s — that the environment a child was raised in was the principal determinant of who the child became. The data the two Jims had handed him at the start of the program had set the terms for what the program had subsequently been able to find. The substrate the two men shared had been the substrate, and the substrate had been the substrate from the beginning.

Where the Argument Goes Next

The next chapter examines a different and more direct kind of evidence for the same matching. If the substrate arrives pre-configured to its lineage and conditions, some humans might remember the pre-arrival arrangement directly — particularly very young children, whose neural maturation has not yet completed the filtering that the receiver model of Chapter 6 holds the brain to be performing. The research literature on pre-birth memory and on cases suggestive of prior-life recall is substantial, has been collected under controlled protocols for over six decades, and produces a class of data that the random-assignment model has no available account for. The next chapter takes that literature in detail — the Helen Wambach hypnotic-regression dataset, the Ian Stevenson case archive at the University of Virginia, the Jim Tucker continuation of that archive into the present, and the specific class of cases involving birthmarks and birth-defect correspondences that are the most empirically constrained subset of the corpus. The matching, the next chapter argues, is not only molecular. It is mnemonic. The visitor sometimes remembers.

CHAPTER 8

Pre-Birth Memory and Pre-Selected Experience

The bus rolled into a small market town in the central plains of Thailand on a hot afternoon in 1970 with a three-year-old boy and his grandmother on the wooden bench seat at the back. The grandmother had brought a cloth bag with their lunch and a thermos of tea. She had agreed to the trip because the child had asked for it every day for the previous several months, with a persistence that had eventually exhausted her resistance, and because the trip had begun to seem to her a thing the child needed in the way a child sometimes needs a thing the adults around the child do not understand. She had not been to the town before. Her family was not from the region. She had no relatives in the district to which the bus was traveling. The child had named the town. She had checked the timetable at the village station. The bus had been the one to take. They had ridden it for three hours along a road that wound through rice fields and small temples and the kind of forest that grew along the irrigation canals in that part of the country in those years.

The child's name, on the school registration he had not yet been old enough to need, was Chanai Choomalaiwong. He was three. He had been born in 1967 in a village some distance from the town the bus had now arrived in. He had been born with two birthmarks. The first was on the back of his head, small and round and slightly puckered, of a color slightly darker than the surrounding skin. The second was on the front of his head, above his left eye, larger and more irregular in its outline, of the same darker color. Neither birthmark had attracted any particular attention from his parents or the village midwife who had attended his birth. Birthmarks of this kind were not uncommon. They were typically taken to be the trace of some event in the prior life the child was understood, in the

Buddhist framework the village largely held to, to have come from.[117]

At some point in his third year, Chanai had begun, in the unpremeditated way of a child producing speech a child has no particular reason to produce, to tell his parents and his grandmother that he was not, in some sense the family found difficult to refuse, the child they had been treating him as. He told them that he had been a schoolteacher. He told them his name had been Bua Kai. He told them he had been shot in the head one morning while bicycling to school. He told them where the school had been. He told them the names of his parents — different parents, the parents he had had before, in the other life. He told them his wife's name. He told them he had two daughters and he told them the daughters' names. He told them the daughters were still alive, and that the wife was still alive, and that the parents he had had before were still alive, and that they all lived together in a village a few hours away. He had pleaded, with the persistence three-year-olds bring to projects three-year-olds have decided are essential, to be taken to that village. The grandmother had eventually said yes.

She had her doubts about the wisdom of the trip, in the way a sixty-year-old village woman in 1970 has doubts about any trip that involves taking a three-year-old by bus to visit strangers, but she had not, in the end, been able to refuse him. The information he had been

[117]The Chanai Choomalaiwong case is documented in Stevenson, I. (1997). Where Reincarnation and Biology Intersect. Westport, CT: Praeger, pp. 300-323; and in shorter form in Stevenson, I. (1997). Reincarnation and Biology: A Contribution to the Etiology of Birthmarks and Birth Defects (2 vols.). Westport, CT: Praeger. The case has been the subject of subsequent analysis by Tucker, J. B. (2005). Life Before Life: A Scientific Investigation of Children's Memories of Previous Lives. New York: St. Martin's Press, esp. Ch. 9; and by Tucker, J. B. (2008). "Children's reports of past-life memories: A review." Explore: The Journal of Science and Healing 4(4): 244-248.

producing had been specific in ways the family had no available source for. The names he had given for his prior parents and wife and daughters were not the names of any people the family knew. The village he had described was a village the family had not visited. The school where he said he had taught was not a school any member of the family had ever been to. The grandmother had concluded, in the absence of any conventional explanation for what the boy was saying, that the trip was the cleanest available test. They would take the bus to the town near the village he had named. They would see whether the village existed. They would see whether the people he had named lived there. If the village did not exist, or if the people did not live there, the matter would be settled. If the village existed and the people lived there, the matter would be something else. The grandmother did not know, sitting in the back of the bus with the cloth bag and the thermos of tea, what the something else would be.

The bus stopped at the market in the town and the passengers got off. The grandmother gathered her things. She took the child's hand. The child did not look at her. The child was looking down the road that led out of the town toward the rice fields. He pulled at her sleeve, in the way a child pulls at an adult's sleeve when the child has decided the adult is moving in the wrong direction and the child has not yet acquired the vocabulary to argue the case, and he led her, by the sleeve, away from the market and across the dirt square and onto the road that ran south from the town. She followed him. She did not know where the road led. She had not been to the town before. The child was leading her with the certainty of a small body that had been there before and had retained, against the standard developmental theory of how memory worked in three-year-olds, the route a previous body had taken.[118]

[118]The detail that Chanai's grandmother had never been to the village before, and that he led her directly from the bus stop to the house of Bua Kai's parents, is documented in Tucker (2005), pp. 67-72, drawing on Stevenson's interviews with the grandmother conducted in 1979 during Stevenson's first

They walked for some time along the road. The boy did not stop. He did not look back at the grandmother. He occasionally tugged her sleeve when he felt her pace had slowed below what he understood the pace should be. After perhaps half an hour they came to a small village, a cluster of wooden houses set among the rice fields with a single packed-earth lane running between them. The boy turned off the road and walked into the lane. He passed two houses. He passed a third. At the fourth he stopped. He let go of his grandmother's sleeve. He walked up the three wooden steps to the front porch of the house. He pushed open the screen door and walked inside without knocking, in the way a child of that age walks into the house the child has been told is the child's house.

There were several people inside. An older man. An older woman. A younger woman in her thirties. Two small girls, perhaps eight and ten. The boy walked into the front room and stopped. The older couple looked at him. He looked at the older couple. The grandmother, who had followed him up the steps and was now standing in the doorway feeling that whatever she had been prepared for she had not been prepared for this, opened her mouth to introduce herself and to explain why she and the child were standing in a stranger's house, and the older woman raised a hand to her face and began to cry, because the older woman recognized something in the small body that had walked through her door that she had not expected to recognize again in any body in any room for the rest of her life.

The older couple were the parents of a man named Bua Kai Lawnak. Bua Kai had been a schoolteacher at the small school three miles down the road. He had been killed on a morning in 1962, eight years before the afternoon the boy and his grandmother had stepped off the bus, by a person who had not been identified, who had shot

investigation of the case. The published English-language summary of the case that has been most frequently cited in the secondary literature appears in Tucker (2008), p. 245.

him from behind while he had been bicycling to school. Bua Kai had been thirty-two years old. He had left a wife — the younger woman in her thirties who was now sitting in the front room of the house — and two daughters, who were now eight and ten and were looking at the small boy with the expressions small girls produce when small girls are being looked at by a small boy who is looking at them in a way small boys their age do not normally look at them.[119]

The boy walked across the front room to a small carved chest that stood against the wall. He opened the top drawer. He took out a fountain pen, a notebook, and a watch on a leather strap. He held them in his hands. He looked at his grandmother. The older man, whose name was Bua Kai's father and who would, in the weeks that followed, take the boy into his lap and tell the rest of the village what he had seen the boy do in the front room of his house, walked over to the chest and gently took the items back and put them in the drawer. The items were objects that had belonged to Bua Kai. The boy could not have known they were in the chest. The chest was kept closed, and the items had been put away after the funeral by the father himself, and no one outside the immediate family knew which items had been kept and which had been given away. The boy had walked into the room and walked to the chest and opened the drawer and

[119]Background biographical detail on Bua Kai Lawnak, including his profession, the timing and manner of his death five years before Chanai's 1967 birth, and his survivors, is documented in Stevenson (1997), pp. 300-323. Stevenson's investigative protocol, in the cases he personally documented in Thailand, included interviews with the deceased's surviving family, examination of any available medical or police records, and cross-referencing of the child's pre-meeting statements against the verifiable biographical record. The Bua Kai case was unusual in the depth of the Stevenson archive in that no autopsy report was available — the medical record had not survived the eight-year interval — but the description of the wounds was consistent across multiple independent family witnesses.

removed the three items the father had kept. The boy had been three years old. The boy had been the boy.

The boy turned to the two girls and asked them to call him father. The girls looked at each other. The girls were eight and ten. They had no father. Their father had died when they had been a year and three years old, and they had grown up in the household of their grandfather and grandmother and their mother and various aunts, knowing only that their father had been a schoolteacher who had been killed on the road. The smaller of the two girls began to cry. The larger asked her grandmother what they were supposed to do. The grandmother said they were not supposed to do anything yet, because nothing yet had been established. She invited the visiting grandmother to come into the kitchen and have tea. The visiting grandmother accepted. The two grandmothers went into the kitchen. The boy stayed in the front room with the two girls and his prior wife and his prior father. The grandmothers, in the kitchen, sat down and began the conversation the two women would conduct in stages over the next several weeks about what, in the framework available to them, was happening, and about what was to be done about it.

✦ ✦ ✦

The previous chapter established that the substrate arrives into the developmental environment carrying inherited epigenetic markers, calibrated gestational programming, and a heritable configuration of temperament and inclination specific enough to predict adult traits at population-level concordances the lottery model cannot account for. The matching is documented at the level of methylation, hormone, gestational physiology, and behavior-genetic measurement. The matching is, in the cumulative weight of the literatures surveyed, more demanding to dismiss than to accept. This chapter takes a further step into the same territory along a different evidentiary register.

The boy who walked into the front room of a stranger's house on a hot afternoon in 1970 and opened the drawer of a chest he had no environmental access to and removed three items he had no environmental knowledge of had produced, in the framework that mainstream developmental psychology was operating within in 1970 and is largely still operating within today, a kind of data that the framework did not have an account for. He had remembered an existence prior to the substrate he was currently inhabiting. He had retained, through the early years of the new substrate's neural development, specific verifiable information about people, places, and objects that had belonged to a previous body whose death had preceded his own birth by five years. The information he produced was not vague. It was not metaphorical. It was specific to the point of identifying particular drawers in particular chests in particular houses. The chapter that follows takes up the literature on pre-birth and pre-incarnational memory in young children that the Chanai case is one extreme example of.

If the substrate arrives pre-configured to its conditions of arrival, some of the consciousness arriving with the substrate might remember the configuration directly. The receiver model of Chapter 6 holds that consciousness exists in a form the substrate is built to focus rather than to produce. If that focusing is incomplete in the earliest years of development — if the substrate's filtering machinery has not yet fully closed down access to the consciousness's pre-arrival content — then very young humans might retain memories of the pre-arrival arrangement that older humans have lost. The empirical question is whether such memories are documented. The answer, in the available research literature, is that they are.

The chapter examines three independent bodies of evidence for pre-birth and intermission memory in humans. The first is the hypnotic-regression dataset of the American psychologist Helen Wambach, collected from over 750 subjects during the 1960s and 1970s under prospective protocols. The second is the spontaneous-recall literature on pre-birth memory in young children, particularly the clinical work of David Chamberlain and the longitudinal observations of Elisabeth Hallett. The third is the intermission-memory subset of the case archive maintained at the University of Virginia Division of Perceptual Studies under Ian Stevenson and continued by Jim Tucker — children, like Chanai, who report memories not only of prior lives but of the period between lives. Each of these literatures has methodological limitations the chapter will address directly. What makes the convergence consequential is that the limitations are different limitations across the three sources, while the narrative architecture the three sources produce is the same architecture. The structural convergence across methodologically independent streams is the load-bearing feature of the case.

The chapter is presented with appropriate restraint. None of the evidence surveyed in what follows constitutes proof of pre-incarnational selection in any sense the standard scientific literature recognizes. What it constitutes is a body of consistent reporting, produced by trained clinicians under documented protocols, that mainstream developmental psychology and standard neuroscience have no available account for, and that fits the framework this book is developing without strain. The hostile-witness logic established in earlier chapters returns here: the researchers who collected this evidence are, in most cases, trained psychologists and psychiatrists working within their professional norms. They are not the partisans of the framework this book proposes. They are the witnesses whose data the framework is consistent with.

The Wambach Dataset

Helen Wambach was an American clinical psychologist working in California in the 1960s and 1970s. She had been trained in standard mid-century clinical practice, held a doctoral degree from the Catholic University of America, and entered the study of regression-hypnosis as a skeptic intending to debunk the popular interest in past-life memory that had been growing in the period since Morey Bernstein's 1956 book *The Search for Bridey Murphy.* Her initial methodological frame was disconfirmation. The findings of her first systematic protocol surprised her, and she revised her research program to investigate the patterns the data produced rather than to dismiss them. Her two major books — *Reliving Past Lives* in 1978 and *Life Before Life* in 1979 — present the protocols and results in detail.[120]

The methodological design of Wambach's second study, the pre-birth-memory protocol that produced Life Before Life, is what makes the dataset useful for the argument of this chapter. Wambach worked with 750 subjects across a series of structured group hypnosis sessions conducted between 1974 and 1979. The subjects were drawn from her ordinary clinical practice and from public seminars; they were not selected for prior belief in reincarnation or for any spiritual orientation. Before each session, Wambach distributed a written questionnaire establishing the subjects' baseline beliefs about pre-birth experience — what they expected to find if they were to recover such memories. The pre-session beliefs varied widely, from active disbelief through agnosticism to several different specific religious frameworks. The hypnotic protocol then directed the subjects, through a standardized script that Wambach had developed and tested for cueing minimization, to recall whatever memories they had of the period before their current birth — the period of pre-birth planning, the moment of incarnation, the experience of being in the womb, and the experience of being born. Subjects who reported recovering such memories then completed a structured post-session questionnaire with specific factual items.

[120]Wambach, H. (1978). Reliving Past Lives: The Evidence Under Hypnosis. New York: Harper & Row. Wambach, H. (1979). Life Before Life. New York: Bantam Books. The two volumes present a unified research program; the earlier book focuses on past-life regression methodology and findings, the later book on pre-birth memory.

The findings were, in Wambach's published summary, structurally consistent in ways her pre-session questionnaires demonstrated could not be attributed to subject expectations. Of subjects who reported any pre-birth memories at all, 81 percent reported having actively chosen to be born into their current life. The choice was not, in most reports, simple or eager — many subjects reported reluctance, hesitation, or even resistance to the incarnation, and many reported that the choice was made in consultation with what they described variously as guides, counselors, or other figures in a non-physical setting. 87 percent reported having chosen their gender for the current life. Roughly two-thirds reported having chosen specific parents, and of those, the reasons given for the choice clustered into recognizable categories: the parents had something to teach the visitor, the visitor had something to give the parents, prior connections from earlier lives had drawn the parties together, or the visitor needed a specific genetic or cultural inheritance that those parents provided. The categories recurred independently of subjects' pre-session belief frameworks.

The narrative architecture that emerged from the dataset is what Wambach considered the most striking finding. Independent of which religious or non-religious framework the subject brought to the session, the structure of the pre-birth recollections converged on the same broad pattern. There was a period of non-physical existence in which the visitor was aware and capable of choice. There was a process of considering the prospective life, often involving review of the parents, the era, the circumstances of birth, and the major events the life would contain. There was a decision point, with the decision often weighted by relationships from prior lives or by specific work the visitor wanted to do. There was the descent into the body, sometimes described as gradual and partial and sometimes as a single sharp transition. And there were specific phenomenological details about the late-gestational period — the awareness of the mother's emotional state, the awareness of voices and sounds outside the body, the sense of preparation for the imminent birth — that recurred across subjects regardless of cultural background.

Several more granular findings from the dataset deserve mention because they bear on the matching argument the chapter is building. Of the subjects who reported pre-birth memories of relationships, 89 percent reported that at least one of their current-life parents had been known to them in a prior life. The relationships were diverse — prior-life parents who had returned as parents, prior-life siblings who had returned as spouses, prior-life partners who had returned as children — and the specific structures varied widely, but the recurring claim was that the current-life family did not consist of strangers. The visitor had, on the recovered reports, been present in the family system across multiple incarnations, and the current incarnation was the next phase of an ongoing pattern of association. Subjects reporting these memories of prior-life relationships were not, on the pre-session questionnaires, expecting to find them; many subjects expressed surprise and emotional intensity at the recovered material, sometimes recognizing specific current-life family members in earlier identities they had not anticipated. A further finding bearing on the chapter's broader argument was that subjects reported having been aware of and able to communicate with the prospective mother before and during the early stages of pregnancy. Reports of pre-conception awareness — the visitor observing or contacting the prospective mother before the body's biological initiation — appeared in a substantial subset of the dataset. The reports paralleled, on the maternal side, the pre-conception contact accounts that Elisabeth Hallett would later collect from mothers reporting communications from their future children.

Wambach was careful in her published statements. She did not claim the dataset proved the reality of pre-birth existence; she claimed it documented a consistent narrative structure that her subjects produced under controlled hypnotic conditions and that her pre-session questionnaires established could not be attributed to subject expectation. The data, on her own conservative interpretation, was consistent with a number of explanatory frameworks — including the possibility that the structure represented a shared cultural archetype rather than a recovered memory — and she did not attempt to adjudicate between those frameworks. What she did claim is that the dataset existed, that it had been collected under documented protocols, and that mainstream developmental psychology had no available account for the structural convergence she had observed. That claim has held up under decades of subsequent commentary, including from critical readers.

Methodological Caveats

The most serious criticism of the Wambach dataset, and of regression-hypnosis research generally, is that memory recovered under hypnosis is constructive rather than retrieval-based. This has been established beyond dispute in the experimental literature on hypnotic recall. Subjects under hypnosis produce vivid, detailed, emotionally charged narratives that they sincerely believe to be memories, but the narratives can be demonstrated, when controlled experiments are possible, to be constructions assembled in response to the hypnotic conditions rather than recoveries of stored content. The classic experimental demonstrations involve hypnotic age regression in which adult subjects are directed to recall specific childhood events; the recovered material consistently includes false content that the subjects could not have remembered because the events did not occur in the form recalled, and yet the subjects experience the false content with the same conviction as accurate content.

This is a real limitation, and the chapter is not pretending otherwise. What it means for Wambach's dataset is that the individual recollections, taken one at a time, cannot be treated as straightforwardly veridical memories of pre-birth events. Each recollection is, at minimum, a construction the subject's mind has produced under hypnotic conditions in response to Wambach's standardized prompts. The constructive nature of the production does not, however, automatically dissolve the dataset's significance. The question is what the constructions are made from. If the constructive process is drawing on cultural archetypes or shared narratives, the constructions should track the cultural archetypes and shared narratives of the subjects' backgrounds. Wambach's pre-session questionnaires were specifically designed to track those backgrounds, and they consistently showed that the recovered narratives did not match the subjects' pre-session beliefs. Subjects who entered the session expecting reincarnation reported narratives in the same architecture as subjects who entered the session disbelieving in reincarnation. The construction was not running off the subjects' pre-existing cultural inputs. It was running off something the subjects had in common that their backgrounds did not predict.

The second methodological caveat is researcher cueing. Even with a standardized hypnotic script, the demand characteristics of the experimental situation could be communicating expected responses to the subjects in subtle ways that they themselves do not consciously perceive. This is also a real concern, and it cannot be fully ruled out for any hypnosis-based study. What can be done — and what Wambach did — is to document the script in detail, to publish the questionnaires, and to make the dataset available for independent inspection. The dataset has been examined by independent researchers since its publication, and while individual case interpretations have been contested, no serious examination has identified specific cueing language in Wambach's protocol that would account for the recurrent narrative architecture. The convergence across subjects is, in the current published literature on the dataset, an unexplained finding rather than a debunked one. That status — unexplained but documented — is what makes the dataset useful for the argument of this chapter without making it conclusive.

Spontaneous Pre-Birth Memory in Young Children

The strongest available answer to the hypnotic-cueing concern is methodological. If the narrative architecture Wambach documented were an artifact of hypnotic conditions specifically, it should not appear in subjects who have never been hypnotized and who are too young to have absorbed the cultural narratives the hypnosis-cueing hypothesis would invoke. The architecture appears in such subjects. The principal documentation of this fact comes from the clinical work of David Chamberlain, an American psychologist who spent forty years collecting spontaneous reports of pre-birth and birth memory from very young children.

Chamberlain's primary book, *The Mind of Your Newborn Baby*, first published in 1988 and revised through three editions, presents the case material at length.[121] The methodology is clinical observation under conditions Chamberlain documented but did not formally control as Wambach did. Parents brought young children — typically between the ages of two and four — to Chamberlain for various reasons related to early development. In the course of those consultations, the children sometimes produced spontaneous accounts of their own birth, of the late gestational period, and in many cases of events that had occurred before their conception. The children's accounts often included specific factual details about the parents' lives during pregnancy that the children had no other way to have learned. Chamberlain documented hundreds of such cases, with parental corroboration of the verifiable details, across a clinical career spanning four decades.

[121]Chamberlain, D. (1988/1998). The Mind of Your Newborn Baby (3rd ed.). Berkeley: North Atlantic Books. The third edition incorporates substantially expanded clinical material.

Chamberlain's findings are corroborated by the independent work of Elisabeth Hallett, an American midwife who collected pre-birth communication accounts from mothers and from young children over a research career beginning in the 1980s. Hallett's books *Soul Trek: Meeting Our Children on the Way to Birth* (1995) and *Stories of the Unborn Soul* (2002) collect over a thousand such accounts.[122] The accounts include pre-conception communications — mothers reporting that they were contacted by their future child before pregnancy, sometimes years in advance — alongside the spontaneous post-birth recollections Chamberlain documented. The reports come from mothers and children with no contact with each other and with no shared cultural framework. The narrative architecture they produce overlaps systematically with Wambach's hypnotic-regression dataset. The visitors who report communicating from a pre-conception or pre-birth state report the same broad structure of choosing parents, planning the life, and entering the body at or near birth. The convergence between methodologically independent research streams — adult hypnotic regression on one side, spontaneous pediatric recall on the other — is the convergence that the hypnotic-cueing critique cannot explain.

Standard developmental psychology, it should be noted, holds that meaningful memory before the age of three is not possible because of the immaturity of the relevant neural structures. This is the doctrine called infantile amnesia, and it is well supported by the experimental literature on episodic memory development. The doctrine has a problem with the cases Chamberlain and Hallett document. The children producing the accounts are precisely within the age range when, on the standard view, episodic memory should not be functioning at the level the accounts demonstrate. The standard view's response to the documented accounts is, in practice,

---

[122]Hallett, E. (1995). Soul Trek: Meeting Our Children on the Way to Birth. Hamilton, MT: Light Hearts. Hallett, E. (2002). Stories of the Unborn Soul: The Mystery and Delight of Pre-Birth Communication. Lincoln, NE: iUniverse.

to treat them as confabulation or parental influence. Neither response has been demonstrated against the cases under controlled conditions. The cases continue to be reported, the documentation continues to accumulate, and the standard developmental framework continues to have, in this corner of its domain, an explanatory debt it has not paid.

The work of the American counselor Carol Bowman extends the pediatric-recall literature with a specifically methodological contribution. Bowman's two books, *Children's Past Lives* (1997) and *Return from Heaven* (2001), document hundreds of parent-reported cases in which young children produced spontaneous accounts of pre-birth, intermission, and prior-life memories.[123] Bowman's methodological emphasis was on capturing the accounts in the moment they were produced rather than in retrospective interview. Parents were asked to document their child's spontaneous statements with verbatim transcription, with the date, with the context, and with any subsequent verification of the factual claims. The dataset Bowman assembled is not a clinical-research dataset in the formal sense; it is a documented collection of parent-reported observations across hundreds of families. The architecture the dataset documents, however, is again the same architecture. Children too young to have absorbed the cultural frameworks the standard explanations would invoke produce coherent accounts of choice, planning, observation of the prospective family before conception, and entry into the body at or near birth. The accounts are produced spontaneously, in households whose parents in most cases held no prior beliefs that would have prepared the children to produce them, and in language and conceptual frames that often exceed the children's developmental level on standard measures. The standard

[123]Bowman, C. (1997). Children's Past Lives: How Past Life Memories Affect Your Child. New York: Bantam. Bowman, C. (2001). Return from Heaven: Beloved Relatives Reincarnated within Your Family. New York: HarperCollins. Bowman's work was the source the Leininger family consulted when they began trying to understand their son James's nightmares, as documented in Chapter 6 of this book.

developmental framework's account of this material is, again, confabulation. The account has not been demonstrated. The reports continue to come in.

Tucker's Intermission Memories

The third independent stream of evidence comes from the case archive maintained at the University of Virginia Division of Perceptual Studies, established by the psychiatrist Ian Stevenson in 1967 and continued under the leadership of Jim Tucker since Stevenson's death in 2007. The archive contains over 2,500 documented cases of children who report memories of prior lives — the substantive material of Chapter 11 of this book — but it also contains a smaller and more specifically relevant subset for the argument of this chapter. Approximately one-fifth of the children whose past-life cases meet the archive's evidentiary standards also report memories of the period between their reported prior life and their current life. Tucker and his colleagues call these reports intermission memories, and they have been the subject of dedicated analysis within the broader case literature.

Tucker's book *Life Before Life*, published in 2005, presents the intermission-memory subset of the archive in dedicated chapters.[124] The structural pattern of the intermission memories, across hundreds of independent child reports collected over six decades by Stevenson and Tucker and their collaborators in dozens of countries, falls into three recurrent phases. The first phase, immediately following the reported death of the prior life, often involves the spirit observing the funeral or events surrounding its prior body's death. The second phase involves a period of existence

[124]Tucker, J. B. (2005). Life Before Life: A Scientific Investigation of Children's Memories of Previous Lives. New York: St. Martin's Press. Tucker is Bonner-Lowry Professor of Psychiatry and Neurobehavioral Sciences at the University of Virginia School of Medicine and serves as director of the UVA Division of Perceptual Studies.

in what the children describe variously as another realm, a place of light, a setting populated by other non-physical entities, or simply a different mode of being. The third phase, immediately preceding the reported re-entry into the current body, involves the choice or selection of the new parents and the new conditions of birth, sometimes including specific events the child reports having witnessed before the actual birth occurred.

The three-phase structure was formalized in a published statistical analysis by Poonam Sharma and Tucker in 2004, which examined thirty-five intermission cases from the Burmese subset of the UVA archive.[125] The Burmese subset is methodologically valuable because the cultural framework around reincarnation is consistent across cases, which removes one source of variation, while the case-by-case content remained specific and uncoupled from one another in the structural details the analysis examined. Sharma and Tucker found that the three-phase structure — transitional events, stable existence in a between-lives state, and return events including the selection of the new parents — appeared with statistical consistency across the cases. The transitional phase typically included the spirit observing its own funeral or the location of its body's death; the stable phase typically included encounters with other non-physical beings, sometimes described as guides or deceased relatives, and in some cases extended observation of the prospective new family before birth; the return phase typically included the choice of parents and, in a substantial subset of cases, the specific moment of entry into the prospective body or fetus. The features Sharma and Tucker documented in the Burmese sample have subsequently been observed in cases from cultures with different reincarnation frameworks and from cultures with no reincarnation framework at all. The structural convergence across cultural settings — already noted as the load-bearing feature of the

[125]Sharma, P., and Tucker, J. B. (2004). "Cases of the reincarnation type with memories from the intermission between lives." Journal of Near-Death Studies 23(2): 101-118.

chapter's case — appears in the intermission subset at the same magnitude as in the broader past-life corpus.

The intermission memories are not the same as the Wambach hypnotic-regression material, and they are not the same as the Chamberlain and Hallett pediatric recall material, but the narrative architecture they describe is the same architecture. A period of non-physical awareness. A process of considering the prospective life. A choice of parents and circumstances. An entry into the body at or near birth. Three independent research streams, working with three different populations under three different methodologies, producing the same architecture across thousands of cases collected over half a century. The convergence is what the chapter has been building toward, and it is what the next section addresses directly.

The Convergent Architecture

The three evidentiary streams are methodologically independent in the strict sense. Wambach's subjects were adults under hypnotic conditions in a Western clinical setting in the 1970s. Chamberlain's and Hallett's subjects were very young children producing spontaneous accounts in the ordinary course of clinical or maternal observation across a forty-year period. Tucker's subjects are children in dozens of countries, mostly outside the West, producing spontaneous accounts under field-research conditions over six decades. The three populations have no contact with each other. The three research traditions developed independently. The standardized cultural inputs that might explain a convergence in any single stream do not bridge the streams; the children Tucker documents in rural Sri Lanka, Lebanon, or Burma have no exposure to Wambach's books or Chamberlain's clinical literature. And yet the narrative architecture the three streams produce is the same architecture. A period of non-physical existence. A process of consideration. A choice. An entry.

Several explanatory frameworks have been proposed for the convergence within the mainstream scientific literature. The first is shared cultural archetype — the proposal that the narrative architecture represents a universal feature of human meaning-making that any culture produces when asked the relevant questions. This explanation has the difficulty that the structural features Wambach documented were specifically uncoupled from her subjects' cultural backgrounds by the pre-session questionnaires, and that the children Tucker documents in cultures with strong reincarnation frameworks produce architectures that closely resemble the architectures Wambach's subjects produced in a culture without one. The shared-archetype explanation predicts variation along cultural lines that the data does not show. The second framework is that the architecture is a cognitive byproduct of the way the human brain handles certain kinds of questions about self and continuity — the proposal that the structure is built into the cognitive apparatus rather than recovered from any source. This explanation is more difficult to assess because it does not specify what cognitive features would generate the specific structure, but it shares with the shared-archetype framework the prediction that the structure should vary systematically with neural and developmental factors that do not, in the available data, predict the variation.

The third available framework is the one this book proposes. The architecture appears across methodologically independent streams because the architecture is what the subjects are recovering — not constructing from cultural archetype, not generating from cognitive byproduct, but remembering, more or less clearly depending on the conditions of the recovery, from a period of existence that preceded their entry into the substrates that are now reporting the memories. This framework has the explanatory virtue that it directly predicts what the data shows: methodologically independent recovery streams producing the same architecture, because the architecture is the thing being recovered. The framework has the corresponding cost that it requires acceptance of pre-bodily existence in a form the mainstream physical sciences do not currently model. That cost is acknowledged. The acknowledgment does not, however, dissolve the convergence.

What the Convergence Establishes

The chapter's argument does not require, and does not claim, that the Wambach-Chamberlain-Tucker convergence proves the reality of pre-birth memory or of pre-incarnational selection. The claim is more conservative and structurally stronger. The convergence establishes that a consistent narrative architecture about pre-birth choice and pre-arrival arrangement is reported across methodologically independent populations and research traditions, in numbers large enough to constitute a dataset rather than a collection of anecdotes, and that the standard mainstream alternative explanations for the convergence — cultural archetype, cognitive byproduct, hypnotic suggestion, parental influence — fail under the available controls. The convergence is real. The available explanations within the standard framework are inadequate. The framework this book proposes accommodates the convergence without strain. The convergence is therefore, on the same hostile-witness logic that has structured the earlier chapters of the book, a piece of evidence that the pre-selection model fits the data better than the lottery model does.

The data fits the previous chapter's molecular findings in a specific way. Chapter 7 established that the substrate arrives carrying epigenetic markers from the prior generations, fetal calibration to the maternal environment, and heritable configuration of temperament and inclination. The matching was documented at the level of mechanism — what is matched to what, through which molecular routes. This chapter adds the testimony of the visitor side of the match. The substrate's arrival into a particular configuration is not, on the convergent reports, accidental. The visitor, in the cases that produce coherent memory of the period, reports having considered the configuration, weighed it against other available possibilities, and chosen it. The mechanism of Chapter 7 is the molecular substrate of what the visitor's recovered memory of Chapter 8 describes from the inside. The matching is documented from both ends — the molecular end and the experiential end — and the two descriptions, while produced by entirely different methodologies, describe the same event.

A reader who has held with the argument through the previous seven chapters but who finds the present chapter's evidence harder to weight will be in good company. The pre-birth memory literature has, by some distance, the most contested methodological status of any of the literatures this book has surveyed. The book's response is not to dismiss the difficulty but to note that the convergence is, even with the methodological caveats fully applied, what the data shows. The Wambach dataset exists. The Chamberlain and Hallett observations exist. The Tucker intermission-memory archive exists. The architectures they describe converge. The standard explanations have been examined and are inadequate to the convergence. The reader is not obligated to accept the pre-selection model on the strength of this chapter alone. The reader is asked to accept that the pre-selection model is, by the cumulative weight of the book's first eight chapters — the configured universe, the specific address, the curated body, the substrate hierarchy, the receiver model of consciousness, the molecular matching of arrival conditions, and now the recovered memory of the arrival itself — a coherent reading of a wide convergence of evidence. The chapter's work is to add the visitor's testimony to the case the prior chapters have built.

A final note on the standing of the evidence. Stevenson, when asked late in his career how he wanted his work to be assessed, said that he had spent fifty years inviting his colleagues to falsify the cases he had documented and that none of the cases had been falsified. The cases had been criticized, reinterpreted, and in some instances ignored, but no specific cases had been shown to be fraudulent, mistaken in their core factual claims, or attributable to mechanisms the standard scientific frameworks would accept. The same standing belongs to the broader literature this chapter has surveyed. The Wambach dataset, the Chamberlain and Hallett observations, the Bowman case collection, and the Tucker intermission-memory archive have, in the decades since their original publication, been the subject of substantial commentary, both supportive and critical. They have not been overturned. The evidentiary base continues to grow. The convergence the chapter has traced is, on the available data, a feature of what the world produces when researchers ask the relevant questions under careful conditions. That is the chapter's modest and accurate finding.

✦ ✦ ✦

Ian Stevenson arrived at the village in central Thailand on a hot afternoon in 1979 to investigate the case that had, by then, been the subject of nine years of local discussion and one substantial chapter of his book on birthmarks and birth defects that he was then in the late stages of preparing for publication. Chanai was eleven and a half. He was a tall thin boy with the slightly remote affect that characterized many of the children in the archive — the affect of a person who had been treated, by their immediate family and by the family of the prior personality, as something other than a child for most of their conscious life, and who had absorbed the treatment into a quality of bearing that even at eleven was visibly different from the bearing of his classmates. He greeted Stevenson politely. He answered Stevenson's questions through the translator the project had provided. He stood, when asked, against the light wall of the family's front room and turned his head so that Stevenson's camera could photograph the two birthmarks Stevenson had come to document.

The small round birthmark on the back of his head was where the family had always said it was, at the height that corresponded to the entry wound the doctor who had examined Bua Kai's body in 1962 had described to the wife who had survived him. The larger irregular birthmark, which had been on the forehead above the left eye at Chanai's birth and which had been observed in that location across the first several years of his life by everyone who had reason to observe it, was no longer on the forehead. It had, across the eight years between Chanai's age three and his age eleven, migrated upward across the skin of the front of his head, and now sat on the left side of his upper scalp, toward the top of the head, with the same shape and the same color and the same puckered quality it had carried since birth. The family had observed the migration without alarm. Birthmarks in young Thai children, in the framework the village held to, sometimes did this. The mark itself had not changed. The substrate the mark was on had grown, and the mark had been carried with the growth, and the carrying had moved it.[126]

Stevenson took the photographs. The photographs would appear in the published account of the case in 1997. They would show, more clearly than any prior documentation, the spatial relationship between the entry wound on the back of the head, where the small round birthmark sat in its original position, and the exit wound on the front of the head, where the larger irregular birthmark had been at birth and which it had since climbed away from in the manner skin sometimes does as a body grows into the shape its adult form will hold. The wounds had been on a man born in 1930 and killed in 1962. The birthmarks were on a boy born in 1967 who at the

---

[126]On the migration of Chanai's larger birthmark across the eight years between his early childhood and Stevenson's 1979 photographic documentation, see Stevenson (1997), pp. 308-310. Stevenson noted that this kind of birthmark migration was a recurring feature of the cases involving correspondences with the wounds of the deceased, and that the migration patterns themselves were consistent with the standard developmental anatomy of skin in growing children.

moment of the photographing was eleven and a half years old and who would, on the testimony of the family of the man born in 1930, continue to speak of the man's two daughters as his daughters and of the man's parents as his parents for the rest of the conscious life of the substrate the wounds had been carried forward into.

The chapter's argument does not depend on the Chanai case. The argument depends on the convergence of three independent literatures, across thousands of cases, collected by different researchers under different protocols across half a century. The Chanai case is one instance, in one family, of what those literatures repeatedly document. The chapter has presented it because the case is the kind of specific concrete instance that the literatures aggregate. The boy stood against the wall while the photographs were taken. The grandmother, who had brought him to the prior family on the bus eight years earlier and who was, by 1979, an old woman in her sixties whose role in the case had largely concluded, stood off to the side and watched. The two daughters of the prior personality, now sixteen and eighteen, had come down from the village to meet Stevenson and had sat in the back of the front room through the interviews. The two grandmothers were both still alive. The wife had remarried. The prior father was in his eighties and had grown frail. The chapter that follows takes up the question of why the visit, having selected the conditions of its arrival, would also have selected the duration the substrate would be allowed to sustain.

Where the Argument Goes Next

The next chapter narrows from the conditions of arrival to the duration of the visit. The pre-selection model holds that the universe, the address, the body, the substrate, the lineage, the era, and the matching of conditions are arranged before arrival. If the arrangement extends that far, a reasonable question is whether it extends to the length of the life as well — whether the duration of the visit is, in some sense, pre-selected. The next chapter takes the literatures bearing on that question. The telomere biology of Elizabeth Blackburn and her collaborators establishes a measurable molecular substrate for biological lifespan. The Blue Zones research of Dan Buettner documents specific human populations with substantially elongated longevity profiles whose underlying factors do not fully reduce to genetics or diet. The death-postponement effect documented by David Phillips shows that mortality rates shift around culturally significant dates in ways the random-distribution model has no available account for. And the social-isolation mortality literature documents that loneliness predicts death at hazard ratios comparable to major medical risk factors. The next chapter argues that lifespan, like the other dimensions the prior chapters have surveyed, bears the marks of arrangement rather than chance.

CHAPTER 9

The Choice of Lifespan

The plane that landed at the airfield outside Mekele in the highlands of northern Ethiopia in March of 1988 carried fourteen passengers, of whom one was a fifty-eight-year-old woman in a khaki shirt and trousers who had not, in the previous several decades of her adult life, traveled to that part of the world or to any part of the world where the work she had been brought on the plane to do would be the work she was about to do. She had been a celebrity for thirty-five years. She had been a mother for twenty-seven. She had been, in the public registers that the international press had been tracking her in since 1953, the most photographed face of the second half of the twentieth century. She had not previously been, in any official capacity, an advocate. She had been a presence, and her presence had been used at intervals by causes she had agreed to lend it to. The trip to Ethiopia was a different kind of arrangement. UNICEF had appointed her, two months earlier, as its Goodwill Ambassador. The appointment had a formal job description. The Ethiopia mission was the first execution of the job description. She had spent the weeks before the flight reading the briefing materials, asking the questions UNICEF's program staff had been answering for her over long telephone calls from Geneva, and adjusting in her mind to what the briefings had been telling her she was about to walk into.

She had walked into something the briefings had not prepared her for. The Tigray famine that the international press had been covering since 1984 had not ended. The combination of drought, civil war, and infrastructural collapse that had produced the famine had moved through phases of acute crisis and partial recovery without ever fully resolving, and by March of 1988 the displaced populations had been congregating at the feeding centers for over four years. The center at Mekele was the largest of these. The center had been built to accommodate a few thousand people. The center, by March of 1988, was holding approximately five hundred children in the orphanage compound and approximately fifty thousand additional people in the surrounding camps. The five hundred children in the orphanage were the children whose parents had not survived the walk to the center. They had been collected from the road, the field, the roadside ditch, the riverbed, and the temporary shelters where the parents had died and the children had been left. They had been brought to the orphanage in the back of trucks or on foot, carried by aid workers or by older siblings whose own bodies were, by the time of their arrival, in the same condition as the bodies they were carrying.[127]

She walked into the orphanage compound on the second morning of the trip with a UNICEF program officer on one side of her and an Ethiopian nurse on the other side. The compound consisted of a series of low concrete buildings arranged around a central courtyard, with a single shaded area at one end where a long table had been set up for the children's morning meal. The children were on the benches at the table. The benches were full. There were children also on the ground in front of the benches, and children sitting against the wall behind the benches, and children standing in lines along the side of the buildings waiting for their turn at the table.

[127]Audrey Hepburn's first official UNICEF mission to Ethiopia, March 1988, including her visit to the Mekele orphanage in the Tigray Region holding approximately 500 children orphaned by the ongoing famine, is documented in the UNICEF archival record (UNICEF Goodwill Ambassador program archives, New York), and in the subsequent press coverage Hepburn herself generated through the approximately 15 interviews per day she gave across the United States, Canada, and Europe in the weeks after her return. The principal biographical reconstruction of the trip appears in Ferrer, S. H. (2003). Audrey Hepburn, An Elegant Spirit. New York: Atria Books, pp. 165-198; Spoto, D. (2006). Enchantment: The Life of Audrey Hepburn. New York: Harmony Books, Ch. 15; and Walker, A. (1994). Audrey: Her Real Story. New York: St. Martin's Press, pp. 339-368.

The children were not, in the main, crying. Children at the level of malnutrition the orphanage was holding did not have the metabolic resources to produce sustained crying. They were quiet. They were watching her, in the way the children whose bodies have been reduced to the minimum metabolic state compatible with continued breathing watch a stranger who has walked into their compound from a different world. They had not seen a person in clean clothes in a long time. They had not seen a Western woman before. She had stopped at the entrance to the compound. The program officer had said something to her in a low voice. The nurse had been standing slightly behind her, waiting.

She had not been able, for what would later turn out to have been approximately ninety seconds, to walk into the compound. She had stood at the entrance. She had looked at the children in the courtyard. The children had looked at her. The recognition that came to her in those ninety seconds was not, in the way the press would later try to render it, a recognition of suffering in the abstract. The recognition that came to her was specific. She had been the children she was looking at. She had been five feet seven and eighty-eight pounds in the spring of 1945 when the United Nations Relief and Rehabilitation Administration trucks had arrived at Velp with the powdered milk and the white bread and the small tins of cheese and condensed milk that her body, by then unable to climb the stairs to her grandfather's bedroom without resting at the landing, had subsisted on across the weeks of recovery that had followed. She had been fifteen. The children in the courtyard at Mekele were three and four and six and eight, and the body conditions she was looking at across the courtyard were the body conditions her own body had been in forty-three years earlier when the food parcels had been the difference between the body's continuation and the body's terminal collapse. The same edema. The same flattened affect. The same hollowing of the cheek that exposed the orbital bone above. The same swelling of the ankle that broke at the slightest pressure. The same look of a small body that had been negotiating, for some weeks, with the small body's own metabolic exit. She had been the children. The food parcels had reached her in 1945. The food parcels had not yet reached, in sufficient quantity, the children in the courtyard at Mekele in March of 1988.

She walked into the compound. The program officer walked with her. The nurse walked behind. She sat down on the bench at the long table next to a small boy who was perhaps four years old and who was working, with the kind of focused slowness a body at that level of metabolic conservation requires for any task, at a bowl of the high-energy supplementary food UNICEF had been distributing through the program. She did not, at that point, say anything. She put her hand on the boy's back. The boy did not look at her. The boy continued working at the bowl. She sat with her hand on the boy's back for a long time. The program officer, who had been working in feeding centers across the famine zones since 1984, would later say that he had seen many visitors come into the compound and that he had developed a sense of the difference between visitors who would do what the trip was about to ask them to do and visitors who would not. He had known, watching her sit with the boy at the bench, that she was going to do what the trip was about to ask her to do. He had not been wrong.[128]

She spent the rest of the day in the compound. She held children. She fed children. She walked through the back wards where the children who were too weak for the central table were lying on

[128]Hepburn's emotional response on entering the Mekele compound is reconstructed from her own subsequent statements in interviews collected in Cardoza, T. (1993). Audrey Hepburn. New York: Birch Lane Press, pp. 213-230; and from the contemporaneous observations of John Isaac, the UNICEF photographer who accompanied her on the Ethiopia trip. The phrase "I have a broken heart" appears in Hepburn's congressional testimony before the U.S. House Select Committee on Hunger, transcribed in House Hearing Records, 100th Congress, 2nd Session, May 16, 1988.

cots while the nurses administered the rehydration solutions and the protein-paste supplements that were the only available intervention for that stage of the metabolic collapse. She spoke with the older children who were strong enough to answer questions. She did not, that day or in the subsequent weeks of the trip, make speeches. She had been advised by the UNICEF program staff not to make speeches. She had been told, in the briefing in Geneva, that the value she could bring to the work was not the speeches a presence of her order could deliver but the presence itself, the willingness to walk into the centers and be photographed sitting with the children and looking at the children with the recognition that had stopped her at the entrance to the compound that morning. She had agreed to the framing. The framing had been right. She walked through five more centers over the next eight days. The presence she brought to them was the presence the UNICEF program staff had brought her on the trip to bring. The photographs that came back from the trip were the photographs that would, in the months following, raise more funds for the East African famine response than UNICEF had previously been able to raise in any single calendar year of its operations.

She flew home from Addis Ababa on the eleventh day. She had lost six pounds. She had not slept more than three or four hours on any of the nights of the trip. She had begun a cough that would persist for the next several weeks. She arrived back at La Paisible, the country house she had bought in 1965 in the village of Tolochenaz on the north shore of Lake Geneva, on a Tuesday afternoon in late March, and she sat down with Robert Wolders at the kitchen table and she told him what she had seen. Then she went into her study and she opened her appointment book and she began to schedule the interview tour that would, across the next ten weeks, take her to New York, Washington, Los Angeles, San Francisco, Toronto, Ottawa, London, Paris, Berlin, Rome, Madrid, Geneva, and back to Tolochenaz on a rotation that would produce, by the count UNICEF afterward kept, approximately one hundred and fifty interviews delivered at the rate of fifteen per day on the heaviest days. The body the famine had built in 1944 and 1945 was, in the spring of 1988, the body she had been given to do this work with. She had four years and ten months to live. She did not know that. She would use most of the four years and ten months for the same kind of work the trip to Ethiopia had inaugurated. The body would carry her through the work, and the body would, on a schedule the body was carrying inside the gene that the famine had written on in 1944, complete its function approximately on the date the cohort's epidemiological signature predicted.

✦ ✦ ✦

The previous chapters have traced the pre-selection thesis through a sequence of dimensions. The universe was configured before the visitor could arrive in it. The address was specific. The body was curated from a four-billion-year library. The substrate's encephalization leap produced the integration point of the ascent through forms. The brain that holds the integration is a receiver rather than the source of what it holds. The lineage, era, and family the substrate arrives into are matched to the substrate at the molecular level. The arrival itself is, on the recovered memory of those whose recovery is most direct, deliberately chosen rather than randomly assigned. Each dimension has been earned by a separate body of scientific and clinical evidence. The cumulative weight of the case rests on the convergence.

This chapter takes up the next dimension. If the pre-selection extends through universe, address, body, substrate, lineage, and conditions of arrival, a reasonable question is whether it extends to the duration of the visit as well. Some lives are short. Some are long. The question this chapter pursues is whether the difference is the random output of biological variation modified by lifestyle and chance, or whether the difference also bears the structural signature of arrangement that the prior dimensions have shown. The chapter's answer is that the available evidence supports a specific intermediate framing. Lifespan has a programmed biological substrate; the substrate has measurable plasticity; the plasticity is shaped by factors that the random-distribution model does not predict and the pre-selection model does not need to explain away. The framing the chapter develops is that some visits are arranged with fixed durations and some are arranged with open contracts, and that the empirical signatures of both arrangements appear in the mortality data.

The chapter takes four bodies of evidence in sequence. The first is the molecular biology of cellular aging — the telomere literature opened by Elizabeth Blackburn and her collaborators, the Hayflick limit on cellular replication that anchors it, and the longevity-gene literature whose most striking single-gene effects have been documented in nematodes by Cynthia Kenyon and in mammals through the FOXO and SIRT pathways. The second is the geographic clustering of extreme longevity in the populations Dan Buettner has named Blue Zones, and the question of what factors beyond genetics and diet explain the clustering. The third is the death-postponement literature documented by David Phillips and his collaborators, in which mortality rates shift around culturally meaningful dates in ways that the random-distribution model has no available account for. The fourth is the social-connection-and-mortality literature, anchored by the Holt-Lunstad meta-analyses and the older clinical trials on social support and survival. Each body of evidence is examined for what it establishes and what it does not. The chapter closes by assembling the four threads into the fixed-duration and open-contract framing.

A note on register before the chapter proceeds. The topic of when and why specific lives end is sensitive, and the chapter does not approach it casually. The pre-selection model does not hold that every person who dies young chose the timing in any sense that would assign blame to the dying or comfort to the bereaved. The model holds that the duration of a visit, like the other dimensions the book has surveyed, is on the available evidence less random than the lottery view assumes — that the timing of departure shows patterns the lottery has no account for and the pre-selection model does. The chapter develops the case with appropriate clinical neutrality. Readers carrying personal grief should know that the chapter does not require their grief to mean any particular thing. The chapter examines the data and reports what the data shows.

The Biological Clock

The most direct biological substrate of human lifespan was identified, in its modern form, by the American cell biologist Leonard Hayflick in 1961. Hayflick demonstrated that normal human cells in culture do not replicate indefinitely; they reach a limit, after approximately forty to sixty cell divisions, beyond which they enter a state of senescence and stop dividing.[129] The finding was contrary to the prevailing view, which held that cells in optimal conditions could replicate indefinitely. The Hayflick limit, as it came to be called, has been confirmed many times since and now constitutes one of the foundational observations in the cell biology of aging. Cells are programmed to stop. The programming is internal to the cell. The programming is identifiable, and the mechanism that implements it was discovered three decades later.

The mechanism is the telomere. Telomeres are repetitive sequences of DNA at the ends of chromosomes that protect the chromosome from degradation and that shorten by a small amount each time the cell divides. When telomeres shorten below a critical threshold, the cell receives an internal signal to enter senescence or

[129]Hayflick, L., and Moorhead, P. S. (1961). "The serial cultivation of human diploid cell strains." Experimental Cell Research 25(3): 585-621. Hayflick's paper was initially rejected by the Journal of Experimental Medicine before being accepted by Experimental Cell Research; the rejection is described in Hayflick, L. (1994). How and Why We Age. New York: Ballantine Books.

apoptosis — programmed cell death. The discovery of telomere structure and the enzyme telomerase that maintains it earned the 2009 Nobel Prize in Physiology or Medicine for the American molecular biologists Elizabeth Blackburn, Carol Greider, and Jack Szostak.[130] Blackburn's subsequent collaboration with the psychologist Elissa Epel demonstrated, in a 2004 paper in the Proceedings of the National Academy of Sciences, that telomere length is not solely determined by genetic inheritance and chronological age. Chronic psychological stress shortens telomeres at measurable rates. Mothers caring for chronically ill children showed telomere shortening equivalent to a decade of additional chronological aging, with the magnitude of the effect proportional to the duration and intensity of the caregiving stress.[131] The biological clock has a programmed substrate and a plasticity layer that responds to the conditions of the life. Both are real.

The plasticity has been documented at the level of single-gene effects in the longevity-gene literature. The American biologist Cynthia Kenyon, working at the University of California at San Francisco in the early 1990s, showed that a single mutation in the daf-2 gene of the nematode worm Caenorhabditis elegans doubled the organism's lifespan.[132] The doubling was achieved by altering the activity of an insulin-like signaling pathway that, when partially suppressed, shifted the organism's resource allocation away from

---

[130]Greider, C. W., and Blackburn, E. H. (1985). "Identification of a specific telomere terminal transferase activity in Tetrahymena extracts." Cell 43(2 Pt 1): 405-413. The Nobel Prize citation, including the role of Jack Szostak's earlier yeast work, appears in the Nobel Foundation press release, October 5, 2009.

[131]Epel, E. S., Blackburn, E. H., Lin, J., Dhabhar, F. S., Adler, N. E., Morrow, J. D., and Cawthon, R. M. (2004). "Accelerated telomere shortening in response to life stress." Proceedings of the National Academy of Sciences 101(49): 17312-17315.

[132]Kenyon, C., Chang, J., Gensch, E., Rudner, A., and Tabtiang, R. (1993). "A C. elegans mutant that lives twice as long as wild type." Nature 366(6454): 461-464.

reproduction and toward somatic maintenance. The daf-2 pathway is conserved across the evolutionary distance from worms to mammals; its mammalian equivalents include the insulin/IGF-1 signaling pathway and the FOXO family of transcription factors, several of whose human variants have been associated, in centenarian-genetics studies, with statistically significant longevity advantages. The genetic substrate of lifespan is modulable. It is not a fixed number assigned at birth; it is a programmed substrate with measurable responsiveness to environmental, behavioral, and stress-related inputs throughout life.

What the biological-clock literature establishes for the argument of this chapter is the first half of the framing. Lifespan has structure. The structure is partly programmed at the cellular level by telomere dynamics, partly programmed at the genetic level by longevity pathways, and partly modulated through the life by stress, behavior, and environment. The programmed substrate sets a band of possible durations within which any individual life is likely to fall, conditional on the substrate's specific configuration. The modulation through life moves the actual duration within that band toward its earlier or later extreme. Neither the programmed substrate nor the modulation, on its own, determines when any specific life ends. Both contribute, in measurable proportions, to the outcome that is observed.

The Blue Zones

The most thoroughly studied populations of extreme longevity are the five regions identified by the National Geographic researcher Dan Buettner as Blue Zones. The term originated with the Belgian demographer Michel Poulain and the Italian physician Gianni Pes, who in 2000 marked the longest-lived villages of Sardinia with blue ink on a research map. Buettner extended the framework to four additional populations: Okinawa in Japan, Nicoya in Costa Rica, Ikaria in Greece, and the Seventh-Day Adventist community of Loma Linda in California.[133] The five Blue Zone populations share an unusual rate of centenarians and supercentenarians, and substantially longer healthspan — the years lived in good health rather than the years lived after major morbidity — relative to comparable populations elsewhere.

The factors Buettner and his collaborators identified as common across the five Blue Zones fall into several recognizable clusters. Plant-based diet with modest caloric intake. Daily natural movement integrated into ordinary life rather than scheduled exercise. Strong family and community networks with frequent intergenerational contact. Membership in religious or spiritual communities with regular practice. And, in the framing that has received the most attention, a culturally specific concept of life-meaning — what the Okinawans call ikigai, what the Nicoyans call plan de vida, what the Seventh-Day Adventists frame in religious-vocational terms. The cluster of factors is well documented. The mechanisms are partly clear: plant-based diet reduces cardiovascular risk, natural movement preserves musculoskeletal function, social connection reduces stress-related mortality. The cluster of factors does not fully explain the magnitude of the longevity advantage observed in the Blue Zones, however, and the residual difference between expected and observed longevity in these populations remains an open question in the demographic and epidemiological literature.

Two features of the Okinawan Blue Zone deserve specific mention because they carry the weight of the social-connection

[133]Buettner, D. (2008). The Blue Zones: Lessons for Living Longer from the People Who've Lived the Longest. Washington, DC: National Geographic Books. The original Poulain-Pes designation appears in Poulain, M., Pes, G. M., Grasland, C., Carru, C., Ferrucci, L., Baggio, G., Franceschi, C., and Deiana, L. (2004). "Identification of a geographic area characterized by extreme longevity in the Sardinia island: The AKEA study." Experimental Gerontology 39(9): 1423-1429.

thread the chapter will return to in the next section. The first is the institution of the moai — a small mutual-support group, typically of five to eight individuals, formed in childhood and maintained for life. Moai members meet regularly across the entire span of adult existence; they provide financial assistance, emotional support, and practical caregiving as members age; and the practice produces a lifelong network of intimate connection that operates as a structural feature of Okinawan society rather than as an individual lifestyle choice. The Okinawan centenarians who live longest are not, on the available data, those with the most favorable genetics or the most stringent diets. They are, with statistical reliability, those most embedded in active moai networks at the time of their hundredth birthday.[134] The second feature is the concept of ikigai itself — a reason for being, often translated as the intersection of what one is good at, what one loves, what one can be paid for, and what the world needs. The empirical observation, documented in the cohort studies of Okinawan and other Japanese populations, is that adults who report a clearly articulated ikigai show significantly reduced all-cause mortality over the subsequent decade, controlling for the standard medical risk factors. The effect is independent of physical health and persists across follow-up periods of twelve and seventeen years.[135]

---

[134]Willcox, B. J., Willcox, D. C., and Suzuki, M. (2001). The Okinawa Program: How the World's Longest-Lived People Achieve Everlasting Health—and How You Can Too. New York: Three Rivers Press. The Okinawa Centenarian Study, which provides the underlying data, has been described in Willcox, D. C., Willcox, B. J., Wang, N. C., He, Q., Rosenbaum, M., and Suzuki, M. (2008). "Life at the extreme limit: Phenotypic characteristics of supercentenarians in Okinawa." Journals of Gerontology, Series A 63(11): 1201-1208.

[135]Sone, T., Nakaya, N., Ohmori, K., Shimazu, T., Higashiguchi, M., Kakizaki, M., Kikuchi, N., Kuriyama, S., and Tsuji, I. (2008). "Sense of life worth living (ikigai) and mortality in Japan: Ohsaki Study." Psychosomatic Medicine 70(6): 709-715. Follow-up: Tanno, K., Sakata, K., Ohsawa, M., Onoda, T., Itai, K., Yaegashi, Y., and Tamakoshi, A. (2009). "Associations of ikigai

The Loma Linda Adventist Health Study, an independent prospective cohort study of approximately 96,000 Seventh-Day Adventists in the United States, provides a complementary line of evidence with a different cultural-religious framing.[136] Adventist men live an average of 7.3 years longer, and Adventist women an average of 4.4 years longer, than comparable Californians outside the religious community. The lifestyle factors associated with the longevity advantage are documented — vegetarian diet in many adherents, abstention from alcohol and tobacco, weekly day of rest, regular community attendance, and the broader integration of religious practice into daily life. As with the Okinawan data, the standard variables account for a substantial fraction of the longevity advantage but not all of it. The residual difference, in both populations, suggests that the integrated coherence of life-meaning, community structure, and daily practice operates as a structural variable in its own right rather than as the simple sum of its identifiable components.

What the Blue Zones data contributes to the argument of this chapter is the observation that extreme longevity clusters geographically and culturally in patterns that are not fully predicted by genetic background, by environmental quality, or by the specific health-behavior variables most commonly invoked. The clustering is real. The factors so far identified account for a portion of the clustering but not all of it. The residual is not metaphysical evidence on its own; it is an open scientific question, currently the subject of active research. What the residual establishes for the framing this

---

as a positive psychological factor with all-cause mortality and cause-specific mortality among middle-aged and elderly Japanese people." Journal of Psychosomatic Research 67(1): 67-75.

[136]Fraser, G. E., and Shavlik, D. J. (2001). "Ten years of life: Is it a matter of choice?" Archives of Internal Medicine 161(13): 1645-1652. Subsequent analyses in the Adventist Health Study 2 series have been published throughout the 2010s by the team at Loma Linda University School of Public Health.

chapter develops is that lifespan is, even within mainstream demographic science, a more clustered and patterned variable than the lottery model would predict. Lives of substantial duration occur disproportionately in specific places, among specific populations, under specific conditions. The pre-selection model does not need to predict the specific Blue Zones to be consistent with their existence; it predicts that the duration of visits would cluster by purpose and by arrangement, which is what the data shows.

The Death-Postponement Effect

A more pointed finding for the pre-selection framing comes from the American sociologist David Phillips, who spent a research career at the University of California at San Diego documenting patterns in mortality timing that mainstream demographic theory had not predicted. Phillips's central observation was that mortality rates, when examined at fine temporal resolution, are not uniformly distributed across the calendar but cluster systematically around culturally significant dates. The mortality of Jewish men, examined in a 1988 paper in The Lancet, dropped measurably in the week before Passover and rose in the week after. The mortality of Chinese women, examined in a 1990 paper in the Journal of the American Medical Association, dropped in the week before the Harvest Moon Festival and rose in the week after. The pattern Phillips called death postponement — the apparent shifting of mortality timing across a culturally meaningful date — appeared in multiple populations under different cultural frameworks, with the magnitude of the shift roughly proportional to the cultural significance of the date in question.[137]

[137]Phillips, D. P., and King, E. W. (1988). "Death takes a holiday: Mortality surrounding major social occasions." The Lancet 332(8613): 728-732. Phillips, D. P., and Smith, D. G. (1990). "Postponement of death until symbolically meaningful occasions." Journal of the American Medical Association 263(14): 1947-1951.

Phillips's findings have been contested in subsequent literature. A 2004 reanalysis by Skala and Freedland, examining the same Chinese Harvest Moon Festival data with refined statistical methods, did not find the postponement effect at the magnitude Phillips reported, and several other replication attempts have produced mixed results.[138] The status of the death-postponement effect in the published literature is therefore mixed — present in some datasets, absent or attenuated in others, with the magnitude depending on the specific population, the specific date, and the statistical method used. The chapter does not treat the Phillips findings as fully settled. What it takes from the literature is the more conservative finding that has held across replications: birthday effects on mortality, in which mortality rates show statistically significant elevations on or near the individual's birthday. The birthday effect is robust and has been replicated under multiple analytic protocols. The interpretation is contested — proposed explanations include increased stress around the date, increased alcohol consumption, and the psychological loading of the birthday as a temporal marker — but the effect itself is documented.

The most rigorous published analysis of the birthday effect is the Swiss study by Vladeta Ajdacic-Gross and colleagues at the University of Zurich, which examined 2.4 million deaths recorded in Switzerland over a forty-year period.[139] The team found that mortality on the day of one's birthday was 6.8 percent higher than on comparison days in the surrounding weeks, with the effect concentrated in older adults and in deaths from cardiovascular causes, falls, and suicide. The magnitude of the effect was large enough to rule out random variation as the explanation, and the

[138]Skala, J. A., and Freedland, K. E. (2004). "Death takes a raincheck." Psychosomatic Medicine 66(3): 382-386.
[139]Ajdacic-Gross, V., Knöpfli, D., Landolt, K., Gostynski, M., Engelter, S. T., Lyrer, P. A., Gutzwiller, F., and Rössler, W. (2012). "Death has a preference for birthdays — An analysis of death time series." Annals of Epidemiology 22(8): 603-606.

statistical analysis controlled for the known confounding factors — seasonal variation, the day-of-week effect, and the possibility of administrative date-of-death recording errors. The birthday effect, in the Ajdacic-Gross dataset, is real, is sizable, and is not attributable to recording artifacts. The interpretation remains contested in detail, but the consistent finding across the largest available analyses is that the timing of human death is partly shaped by the calendar of the dying person's life history.

The interpretation this chapter takes is structural rather than strongly causal. The death-postponement and birthday-effect literatures, taken together, establish that mortality timing is not uniformly random but is influenced, at measurable levels, by culturally and personally meaningful dates. Whether the mechanism is conscious resistance, unconscious physiological readiness, or some other route, the timing of human death is, on the data, partly shaped by what the dying find meaningful. This finding is consistent with the pre-selection model in a specific way. If the duration of a visit is arranged before arrival, with culturally and personally meaningful dates having significance in the arrangement, then mortality timing should show clustering around such dates. The clustering is documented. The mechanism is, at present, not fully understood within the standard demographic framework. The pre-selection framing does not require the mechanism to be understood within that framework to be consistent with the clustering. It expects the clustering.

Social Connection and Survival

The fourth body of evidence for the structural patterning of lifespan comes from the social-connection literature. The relationship between social ties and mortality has been documented across multiple research traditions for over four decades, but the most rigorous synthesis is the meta-analytic work of the American psychologist Julianne Holt-Lunstad and her collaborators.[140] Holt-Lunstad's 2010 meta-analysis, pooling data from 148 prospective studies covering over 308,000 participants followed for an average of 7.5 years, found that individuals with stronger social relationships had a 50 percent higher likelihood of survival relative to those with weaker relationships. The hazard ratio is comparable in magnitude to the mortality effect of smoking fifteen cigarettes per day, and exceeds the mortality effect of obesity, physical inactivity, and exposure to air pollution. The finding has been replicated across studies, across populations, and across methodologies. Social isolation is, on the meta-analytic evidence, among the most consequential single predictors of mortality currently identified in the epidemiological literature.

The clinical-trial evidence on social support and survival is more contested. The most discussed clinical trial in the field is the 1989 study by the Stanford psychiatrist David Spiegel, published in The Lancet, which followed eighty-six women with metastatic breast cancer randomized to standard care or to standard care plus weekly group support sessions. The supported group survived, on average, eighteen months longer than the standard-care group.[141] The 1989 finding generated substantial follow-up research, including a

[140]Holt-Lunstad, J., Smith, T. B., and Layton, J. B. (2010). "Social relationships and mortality risk: A meta-analytic review." PLoS Medicine 7(7): e1000316. Subsequent extension: Holt-Lunstad, J., Smith, T. B., Baker, M., Harris, T., and Stephenson, D. (2015). "Loneliness and social isolation as risk factors for mortality: A meta-analytic review." Perspectives on Psychological Science 10(2): 227-237.

[141]Spiegel, D., Bloom, J. R., Kraemer, H. C., and Gottheil, E. (1989). "Effect of psychosocial treatment on survival of patients with metastatic breast cancer." The Lancet 334(8668): 888-891. Replication trial: Goodwin, P. J., Leszcz, M., Ennis, M., Koopmans, J., Vincent, L., Guther, H., Drysdale, E., Hundleby, M., Chochinov, H. M., Navarro, M., Speca, M., and Hunter, J. (2001). "The effect of group psychosocial support on survival in metastatic breast cancer." New England Journal of Medicine 345(24): 1719-1726.

2001 New England Journal of Medicine multi-site replication trial led by Pamela Goodwin that did not reproduce the survival advantage at the original magnitude. The interpretation of the contested replication remains active in the clinical literature. What the 1989 trial and the subsequent meta-analytic work jointly establish is that the relationship between social connection and mortality is real, that the magnitude of the relationship is substantial, and that the specific clinical-intervention versions of the finding are sensitive to the design of the intervention in ways that complicate translation into standard medical practice.

For the framing of this chapter, the social-connection literature contributes a structural variable that the standard medical models of mortality do not centrally treat. Mortality is not, on the available evidence, primarily a function of biological substrate plus health behavior. The substrate matters and the behavior matters, but a third variable — the density and quality of the dying person's connection to others — exerts an independent effect of comparable magnitude. The pre-selection model accommodates this finding directly. If the visit is arranged with specific relationships as part of its purpose, the visit's duration should track the integrity of those relationships. The empirical fact that mortality clusters around social-connection variables at the magnitudes Holt-Lunstad has documented is the operational signature of that arrangement in the available data.

The most striking historical natural experiment in the social-connection literature is the case of Roseto, Pennsylvania, a small town settled in the late nineteenth century by immigrants from a single village in southern Italy. The American physician Stewart Wolf and the sociologist John Bruhn conducted a fifty-year study of Roseto beginning in the early 1960s, motivated by the observation that the town's mortality from heart attack was approximately half the rate of neighboring towns with comparable demographic and economic profiles.[142] The Rosetans, on every standard medical risk-factor measure, should have been at elevated cardiovascular risk: their diet was high in fat, cured meats, and full-fat cheeses; many residents smoked; physical exercise was modest by mid-century American standards; and the town's industries — quarrying and garment work — were stressful and physically demanding. The mortality data, however, was anomalously favorable. What distinguished Roseto from its neighboring towns was not biology or behavior. It was social structure. Three-generation households were the norm. Mutual aid was strong. Communal celebrations were frequent and integrative across class. Wealth differences were modulated by social custom to avoid visible status display. The townspeople knew one another's lives in detail and provided practical assistance through illness, financial reversal, and family crisis. Wolf and Bruhn's fifty-year follow-up documented what happened when the social structure changed. Beginning in the 1960s, as the town modernized and the second and third generations dispersed into the broader American suburban pattern, the cardiovascular mortality rate rose to match neighboring towns. The diet had not appreciably changed. The smoking rate had not appreciably changed. The social structure had changed. The Roseto Effect, as it came to be called, is one of the cleanest available demonstrations that social connection operates as a structural determinant of mortality at magnitudes comparable to or exceeding any individual health-behavior variable.

The neurobiological substrate of the social-connection effect has been examined most directly by the late American social

[142]Wolf, S., and Bruhn, J. G. (1993). The Power of Clan: The Influence of Human Relationships on Heart Disease. New Brunswick, NJ: Transaction Publishers. The fifty-year longitudinal study findings on the dissolution of the Roseto effect as the town modernized are summarized in Egolf, B., Lasker, J., Wolf, S., and Potvin, L. (1992). "The Roseto effect: A 50-year comparison of mortality rates." American Journal of Public Health 82(8): 1089-1092.

neuroscientist John Cacioppo, whose research program at the University of Chicago documented the physiological consequences of chronic loneliness across two decades of investigation.[143] Cacioppo and his collaborators showed that perceived social isolation — the subjective experience of loneliness, independent of the objective number of social contacts — produces measurable elevations in inflammatory markers, in glucocorticoid resistance, in cardiovascular reactivity, and in indicators of accelerated cellular aging including telomere shortening of the kind Blackburn and Epel had documented from a different angle. Loneliness, on Cacioppo's data, operates through the same molecular substrate as chronic stress and through several additional routes specifically associated with social-evaluative threat. The substrate of social connection is not an abstract psychosocial variable. It is a measurable input to the body's stress-response and inflammation systems, with consequences for the molecular biology that the chapter's first section identified as the substrate of programmed lifespan. Social connection modulates the telomere clock. The pathways are documented.

Fixed Duration and Open Contract

The four bodies of evidence assembled above support an intermediate framing of the pre-selection of lifespan. Lifespan is neither a fixed number assigned at conception nor an arbitrary product of biological and circumstantial chance. It is, on the cumulative evidence, a planned duration with a band of acceptable individual modulation, and the modulation is shaped by factors that the standard framework has identified but that the standard framework does not centrally weight.

[143] Cacioppo, J. T., and Patrick, W. (2008). Loneliness: Human Nature and the Need for Social Connection. New York: W. W. Norton. Foundational paper: Cacioppo, J. T., and Hawkley, L. C. (2003). "Social isolation and health, with an emphasis on underlying mechanisms." Perspectives in Biology and Medicine 46(3 Suppl): S39-S52.

The framing this book develops uses two terms. A fixed-duration visit is a visit whose planned length is narrow, in which the timing of departure is arranged at or near a specific point regardless of the individual modulation that would otherwise widen or narrow the window. An open-contract visit is a visit whose planned length is broader, in which the individual modulation — health, relationships, sense of purpose, will — is the principal determinant of the actual duration within a wider window. Most lives, on the framing, are open-contract visits within fairly broad windows. Some are fixed-duration visits whose timing reflects the specific work the visit was assembled to do.

Within this framing, several features of the mortality data become intelligible that the lottery model has trouble accounting for. The death-postponement clustering around culturally meaningful dates is consistent with an open-contract structure in which the dying have, at the margin, the capacity to bring or postpone the departure relative to events they find meaningful. The Blue Zones clustering of extreme longevity is consistent with the existence of populations whose individual visits are disproportionately open-contract and whose conditions of life sustain the modulation toward the upper end of the window. The social-connection mortality effect is consistent with relationship-integrity being a structural determinant of the modulation rather than a peripheral lifestyle factor. The telomere-shortening response to chronic stress is consistent with the modulation being implemented biologically through the same molecular substrate that calibrates the planned duration. None of these consistencies proves the framing. Each removes an explanatory difficulty that the lottery model carries unresolved.

The fixed-duration cases are the harder topic, and the chapter treats them with appropriate restraint. Visits that end early in childhood, in adolescence, or in young adulthood occur. The pre-selection model does not require any specific interpretation of any individual case. What the model holds is that the cumulative distribution of such cases, considered as a class, shows patterns that the lottery model does not predict. Children who die very young show, in the cross-cultural reincarnation literature surveyed in Chapter 8 and to be developed in Chapter 11, a disproportionate rate of return to closely related families within a short interval — a pattern that has been documented in Stevenson's archive and that is structurally consistent with the fixed-duration framing the present chapter has developed. The brief life is not random. It is, on the pre-selection model, a specific kind of visit whose duration is arranged in advance, whose purpose may be narrower than that of an extended visit, and whose impact on the family system is part of what the visit was assembled to do. None of this consoles. None of it is offered as consolation. It is offered as the chapter's most conservative reading of the available data on the duration of human lives.

A note to readers carrying the loss of a brief life. The framing this chapter develops does not claim to explain any specific death, and the book's voice would betray itself if it tried to. What the framing does is reject two errors the lottery model invites the bereaved to adopt. The first error is the conclusion that the death was meaningless — that the visit, having ended too soon, ended without purpose, leaving only the random arithmetic of biological accident. The pre-selection framing rejects this conclusion. The second error is the conclusion that the bereaved are responsible for the death — that something they did or did not do produced the timing they did not want. The pre-selection framing equally rejects this conclusion. The brief visit, on the framing, is a visit whose duration was part of what it was. The reasons for any specific arrangement are not knowable from the bereaved's position, and the book makes no attempt to specify them. The framing offers what it can: the recognition that the visit was a visit, that its duration is part of what it was assembled to be, and that the meaning of the visit is not exhausted by its length. That is, at the chapter's most honest, what the data the chapter has surveyed supports.

What This Establishes

The chapter's load is modest by the standards of the book's earlier chapters. It is to establish that lifespan, like the other dimensions the prior chapters have surveyed, shows empirical signatures of patterning that the lottery model has not accounted for and that the pre-selection model accommodates without strain. The molecular biology of cellular aging shows a programmed substrate with measurable plasticity. The Blue Zones data shows that extreme longevity clusters by geography, culture, and life-meaning in patterns whose underlying factors do not fully reduce to identified variables. The death-postponement and birthday-effect literatures show that mortality timing is partly shaped by personally and culturally meaningful dates. The social-connection literature shows that mortality is structurally influenced by the integrity of the dying person's relationships at magnitudes comparable to major medical risk factors. Each of these findings is consistent with mainstream science in its own register. The conjunction of the findings, taken together, supports a framing in which lifespan is partly planned and partly modulable — fixed-duration for some visits and open-contract for others — rather than randomly distributed within the biological band.

The framing also closes the second part of the book. The four chapters of Part Two have established the blueprint of the pre-selection: the ascent through forms, the receiver model of consciousness, the matching of lineage and conditions of arrival, the recovered memory of the arrangement, and now the patterning of duration. The visit's universe was configured, its address was specific, its body was curated, its substrate was tuned for integration, its consciousness predates the substrate, its conditions of arrival are matched at the molecular level, its arrival is reported as chosen, and its duration shows the empirical signatures of arrangement. The blueprint is, in the cumulative weight of the evidence the book has surveyed, more demanding to dismiss than to provisionally accept.

The blueprint is not, on the book's framing, a deterministic script. The visitor is not a passenger whose entire experience is fixed at the moment of departure. Each of the dimensions the book has surveyed is a structural parameter within which the visit operates with real agency. The universe is configured but the visit acts within it. The body is curated but the choices the body makes are the visitor's. The lineage is matched but the relationships are lived. The duration is arranged but the modulation through life — the stress, the connection, the meaning, the will — is the visitor's contribution to where within the arranged band the actual departure falls. The blueprint sets the conditions of the visit. The visit, within the conditions, is what the visitor does. Part Three of the book takes up the evidence that this is the structure the data supports.

✦ ✦ ✦

The bedroom at La Paisible faced east across the orchard and the gravel drive toward the village of Tolochenaz and, beyond Tolochenaz, the low waters of Lake Geneva that on clear mornings reflected the Alps along the southern shore. She had bought the house in 1965, three years after the birth of her first son and during the long marriage to Mel Ferrer that would end in 1968. She had chosen the village for the schools and for the privacy and for the orchards and for the kind of slow steady life she had been trying, throughout her acting career, to construct around the parts of her that the acting career had been organizing in directions she had not wanted to be organized. She had kept the house through the second marriage to Andrea Dotti and through the second motherhood and through the long quiet decade after the films had largely stopped. By the time Robert Wolders had moved in with her in 1981 the house had been, for sixteen years, the place that was hers. She had named it La Paisible. The name meant the peaceful one.

She had returned to La Paisible from Cedars-Sinai Medical Center in Los Angeles on the eighteenth of December, 1992, on a private jet that Hubert de Givenchy had arranged for her when it had become clear that her body would not, in the condition the surgeons had returned her to after the laparoscopy in early November, sustain a commercial flight. Givenchy had filled the cabin with white flowers. The jet had taken her, with Robert Wolders and the two sons, from Los Angeles to Geneva in fourteen hours. The car had driven her from the Geneva airport to Tolochenaz in another forty minutes. She had walked into the house. She had sat down in the chair by the window in the bedroom. She had slept for several hours. She had then, in the way of a person who had been informed that the most that could be done had been done and that the time remaining to her was on the order of weeks, opened her appointment book and begun to organize what remained of the appointment book around the work she still wanted to complete.[144]

The two sons came in to see her every day across the next four weeks. Sean was thirty-two. Luca was twenty-two. They had grown up in the house. They had grown up around her in the long quiet years that had followed the Hollywood career, and they had

[144]On Audrey Hepburn's diagnosis in November 1992 with pseudomyxoma peritonei, the surgery at Cedars-Sinai, Hubert de Givenchy's private jet arrangement, and the return to La Paisible: see Ferrer (2003), pp. 245-272; Spoto (2006), Ch. 16; Walker (1994), pp. 360-372. On the specific cancer — a rare appendiceal malignancy in the pseudomyxoma peritonei family, statistically over-represented in cohorts with prenatal nutritional deprivation — see Sugarbaker, P. H. (2006). "New standard of care for appendiceal epithelial neoplasms and pseudomyxoma peritonei syndrome?" Lancet Oncology 7(1): 69-76; and the Dutch Famine Birth Cohort follow-up by Lumey, L. H., Stein, A. D., Kahn, H. S., and Romijn, J. A. (2009). "Lipid profiles in middle-aged men and women after famine exposure during gestation: The Dutch Hunger Winter Families Study." American Journal of Clinical Nutrition 89(6): 1737-1743. Robert Wolders's recollection that "the doctors gave her three months, she lived three months" appears in his interview with Cardoza, T. (1993). Audrey Hepburn. New York: Birch Lane Press, p. 244.

been at her bedside in Los Angeles during the surgery and the chemotherapy that had not, in the end, accomplished what the chemotherapy had been ordered to accomplish. The cancer was a rare appendiceal malignancy in the family the pathologists called pseudomyxoma peritonei. It had been growing for, on the surgeons' best estimate from the laparoscopy findings, between four and eight years. By the time it had produced the abdominal pain she had begun to feel on the Somalia trip the previous September, it had spread through the lining of the peritoneal cavity in the gelatinous coating the disease characteristically produced, and the only available intervention was the cytoreductive surgery the team in Los Angeles had attempted in November and had, in the second exploratory operation, concluded could not be completed. The disease was, in the Dutch Famine Birth Cohort epidemiological literature that the Leiden group had begun publishing in the late 1990s and that would mature substantially over the decade after her death, statistically over-represented in women who had been in the third trimester of fetal development during the Hunger Winter of 1944 and 1945. She had been fifteen during that winter, not in utero, but the cohort signature she carried in the methylation pattern of the insulin-like growth factor 2 gene had been the same signature the cohort women carried, written into the body by the same five-month window of caloric restriction, and the body the famine had built was completing its work on a schedule the cohort signature predicted within a band of biological variation that the epidemiological data would, in subsequent decades, document with statistical precision.

On the last Christmas she had spent in the house, eleven days after returning from Los Angeles, she had asked a friend who had been visiting to go into Lausanne and buy three winter coats for her. The coats were for Givenchy and for Sean and for Robert Wolders. She had wrapped them herself. She had given them out on Christmas morning in the living room. She had told the three men, when they had unwrapped the coats, that they were to think of her when they wore them. She had told Wolders, after the others had left and they had gone up to bed, that it had been the most beautiful Christmas she had ever had. He had not, that night, known what to say to her. He had said, in the years afterward when he was asked about that Christmas and about the last weeks at La Paisible, that the thing she had been doing in the final weeks had not been the dying of a person who was being taken from a life she had not finished. The thing she had been doing had been the completion, in the small attentive ways she had organized the days around — the coats, the letters to the UNICEF country offices, the calls to the families of the children she had met on the field missions, the conversations with the sons about the disposition of the house and the practical matters that would follow — of the work the visit had been arranged to do.[145]

She died at her home in the evening of January 20, 1993, with Robert Wolders and the two sons in the room. She was sixty-three years old. The funeral was held at the village church in Tolochenaz four days later. The priest who conducted the service was the same priest who had married her to Mel Ferrer at the same church in 1954 and who had baptized Sean at the same church in 1960. The interment was at the village cemetery on the same afternoon. The visit had been four years and ten months from the orphanage in Mekele. The visit had been sixty-three years and eight months from her birth in Brussels in May of 1929. The work the visit had been assembled to do had been, in the cumulative weight of the field missions and the interviews and the testimony to the United States Congress and the speeches to the General Assembly and the photographs of her sitting with the children at the long tables in the centers, completed. The body the famine had built had carried her

[145]Robert Wolders's reflections on the last Christmas at La Paisible, including the three winter coats Hepburn arranged for Givenchy, Sean, and Wolders himself, are documented in Wolders's interviews with the Audrey Hepburn Children's Fund archive and in Ferrer (2003), pp. 268-272. The coats remain in the collection of the Hepburn estate.

through it. The cohort signature had, on the schedule the cohort signature was carrying, completed the body's function. The peaceful one had been the place she had chosen to finish in. She had finished there. The chapter that follows, opening the third part of the book, takes up the question of how the visit's purpose is held across the substrate that carries it, and how the substrate transmits what the visit has done to the substrates that will follow.

Where the Argument Goes Next

Part Three of the book turns from the blueprint to the evidence in its most concentrated form. The next chapter examines what is now called epigenetic memory — the molecular substrate by which experience encoded in one generation is carried into the next, and the broader phenomenon of memory in matter that the Yehuda and Dias-Ressler findings in Chapter 7 introduced and that the present chapter has touched in the telomere work. The chapter after that turns to the cross-substrate memory documented in the Stevenson and Tucker case archive — the over 2,500 children who have reported specific, verifiable memories of prior lives, including the birthmark-and-birth-defect correspondence subset that is the most empirically constrained portion of the corpus. The chapter after that turns to the quantum substrate underneath the receiver model — the experimental literature on quantum coherence in biological systems, from avian magnetoreception to photosynthesis to the microtubule findings of Chapter 6 — that grounds the receiver model in measured physics rather than speculation. Part Three is the cumulative evidence the rest of the book's argument has been preparing the reader to weigh.

CHAPTER 10

Memory in Matter

The doctor who examined the body of Bua Kai Lawnak on a morning in the early summer of 1962 was a general practitioner who served the cluster of small villages along the dirt road that connected them, in the central plains region of Thailand the schoolteacher had been bicycling toward when he had been shot. The doctor had come out from the dispensary in the larger town to the south on the back of a motor scooter driven by a younger man who worked as a courier in the district, with his medical bag strapped to the rear rack and a sheaf of paper forms folded into the inside pocket of his shirt. He had been called by the headman of Bua Kai's village, who had been called by the boy who had found the body on the road. The body was on the road, on its side, with the bicycle a short distance away and the cap that had been on the head several meters further along where it had come to rest. A small group of villagers had gathered. The widow had been brought down from the house. The widow was thirty-one. She had a one-year-old and a three-year-old at the house with her mother-in-law. She was not, by the time the doctor arrived, making any sound. She was standing in the way the women of that province stood when they had been told a thing they had not expected to be told that morning.

The doctor knelt by the body. He turned the body, with the help of the headman, onto its back. He examined the wound on the back of the head. It was a small round wound, perhaps eight or nine millimeters in diameter, with the kind of clean entry margins that gunpowder of any sufficient velocity produced when it passed through human bone at close to perpendicular. He examined the wound on the forehead, above and slightly to the right of the left eye. It was larger, perhaps two centimeters across at the widest, with the irregular outline and the protruding fragments of cranial bone that an exit wound produced when the projectile that had entered the smaller wound on the other side of the head had carried, by the time it had exited, both its remaining kinetic energy and a substantial portion of the cortical tissue it had passed through. The doctor had been a general practitioner in the district for eleven years. He had examined several bodies in similar conditions over those years. The pattern of the two wounds did not, on his examination, admit of any ambiguity as to which was the entry and which was the exit, or as to the position the shooter had been in relative to the bicycle.[146]

He filled in the form. He noted the date and the location and the approximate time of death, which the body's temperature and the rigor of the limbs suggested was a few hours earlier. He noted the manner of death, which was a single gunshot from behind at a range of perhaps two to five meters. He noted the two wounds. He did not note any other findings of significance, because there were no other findings of significance. The body was the body of a thirty-two-year-old man in otherwise unremarkable health who had been bicycling along a road on the way to the school where he had taught for the previous nine years, and who had been killed by a single projectile fired by a person whose identity he did not, on his examination of the

[146]On the medical examination of Bua Kai Lawnak's body in central Thailand in 1962, including the description of the entry and exit wounds, see Stevenson, I. (1997). Where Reincarnation and Biology Intersect. Westport, CT: Praeger, pp. 300-323. No autopsy report was preserved, but Stevenson conducted interviews with multiple family members in 1979 — including Bua Kai's widow, his parents, and his surviving brothers — each of whom independently described the doctor's findings on the morning of the death. The description of the entry wound as small and round on the back of the head, and of the exit wound as larger and irregular on the forehead, is consistent across the witness accounts and was the basis on which Stevenson, when he subsequently photographed Chanai Choomalaiwong's birthmarks in 1979, identified the spatial correspondence between the marks on the boy and the wounds on the deceased man documented in the family's collective memory.

body, have any way to determine. He folded the form, put it back in his shirt pocket, and stood up. He spoke briefly with the headman about the arrangements for the funeral. The body would be wrapped and carried to the house. The widow would receive the body at the house. The wat would conduct the cremation rites two days later. The family had property and standing in the village; the funeral would not be hasty. The doctor walked over to where the widow was standing and offered her, in the few words his medical training had given him for the occasion, what he had to offer. She nodded. She did not look at him. She walked over to the body. She sat down on the ground beside the body. She put her hand on the body's chest. She remained there until the men from the village came with the wrapping and lifted the body and carried it up the road toward the house.

The body had been a body. The body had been the substrate of a life that had begun thirty-two years earlier in another village in the same district and had moved through childhood and adolescence and the teacher's college in the provincial capital and the early teaching positions and the marriage in 1958 and the birth of the first daughter in 1959 and the birth of the second in 1961 and the morning rides on the bicycle to the school three miles down the road for the previous nine years. The body had been the integration point of all of that. The body had carried the cellular signatures of every infection the body had survived, every meal the body had eaten, every emotion the body had registered, every relationship the body had built across the thirty-two years the body had been the platform of the life. The body had been, on every measure the chapter that follows will take up, an archive of the life it had been the substrate of. And the body, on the road that morning, had stopped being the integration point. The body remained. The archive remained. The life that had been integrating the archive moment by moment had, in the four or five minutes between the projectile and the doctor's arrival, departed. The substrate had returned to what the substrate was when no longer integrated. The chapter that follows takes up the substrate.

✦ ✦ ✦

Part Two of this book closed with a framing that connected the receiver model of consciousness — the brain as antenna for a signal it does not generate — to the dimensions of the visit that are arranged before arrival. Universe, address, body, substrate, lineage, conditions, and duration: each is documented, at separate evidentiary registers, as more configured than the lottery view assumes. The visitor that arrives into this configuration arrives as a signal the substrate is tuned to receive. The substrate that receives the signal is the body and brain assembled from the four-billion-year library of biological design surveyed in earlier chapters. The configuration and the receiver were the burden of Part Two.

Part Three turns to the body of evidence that gives the receiver model its operational substance. If consciousness is received rather than produced, the brain is not the only thing the visitor inhabits, and the brain is not the only thing that stores what the visitor learns and feels. The body itself — at the cellular level, at the molecular level, at the level of fascia and viscera and the millisecond-by-millisecond regulation of the autonomic system — is, on the available scientific evidence, a substrate of memory in its own right. The brain is, on the receiver model, the integration point of the body's memory. The body's memory is the substrate the brain integrates. The chapter takes the case for memory in matter as the first of the three evidence chapters that constitute Part Three. The chapter that follows takes up memory across lifetimes. The chapter after that takes up the quantum substrate underlying both.

The opening of Chapter 1 contained a phrase the chapter promised to earn at length later in the book. The phrase was that the cell knows and the body remembers, and that the remembering extends further back than the present life. The first half of the phrase is the load of the present chapter. The cell knows. The body remembers. Within the single lifetime, the substrate of memory extends well beyond the neural pathways that the standard introductory neuroscience textbooks identify as the seat of memory storage. The body is not the brain's vehicle. The body is the brain's environment, the brain's substrate, and the brain's distributed archive. The brain, on the available evidence, is the part of the body specialized for the integration of memory the rest of the body is storing in parallel.

The chapter takes the case in five parts. The first part is the cellular biology of memory across mitotic division — the somatic epigenetic mechanisms that allow a fertilized egg to differentiate into the hundreds of distinct cell types of the adult body while every cell retains an identical genome. The second part is the molecular biology of long-term memory in the synapse, anchored by the Nobel-Prize-winning work of Eric Kandel on the marine snail Aplysia californica, and extended by Kandel's later finding that long-term memory storage involves a prion-like protein mechanism that had been thought to belong to a different domain of biology entirely. The third part is the distributed-neuropeptide architecture documented by Candace Pert in the 1980s and 1990s, in which the chemical messengers of the nervous system are present throughout the body and operate as a system that is, in Pert's phrase, the body's molecular substrate of emotion. The fourth part is the trauma-medicine literature anchored by Bessel van der Kolk, in which the somatic storage of traumatic memory is established not as theoretical possibility but as the standard clinical framework in modern trauma care. The fifth part assembles the threads into the body-as-archive thesis that the next chapter then extends across substrates.

Cellular Memory Across Mitotic Division

The most basic instance of memory in matter is the one every reader carries in every cell of the body. The fertilized egg from which the body grew was a single cell with a single genome. That genome was copied, with high fidelity, into every one of the approximately thirty trillion cells that constitute the adult human body. Every cell in the body has the same DNA sequence. And yet the cells are not the same. The cells of the liver perform liver functions and not heart functions. The cells of the heart contract on cardiac timing and do not produce bile or filter toxins. The cells of the cortex form neural networks and do not secrete digestive enzymes. Every cell carries the full library of genetic instructions, but each cell reads only the portions of the library relevant to its specific role and ignores the rest. The capacity to read the right portions and ignore the others is a memory the cell maintains across every division it undergoes, for the entire duration of the life. This is cellular memory in its most fundamental form.

The mechanism of cellular memory across mitotic division is the same epigenetic machinery that Chapter 7 introduced at the transgenerational level. DNA methylation marks specific positions on the genome that are not to be read in the current cell type. Histone modifications open or close regions of the chromosomes to transcription. Non-coding RNA species reinforce the silencing of regions that need to remain quiet for the cell's role to be maintained. The regulatory annotation that distinguishes a liver cell from a neuron is the methylation pattern, the histone state, and the non-coding RNA environment specific to each cell type. When a liver cell divides, it does not produce a generic daughter cell that has to relearn that it is a liver cell. It produces a daughter cell whose epigenetic annotation matches the parent's, allowing the daughter to continue being a liver cell without re-reading the entire genome to decide which cell type it should be. The annotation is inherited, faithfully and reliably, across every cell division the lineage undergoes, for the entire life of the organism.[147]

The specificity of the cellular-memory mechanism was given a striking demonstration by the Japanese cell biologist Shinya Yamanaka, whose 2006 paper in Cell established that a small set of transcription factors — now called the Yamanaka factors — could reprogram an adult differentiated cell back to a pluripotent state resembling the original embryonic stem cell from which all the body's tissues had derived.[148] The implication of the Yamanaka work is that cellular memory is, in the most literal sense, both stable and reversible. The adult cell carries a memory of having been a skin fibroblast, a memory that is faithfully transmitted across its mitotic divisions through ordinary life. The memory can be erased — overwritten by a defined biochemical intervention — and the cell can be returned to a state of openness in which it could become any cell type the body contains. The intervention does not change the cell's DNA. It changes the cell's epigenetic annotation. The annotation is the memory. The Yamanaka demonstration is the cleanest available proof that what makes a cell what it is, in any given moment of life, is not its sequence of inherited base pairs but the inheritance of regulatory marks that say which base pairs to read.

Cellular memory across mitotic division is the foundational case of memory in matter. The mechanism is well-documented, the field that studies it is enormous and growing, and the practical

[147]The standard reference for the cellular-memory mechanism is Allis, C. D., Caparros, M.-L., Jenuwein, T., and Reinberg, D. (eds.) (2015). Epigenetics (2nd ed.). Cold Spring Harbor: Cold Spring Harbor Laboratory Press. For an accessible synthesis: Carey, N. (2012). The Epigenetics Revolution. New York: Columbia University Press.

[148]Takahashi, K., and Yamanaka, S. (2006). "Induction of pluripotent stem cells from mouse embryonic and adult fibroblast cultures by defined factors." Cell 126(4): 663-676. Yamanaka was awarded the 2012 Nobel Prize in Physiology or Medicine for this and related discoveries.

applications — regenerative medicine, cancer therapeutics, age-reversal research — are among the most active areas of contemporary biology. The case is established. The cell remembers what it is. The body that the cells compose remembers what it is by the integration of every cell's individual memory of itself. The substrate of memory begins below the level of the neuron and extends throughout every tissue the body contains.

The Molecular Biology of Memory: Kandel and Aplysia

The most extensively documented case of memory in matter at the level of the individual neuron is the work the Austrian-American neuroscientist Eric Kandel has built across a research career of more than six decades, much of it conducted on the marine sea slug Aplysia californica. Kandel was awarded the 2000 Nobel Prize in Physiology or Medicine for the discoveries that made the molecular basis of memory storage available to direct experimental investigation.[149] Aplysia was chosen as a model organism for the practical reason that its nervous system contains roughly 20,000 unusually large neurons, several thousand times fewer than the human brain but functionally similar at the cellular and molecular level. The simplicity of the Aplysia nervous system allowed Kandel's group to identify specific neurons whose individual changes during learning could be directly measured.

The findings that emerged across Kandel's program established the molecular machinery of memory storage with a precision that earlier eras of neuroscience had only theoretical access to. Short-term memory, lasting minutes to hours, was shown to involve modifications of existing synaptic proteins — the chemical

[149]Kandel, E. R. (2001). "The molecular biology of memory storage: A dialogue between genes and synapses." Science 294(5544): 1030-1038. Book-length treatment: Kandel, E. R. (2006). In Search of Memory: The Emergence of a New Science of Mind. New York: W. W. Norton.

signaling apparatus between neurons — without changes to the underlying genome. Long-term memory, lasting days to a lifetime, was shown to require the activation of specific transcription factors, particularly a protein called CREB, which travels to the cell nucleus and turns on the production of new proteins that physically remodel the synapse. The physical change at the synapse — new connections, strengthened connections, sometimes the growth of new synaptic terminals altogether — is the substrate of long-term memory. The memory is not stored as an abstract information pattern in some non-physical register. It is stored as the physical structure of the neuron's connections to other neurons. Memory, on Kandel's work, is anatomical.

Kandel's later work, conducted with his postdoctoral collaborator Kausik Si, added a finding that few in the field had anticipated. The maintenance of long-term memory — its persistence across years and decades, despite the constant turnover of the molecules that constitute the synapse — appears to involve a class of proteins that behave like prions. Prions are infectious protein particles that maintain a self-perpetuating conformational state, most famously in the context of neurodegenerative diseases such as Creutzfeldt-Jakob and bovine spongiform encephalopathy. Si and Kandel's 2003 paper in Cell demonstrated that an Aplysia synaptic protein called CPEB exists in two stable conformational states, and that activity at the synapse can drive CPEB from its inactive form into a prion-like aggregating form that self-perpetuates locally at the synapse.[150] The aggregated form of CPEB sustains the synaptic modification long after the original signaling event has ended. The memory persists because the protein's conformational state is self-

[150] Si, K., Lindquist, S., and Kandel, E. R. (2003). "A neuronal isoform of the Aplysia CPEB has prion-like properties." Cell 115(7): 879-891. Mammalian extension: Stephan, J. S., Fioriti, L., Lamba, N., Colnaghi, L., Karl, K., Derkatch, I. L., and Kandel, E. R. (2015). "The CPEB3 protein is a functional prion that interacts with the actin cytoskeleton." Cell Reports 11(11): 1772-1785.

reinforcing, in a manner mechanically analogous to the self-reinforcing protein aggregation of prion disease but in service of a normal biological function rather than a pathological one. Memory, on this finding, is not only anatomical. It is a specific kind of stable protein state that the synapse maintains by its own internal dynamics.

What the Kandel program contributes to the argument of this chapter is the documentation of memory as a property of molecular matter. The synapse is matter. The CPEB protein is matter. The conformational state that sustains the memory is a state of that matter. And the persistence of the memory across years is the persistence of the state. There is nothing magical or non-physical about the substrate. The mechanism is biochemistry. What is striking is that the biochemistry is, in its operation, a system whose function is precisely the preservation of past information across time. The matter is doing what matter is not, on the standard introductory account, supposed to be able to do. The matter is remembering. And the matter is doing this in service of an organism whose overall capacity to remember extends, as the rest of this chapter will argue, far beyond the synapses that the Kandel program has so beautifully characterized.

Distributed Memory: Pert and the Neuropeptide System

The American neuroscientist Candace Pert spent her research career, much of it at the National Institute of Mental Health, characterizing a class of chemical messengers called neuropeptides — short protein sequences that act as signals between cells and that, at the time her work began in the early 1970s, had been thought to belong primarily to the brain. Pert's first major discovery, made as a graduate student in the laboratory of Solomon Snyder at Johns Hopkins in 1972, was the opiate receptor — the cellular target of morphine and heroin and, as was subsequently established, of the body's own endogenous opioid peptides, the endorphins.[151] The discovery of the opiate receptor and its endogenous ligands established that the brain produces its own opioid compounds and uses them in normal signaling — a finding that revolutionized the understanding of pain modulation, addiction, and the neurochemistry of reward.

What Pert's subsequent program revealed was that the neuropeptide system was not confined to the brain. The receptors for neuropeptides — the cellular machinery that detects and responds to the chemical signals — appeared throughout the body. They were present on cells of the immune system, on cells of the digestive tract, on cells of the kidneys and reproductive organs, on cells of the heart and blood vessels. The same chemical signals that the brain used for emotional and cognitive regulation were operating in tissues throughout the body, with the same precision of receptor binding and the same downstream cellular consequences. The body, on Pert's work, was not a vehicle that the brain inhabited. The body was a distributed network of communication operating through chemical signals that the brain participated in but did not exclusively control.

Pert's broader framing of the implications, published in her 1997 book Molecules of Emotion, was that the molecules her research had characterized — neuropeptides and their receptors — were the physical substrate of what is ordinarily called emotion, and that the substrate was distributed throughout the body rather than localized in the brain.[152] Emotions, on this framing, are not

[151]Pert, C. B., and Snyder, S. H. (1973). "Opiate receptor: Demonstration in nervous tissue." Science 179(4077): 1011-1014. The discovery was made in 1972 when Pert was a graduate student in Snyder's laboratory at Johns Hopkins; the credit allocation later became a source of public controversy, with Pert ultimately not included in the 1978 Lasker Award given to Snyder for the work.

[152]Pert, C. B. (1997). Molecules of Emotion: Why You Feel the Way You Feel. New York: Scribner. The book combines Pert's autobiographical account of the opiate-receptor discovery with

phenomena that occur in the brain and have consequences for the body. Emotions are bodily events in which the brain participates as one node among many. The gut, the heart, the immune system, the reproductive organs — each of these is a site of neuropeptide activity that constitutes part of the body's overall emotional state at any given moment. The body knows what the body is feeling. The brain integrates and reports on the knowing. The distinction is consequential, and Pert was careful to ground it in the experimental data on receptor distribution she and her collaborators had accumulated over two decades.

Pert's findings have been substantially extended by subsequent work on the enteric nervous system — the dense neural network surrounding the digestive tract, which contains approximately 500 million neurons, roughly the same as the spinal cord. The American gastroenterologist Michael Gershon, whose 1998 book The Second Brain synthesized the field, established that the enteric nervous system operates with substantial autonomy from the central nervous system, generates its own neurotransmitters at concentrations exceeding those of the brain itself in the case of serotonin, and communicates with the brain in both directions through the vagus nerve and the bloodstream.[153] The gut, in the modern framing, is a substrate of distributed cognition that participates in the body's overall regulatory and emotional state in ways the older brain-centric model did not anticipate. The phrase that runs throughout the popular and scientific literature on this subject — that the gut has a brain of its own — is not metaphor. It is description.

---

the broader scientific framework of neuropeptide signaling she developed in the 1980s and 1990s.

[153]Gershon, M. D. (1998). The Second Brain: A Groundbreaking New Understanding of Nervous Disorders of the Stomach and Intestine. New York: Harper Perennial. Foundational research: Gershon, M. D. (1981). "The enteric nervous system." Annual Review of Neuroscience 4: 227-272.

An analogous and equally consequential body of work has been developed in neurocardiology, the field that studies the neural architecture intrinsic to the heart itself. The Canadian physiologist J. Andrew Armour, whose 1991 paper introduced the term intrinsic cardiac nervous system to describe the densely innervated neural network embedded in the heart muscle and surrounding tissues, established that the heart contains approximately 40,000 sensory neurons of its own — enough to constitute what Armour himself referred to as the little brain on the heart.[154] The intrinsic cardiac nervous system operates with substantial autonomy from the central nervous system, processes sensory information from the cardiovascular system, and communicates bidirectionally with the brain through the vagus nerve. The information flow from the heart to the brain, on the available measurements, exceeds the information flow from the brain to the heart, both in volume and in the regulatory consequences for the body's overall state. The heart is, on this picture, not the brain's pump. The heart is a sensing, processing, and signaling organ that participates in the body's distributed cognition at the same level as the gut and the brain. The body's substrate of memory and processing is, on the convergence of the neurocardiology and neurogastroenterology literatures, distributed across at least three major neural centers, each of which carries the information specific to its role and contributes to the integrated experience the body reports as conscious life.

---

[154] Armour, J. A. (1991). "Anatomy and function of the intrathoracic neurons regulating the mammalian heart." In I. H. Zucker and J. P. Gilmore (eds.), Reflex Control of the Circulation (pp. 1-37). Boca Raton: CRC Press. Synthesis: Armour, J. A. (2008). "Potential clinical relevance of the 'little brain' on the mammalian heart." Experimental Physiology 93(2): 165-176.

Trauma Memory in the Body: Bessel van der Kolk

The clinical synthesis of the somatic-memory literature that has had the broadest impact in contemporary medicine and psychology is the work of the Dutch-American psychiatrist Bessel van der Kolk, whose 2014 book The Body Keeps the Score has become the standard reference in trauma medicine and one of the best-selling works of medical nonfiction of the twenty-first century.[155] Van der Kolk's central clinical observation, developed across forty years of practice and supported by an enormous body of research on post-traumatic stress disorder, is that traumatic experience is encoded not primarily in conscious narrative memory but in the body's autonomic regulation, in the musculoskeletal patterns of held tension, in the breathing rhythms, and in the cellular and neurochemical signatures of chronic physiological alarm. The trauma victim, on van der Kolk's framing, is not a person whose mind has stored a difficult memory. The trauma victim is a person whose body is continuing to respond to a threat that has ended, because the body's regulatory systems have been remodeled by the trauma in ways that conscious recollection of the event does not, on its own, reverse.

The clinical consequence of van der Kolk's framing is that the standard talk-therapy approach to trauma — recall the event, process it cognitively, integrate the meaning — is necessary but not sufficient. The body's regulatory patterns require interventions that operate at the body's level: breath work, somatic-experiencing therapy, yoga, eye-movement desensitization and reprocessing, and in some cases pharmacologic interventions targeting the autonomic system directly. The interventions are not adjuncts to the real

[155]van der Kolk, B. (2014). The Body Keeps the Score: Brain, Mind, and Body in the Healing of Trauma. New York: Viking. Foundational research: van der Kolk, B. (1994). "The body keeps the score: Memory and the evolving psychobiology of posttraumatic stress." Harvard Review of Psychiatry 1(5): 253-265.

treatment, which would be the cognitive work. The interventions are the real treatment, because the location of the trauma's storage is the location at which the trauma must be addressed. The brain participates in the storage, but the brain is not the location. The location is the body.

What van der Kolk's clinical framework establishes for the argument of this chapter is that somatic memory is not theoretical. It is the operating substrate of modern trauma medicine. The clinicians treating post-traumatic stress disorder, dissociative disorders, and complex developmental trauma in major teaching hospitals and academic medical centers in North America and Europe operate within a framework that assumes the body holds memory in a substantively different way than the brain holds it, and that the body's memory has clinical consequences distinct from the brain's. The framework is not fringe. It is the mainstream clinical synthesis of fifty years of research on the neurobiology of stress and trauma. The body keeps the score. The phrase that names the book has, since the book's publication, entered the standard vocabulary of medical and psychological practice.

The neurophysiological substrate that van der Kolk's clinical framework operates through has been most systematically developed by the American behavioral neuroscientist Stephen Porges, whose polyvagal theory provides the autonomic-nervous-system architecture of the trauma-medicine framework.[156] Porges's framework identifies three distinct branches of the autonomic nervous system, each evolved at a different stage of vertebrate history and each associated with a different mode of regulation in

[156]Porges, S. W. (2011). The Polyvagal Theory: Neurophysiological Foundations of Emotions, Attachment, Communication, and Self-Regulation. New York: W. W. Norton. Original theoretical paper: Porges, S. W. (1995). "Orienting in a defensive world: Mammalian modifications of our evolutionary heritage. A polyvagal theory." Psychophysiology 32(4): 301-318.

response to safety, threat, or overwhelming danger. The most recently evolved branch, the ventral vagal complex, supports social engagement, calm regulation, and the integration that ordinary waking consciousness depends on. The older sympathetic branch supports the fight-or-flight response. The most ancient branch, the dorsal vagal complex, supports a freeze or shutdown response that mammals share with their reptilian ancestors. The clinical consequence is that trauma is, in Porges's framing, the body settling into a chronic activation of the older branches and a chronic inhibition of the most recent. The body's substrate of memory and regulation is thereby locked into a state that the conscious mind, operating from a substrate whose connection to the integrated regulatory system has been compromised, cannot reach by ordinary willing alone. The trauma-medicine interventions van der Kolk and others have developed work precisely at the level Porges has characterized: they re-engage the ventral vagal complex through breath, posture, social attunement, and rhythmic activity, allowing the body's regulatory hierarchy to re-stabilize from the bottom up. The mechanism is not theoretical. It is the operational basis of contemporary trauma treatment.

The mechanisms van der Kolk's framework operates through include several of the substrates this chapter has surveyed. The hypothalamic-pituitary-adrenal axis dysregulation that Chapter 7 introduced in the context of the Yehuda Holocaust-cohort studies is, in the trauma medicine framework, the same substrate that holds bodily trauma memory within a single lifetime. The neuropeptide system that Pert characterized is the chemical signaling architecture through which the body's distributed emotional regulation operates. The cellular epigenetic mechanisms that the chapter's first section described are the substrates of the long-term durability of the trauma's bodily traces. These are not separate findings. They are the same body of biological work, looked at from different methodological angles. The trauma medicine literature is the clinical translation of what the cellular biology, the neuropeptide biology, and the epigenetic biology have separately documented: the body, at the level of its operating molecular and cellular systems, holds the durable record of what the body has lived through.

The Lipton Synthesis

A popular synthesis of the cellular-biology side of this material, widely read in the years since its publication and widely discussed in clinical and integrative-medicine settings, is the American cell biologist Bruce Lipton's 2005 book The Biology of Belief.[157] Lipton's central framing is that the cell membrane, not the cell nucleus, is the cell's primary regulatory center, and that the proteins embedded in the membrane — the receptors and effectors that detect and respond to environmental signals — operate as the cell's interface with the conditions it inhabits. The framing is not original to Lipton in its biochemical content; the cell-signaling literature on membrane receptors is enormous and well-established, and the standard cell-biology textbooks treat membrane receptor biology as foundational. What Lipton added to the textbook material was a particular interpretive frame in which the cell's responsiveness to its environment was treated as evidence that consciousness or belief operates on the body at the cellular level.

The book's reception has been mixed. The cell-signaling biology Lipton describes is correctly summarized, in its essentials, from the mainstream literature, and the framework of cell-membrane responsiveness to environmental signals is well-supported by the

[157]Lipton, B. H. (2005). The Biology of Belief: Unleashing the Power of Consciousness, Matter and Miracles. Santa Rosa, CA: Mountain of Love/Elite Books. The book has been the subject of substantial discussion in the integrative-medicine community since its publication, with the cell-signaling biology generally accepted as a correct summary of mainstream work and the broader interpretive framing remaining the subject of active debate.

standard textbooks. The framing that the cell's responsiveness constitutes evidence for consciousness as a causal factor at the cellular level is, by contrast, controversial within the scientific community and has not been accepted into the mainstream framework. The chapter's working stance is to cite the cell-signaling biology Lipton accurately summarizes while remaining agnostic about the strong interpretive frame he places on it. Readers interested in the cellular biology in detail are best served by the primary literature in the standard textbooks. Readers interested in Lipton's accessible synthesis can read his book with the understanding that its scientific content is the mainstream cell-signaling literature and its interpretive framing is the author's own.

What the cell-signaling biology contributes to the argument of this chapter is independent of Lipton's interpretive frame. The cell is responsive to its environment through membrane receptor mechanisms that detect signals — hormonal, chemical, mechanical — and translate them into changes in the cell's internal state. The internal state changes include alterations in gene expression, in protein production, in metabolic activity, and in the cell's structural configuration. The responses are documented at the molecular level and are the standard subject matter of contemporary cell biology. What the documentation establishes for the chapter's broader argument is that the cell is not an isolated unit reading from its fixed genetic program. The cell is, throughout the duration of its existence, integrating signals from the body it belongs to, the systems it serves, and the larger environment those systems operate in. The integration is the cell's contribution to the body's overall memory. The body's overall memory is the integration of the integrations occurring at every cell of every tissue, continuously, for the entire duration of the life.

The Body as Archive

Pulling the threads together. The cellular biology of memory across mitotic division establishes that every cell carries the inheritance of its lineage-specific epigenetic annotation for the entire life of the organism, and that the annotation is responsive to conditions the cell encounters in ways that modify the inheritance. The molecular biology of synaptic memory establishes that the neurons of the brain encode their long-term memory in the physical structure of their synaptic connections, in protein conformational states that self-perpetuate, in changes that are anatomically real. The neuropeptide biology of Pert and the enteric-nervous-system biology of Gershon establish that the molecular signaling of emotion and regulation operates throughout the body's organs and tissues, not only in the brain. The trauma medicine of van der Kolk establishes that the body's autonomic, musculoskeletal, and visceral regulation holds the durable record of the body's lived experience in a way that conscious recall does not, on its own, exhaust.

The body, on the convergence of these literatures, is an archive. Every tissue of the body holds, at the molecular and cellular level, a record of the conditions the body has lived through, the chemical signals the body has received, the structural patterns the body has assembled in response, and the regulatory states the body has settled into. The brain is part of this archive — the part most specialized for integration, for cross-tissue communication, and for the production of the integrated narrative the body refers to as its life — but the brain is not the only part of the archive, and the brain is not the location at which the archive is centrally stored. The archive is distributed across the body's substrates. The brain is the index. The body is the library.

This framing has an immediate consequence for the broader argument of the book. The receiver model of Chapter 6 held that consciousness is a signal the brain is configured to receive rather than produce. The substrate the brain receives the signal through is the rest of the body, which carries the memory at the molecular and cellular level. The conscious experience of any moment is the integration, by the brain, of the body's distributed memory with the signal the brain is tuned to. The conscious self the visitor inhabits in any given moment is not the brain's output. It is the brain's report on the integration the body is performing across all its substrates simultaneously. The body, on this framing, is what the visit feels like from the inside. The body's memory is what the visit accumulates as it proceeds.

What This Establishes

The case for memory in matter, within the single lifetime, is the case Part Three's first chapter has assembled. The science is mainstream. The substrates are documented. The mechanisms are well-characterized. The clinical translation, in trauma medicine and regenerative biology, is among the most active areas of contemporary research. The case is not contested in its essentials by anyone working in the relevant fields. Memory is, within the single lifetime, a property of matter in ways the older brain-centric framework had not anticipated.

The chapter's contribution to the book's larger argument is to establish the substrate. If matter holds memory within a single lifetime, the question the next chapter takes up is whether matter holds memory across lifetimes — whether the molecular and cellular substrates the present chapter has characterized are the only substrates by which memory is preserved, or whether other substrates, operating in registers the present chapter has not engaged, carry information across the boundary that biological death represents. The present chapter has not made any claim about that further question. It has established that the within-lifetime case for memory in matter is sufficient on its own to dissolve the older intuition that memory is the brain's exclusive function and that the body is the brain's vehicle. Memory is the body's function. The brain integrates and indexes. The body remembers. The promise of Chapter 1 has, at this point in the book, been earned for the within-lifetime case. The next chapter examines whether the same claim can be supported across substrates.

Before that examination begins, it is worth registering what the within-lifetime case has cumulatively accomplished. Every cell in the reader's body, at this moment, carries methylation marks that record some portion of what the body has lived through. Every synapse in the reader's brain holds the conformational state of CPEB-class proteins that maintain the long-term memories the synapse stores. The reader's gut, heart, and immune system are processing neuropeptide signals that constitute the body's distributed emotional and regulatory state. The reader's autonomic nervous system holds, in its current regulatory configuration, the bodily inheritance of every significant experience the reader has had, particularly those experienced under conditions of safety, threat, or overwhelming danger that have shaped the body's settled patterns. None of this is metaphor. All of it is documented mainstream science. The body the reader inhabits is, at the molecular and cellular level, a living archive of the visit so far. The visit continues. The archive continues to fill.

✦ ✦ ✦

Five years and four months after the doctor folded the form for the body of Bua Kai Lawnak on the road in the central plains and put the form back into his shirt pocket and the men from the village wrapped the body and carried it to the house, a woman in another village in the same district who had not known the dead man and who would not, until her son was three years old, know that her son had any connection to him, gave birth in the back room of her family's wooden house on an October morning in 1967 to the second of the four children she would have over the course of her marriage.

The labor had been her second; she had delivered a daughter two years earlier in the same room with the same midwife, and the second delivery went, as second deliveries often go, more quickly. The child was a boy. The midwife wiped him, weighed him on the small scale the dispensary had loaned to the village, and laid him on his mother's chest. The mother looked at him. He was, in the way the firstborn moments of any infant are, the firstborn moment of an infant. The mother was tired. The midwife, after she had washed the boy and wrapped him in the cloth the mother's mother had prepared the previous week, pointed out to the mother two small irregularities on the child's head that the mother had not, in the first moments of looking at him, registered.

The first was a small round dark mark on the back of the boy's head, perhaps eight or nine millimeters in diameter, with the slightly puckered quality the midwife had seen on other newborns over the years and that the village understood as the trace, in the body of the infant, of some event in the prior life the infant was understood, in the Buddhist framework the village largely held to, to have come from. The second was a larger irregular dark mark on the forehead above the boy's left eye, perhaps two centimeters across at its widest, with the same puckered quality and the same darker color as the smaller mark on the back of the head. The midwife had seen the pattern before. She did not comment on what the pattern might indicate. She told the mother that the marks were birthmarks, that birthmarks of various kinds were common in infants of the district, and that the mother should not concern herself with them. The mother nodded. She looked at the marks. She did not, that morning, register any particular significance in them. She was thirty and she was tired and she had a daughter of two in the front room of the house and a husband who would come back from the fields in the late afternoon and a body of her own that the second delivery had asked a great deal of. She held the boy. She rested. The marks were marks. The boy was the boy. The boy was hers.[158]

The body the boy had been given to inhabit was a body whose cells, in the moment the midwife was holding the boy under the lamp in the back room, were assembling the molecular and cellular memory the chapter that has just closed has documented. The methylation patterns on every cell's genome were establishing the lineage-specific regulatory annotation that would persist across every cell division the body would subsequently undergo. The synaptic machinery in the developing brain was preparing the CPEB-class proteins that would, across the next eight decades if the body sustained that long, hold the conformational states of every long-term memory the body's nervous system would acquire. The neuropeptide receptors throughout the gut and heart and immune system were in their final stages of fetal differentiation. The autonomic regulation was, in the way the autonomic regulation of any newborn is, calibrated for the world the body was about to enter and not yet for the world the body had not yet encountered. The body the boy had been given was, by every measure the present chapter has assembled, an empty archive ready to begin its filling. The within-lifetime substrate of memory in matter was, at the moment the boy lay against his mother's chest in the back room, ready for the visit the boy was now beginning. The within-lifetime substrate was

[158]On the standard role of birthmarks in the Buddhist framework of central Thai villages of the mid-twentieth century, and the routine acceptance of small irregular marks at birth as traces of prior-life events: Stevenson, I. (1997). Where Reincarnation and Biology Intersect. Westport, CT: Praeger, pp. 1-30; and Stevenson, I. (1983). Cases of the Reincarnation Type, Vol. IV: Twelve Cases in Thailand and Burma. Charlottesville, VA: University Press of Virginia. The midwife's framing — that birthmarks are common, that they indicate prior-life events, and that the mother should not concern herself with them — was the standard cultural attitude in the region at the time and is documented across many of Stevenson's Thai case interviews.

not, however, the only substrate the body the boy had been given was carrying. The two marks the midwife had pointed out were the surface of a substrate the present chapter has not engaged. The substrate of cross-lifetime memory in matter is the substrate the next chapter takes up.

Where the Argument Goes Next

The next chapter narrows from the within-lifetime substrate of memory in matter to the cross-substrate question — whether matter holds memory across the boundary biological death represents, and whether the molecular and cellular machinery the present chapter has documented is the only machinery by which memory is preserved. The chapter takes the case in three parts. The first is the documented case archive of children who report memories of prior lives, anchored at the University of Virginia Division of Perceptual Studies under Ian Stevenson from 1967 to 2007 and under Jim Tucker from 2007 to the present. The second is the specific subset of cases in which the child's body bears birthmarks or birth defects in correspondence with the wounds or anatomical features of the prior personality — the subset of which the Chanai-Bua Kai case the present chapter has bookended is one of approximately three hundred documented examples. The third is the methodological literature surrounding the archive, including the case-strength scale Tucker developed to standardize the evaluation of new cases against the archive's evidentiary criteria. The next chapter is the closest the book comes to taking up the question of what continues, in concrete documented terms, across the substrate boundary the present chapter has stopped at.

CHAPTER 11

Memory Across Lifetimes

The function room at the hotel in San Diego where the surviving veterans of the USS Natoma Bay had gathered for their annual reunion in the second week of September of 2002 was on the second floor at the end of a long carpeted corridor, with a row of windows looking out over the parking structure and a folding table against the far wall on which the association historian had arranged the photographs, the typewritten casualty rosters, the unit citations, and the small handful of personal artifacts that the dwindling membership of the ship's crew had brought along to the reunion each year since the reunions had begun in the mid-1980s. The men in the room were in their seventies and eighties. They had come in from Texas and Arizona and Indiana and Pennsylvania and Florida, in the way the men of their generation came in to their annual reunions in those years, with their wives at their elbows and their grandchildren on the family-album pages they had brought to show one another and their own bodies in the late stages of the lives that had begun, for most of them, in the small American towns of the 1920s and 1930s and had passed through the same Pacific carrier deployments of 1943 to 1946 that had bound them to one another for the sixty intervening years.

Bruce Leininger had arrived on the Friday afternoon, alone, in the rented sedan he had picked up at the San Diego airport that morning. He was forty-seven. He had not previously attended a military reunion of any kind. He had spent the previous two years, since the May night in 2000 his wife had heard their son screaming about the airplane crash, assembling the documentary record that he had walked into the function room that afternoon to compare against the recollections of the men who had been on the ship the son had named. He had with him a notebook with his son's specific statements listed in the order they had been produced, with the dates the statements had been recorded, with cross-references to the contemporary published material he had been able to consult. He had with him a manila folder of typewritten correspondence with the Natoma Bay Association's historian, John DeWitt, who had been responsible for the photograph and document display at the front of the room. He had with him, in the way a forty-seven-year-old human-resources executive arrives at the place a forty-seven-year-old human-resources executive has arrived at by following the evidence the executive's two-year investigation has been compiling, the expectation that the afternoon would resolve what the previous two years had progressively been unable to resolve, in whichever direction the resolution would take.[159]

He had introduced himself at the registration table to the association president and to John DeWitt and to the small handful of veterans who had been standing near the entrance. He had explained, in the brief careful sentences he had rehearsed for the introduction, that his interest in the ship was a research interest related to the history of the Pacific carrier operations, and that he was specifically interested in the unit records of VC-81, the composite squadron that had operated off the Natoma Bay through the Mariana, Leyte, Luzon, and Iwo Jima campaigns. He had not, at the introduction,

[159]The Natoma Bay Association reunion in San Diego on September 8-10, 2002, and Bruce Leininger's attendance is documented in Matlock, J. G. (2022). "Clarifying muddied waters, Part 1: A secure timeline for the James Leininger case." Journal of Scientific Exploration 36(1): 100-130, which presents the timeline established from contemporaneous primary documents and ABC News video footage from late 2002. Bruce's specific discoveries at the reunion — the identification of James M. Huston Jr. as the only Natoma Bay pilot killed during the Iwo Jima operation, the connection to the Chichi Jima mission of March 3, 1945, and the acquisition of war diary records of Huston's VC-81 squadron — are documented in Leininger, B., Leininger, A., and Gross, K. (2009). Soul Survivor: The Reincarnation of a World War II Fighter Pilot. New York: Grand Central Publishing, Chs. 8-10.

mentioned his son. He had not, at the introduction, mentioned the two-year investigation. He had presented himself as a researcher. The veterans, who had been receiving researchers of varying degrees of seriousness at their reunions for the better part of a decade, had welcomed him with the courtesy of men who had served on a ship most of the country did not remember and who were grateful for the attention of anyone who had bothered to look.

He walked over to the display table. The photographs were arranged in chronological order along the length of the table, beginning with the ship's commissioning in October of 1943 at Vancouver, Washington, and ending with the decommissioning in May of 1946 at Bremerton. The typewritten casualty roster was at the center of the display. The roster listed the eighteen members of the ship's crew killed in action during the deployment, in the order of the dates of their deaths, with the operational engagement in which each had been killed listed in the right-hand column. Bruce read down the list. He read it three times. The eighteenth name on the list — the last addition to the casualty record before the ship's withdrawal from the Iwo Jima support operation in March of 1945 — was the name James M. Huston, Jr., listed with the rank Lieutenant Junior Grade, with the date March 3, 1945, with the operational entry Chichi Jima Strike, and with the laconic notation in the third column: Aircraft lost, harbor strike, body not recovered.[160]

[160]On the operational record of Lt. (j.g.) James M. Huston Jr. of VC-81 squadron, USS Natoma Bay, and the Chichi Jima mission of March 3, 1945 in which his FM2 Wildcat aircraft was shot down by Japanese anti-aircraft fire during a harbor strike, see the Natoma Bay Association's casualty records as compiled by John DeWitt, and the contemporaneous after-action reports of VC-81 archived at the Naval History and Heritage Command. The detailed treatment in Soul Survivor (Leininger, Leininger, and Gross, 2009), Ch. 9, includes the eyewitness accounts of Bob Greenwalt, Jack Durham, and John Provost, who saw Huston's plane hit and go down.

Bruce did not, in the moment of reading the entry the third time, say anything to the men around the display table. He stood at the table for some minutes longer. He noticed that the photograph on the corner of the display, slightly faded and labeled in the same typewritten font as the roster, was the official Navy portrait of a young pilot in flight gear, taken in February of 1945 according to the date stamped in the lower-right corner of the print. The pilot's face was the face of a serious twenty-one-year-old in the way the faces of twenty-one-year-old pilots in February of 1945 were the faces of serious twenty-one-year-olds, with the slight crookedness of mouth that came from being asked to smile by the photographer when the photographer had been the only thing standing between the pilot and the next deployment. The name typed under the photograph was the name James M. Huston, Jr. The face under the name was a face Bruce had seen, in altered form and at smaller scale, every morning for the previous four years and seven months across the breakfast table in Lafayette. Bruce stood at the table. The men around the table continued the conversations they had been having. Bruce looked at the photograph. The face looked back at him.

He spent the next forty-eight hours at the reunion. He spoke with the veterans who had known Huston — Bob Greenwalt, who had been Huston's wingman on the morning of the Chichi Jima strike; Jack Durham, who had flown the aircraft on Huston's left side; John Provost, who had been in the second flight position behind Huston when the anti-aircraft round had hit the engine of Huston's FM2 Wildcat at approximately 0850 hours local time and the aircraft had spiraled down into the harbor of Futami Ko without recovery. He took the war diary of VC-81 that John DeWitt had brought to the reunion and made copies of the pages that covered the period from the squadron's deployment in October of 1944 to the morning Huston had been killed. He took the photographs that the association had been able to provide. He took the rosters of the squadron's personnel and the names and contact information of the veterans whose memories he would, across the following months, conduct interviews with by telephone from his office in Lafayette. He flew home on the Sunday afternoon. The plane took off from San Diego and flew east across the southwest into the night. Bruce sat in the window seat. He did not work on the notebook he had brought with him. He looked out the window. The plane was over Texas when the lights of Dallas came into view below. He had been the boy's father since the boy had been born in 1998. He was, by the time the plane had passed Dallas and was beginning the descent toward New Orleans, beginning to register what he had also been.[161]

✦ ✦ ✦

The previous chapter established that memory, within a single lifetime, is a property of matter. The cell remembers what it is. The synapse holds long-term content in prion-like protein states. The neuropeptide system carries the body's distributed regulatory and emotional state. The enteric and intrinsic cardiac nervous systems process and store information at the level of their own neural populations. The body's autonomic configuration encodes the durable trace of what the body has lived through. The brain integrates and indexes the distributed archive. The body remembers.

[161]On the timeline of Bruce Leininger's continued documentation in the months after the September 2002 reunion, including the receipt of nine rolls of microfilmed records from John DeWitt in December 2002 and the war diary of VC-81 in late September 2002, see Matlock (2022), pp. 109-115. The case file is unusual within the UVA archive for the extensive contemporaneous documentation Bruce produced — notebooks, dated internet print-outs, correspondence with the Natoma Bay Association, and the ABC News video footage from late 2002 — which Jim Tucker has cited as providing methodologically strong evidence that James's specific statements (including the name 'Natoma' and the identification of Jack Larsen) preceded any contact the family had with information about Huston or the ship.

This is the within-lifetime case. The case is mainstream, well-documented, and clinically operational.

The present chapter takes up the further question that the within-lifetime case raises but does not answer. If memory is a property of matter within a single life, is matter the only substrate by which memory is preserved? When the body's matter disassembles at biological death, does the memory disassemble with it, or does some component of what the body held continue to exist in a substrate the body's matter did not exhaust? The receiver model of Chapter 6 already implied that consciousness exists in a form the substrate does not generate. The present chapter examines the most extensively documented empirical literature on whether memory, like consciousness, persists across the substrate transition that death represents.

The literature has a single principal source — the case archive maintained at the University of Virginia Division of Perceptual Studies, established by the American psychiatrist Ian Stevenson in 1967 and continued under the direction of the American child psychiatrist Jim Tucker since Stevenson's death in 2007. The archive now contains over 2,500 documented cases of young children who have produced specific, verifiable memories of prior lives, collected under controlled investigative protocols across six decades and over a dozen countries. The archive is the largest single body of empirical research on the question of cross-substrate memory in the scientific literature. It is not the only such body, and it is not without methodological complications, but it is the case material that any serious examination of the question must engage. The chapter examines the archive in detail, with particular attention to the methodological protocols Stevenson developed, the specific subsets of the case material that are most empirically constrained, and the falsification posture that Stevenson maintained across his career.

The chapter is presented with the same evidentiary restraint as Chapter 8. The cases the archive documents are, individually, anecdotal in the technical sense — each case is a particular child's particular reports about a particular prior life. The strength of the corpus is not in any single case but in the structural features that recur across the corpus as a whole, in the cross-cultural consistency of the recurring features, and in the falsification posture under which the corpus has been collected. The chapter examines what the corpus establishes, what it does not, and what the standard mainstream alternatives have offered as alternative explanations. The chapter's working conclusion is that the standard alternatives, examined against the actual contents of the published archive, are inadequate to the corpus, and that the framework this book has been developing — in which the substrate the visitor inhabits is not the only carrier of what the visitor brings — fits the corpus without strain.

The UVA Archive

Ian Stevenson was a Canadian-born American psychiatrist who served as chairman of the Department of Psychiatry at the University of Virginia School of Medicine from 1957 to 1967. In 1967, he stepped down from the chairmanship to found what was then called the Division of Personality Studies, now the Division of Perceptual Studies, with funding from the Chester Carlson estate. Stevenson had become interested in the cross-cultural literature on cases of children who reported memories of prior lives, and he intended to investigate the cases under the protocols he had developed for clinical psychiatric research. His first major publication on the subject, the 1966 monograph Twenty Cases Suggestive of Reincarnation, presented the methodological protocol and twenty case studies that exemplified what would, over the subsequent four decades, become the standard format of the field.[162]

[162]Stevenson, I. (1966/1974). Twenty Cases Suggestive of Reincarnation (2nd ed., revised and enlarged). Charlottesville:

Stevenson's investigative protocol established the standards that the subsequent literature has followed. When a case was reported to the Division — typically through a network of correspondents who had been briefed on the relevant indicators — a senior investigator traveled to the location of the case and conducted detailed interviews with the subject child, with the parents and other relatives of the subject, with any neighbors or witnesses who had observed the subject's statements, and, when the prior personality the child claimed to remember had been identified, with the family and acquaintances of the prior personality. The interviews were conducted in the local language with the assistance of trained translators. The investigators documented the specific factual claims the child had made, the order in which they had been made, the witnesses who had heard them, and the dates at which they had been recorded. The verification process then compared the child's statements against the records and testimony of the prior personality's family — birth certificates, death certificates, occupational records, residential records, photographs, and the personal recollections of those who had known the deceased.

The protocol was specifically designed to minimize the most obvious alternative explanations. To rule out fraud, Stevenson and his investigators worked, whenever possible, to document the child's statements before the families could meet — meaning that the child's specific factual claims were on record before any opportunity for cross-contamination. To rule out cryptomnesia (the unconscious recall of information the child had encountered through ordinary channels), the investigators traced the geographic, social, and informational distance between the child's environment and the prior personality's environment, and discarded or downgraded cases in which the distance was small enough that ordinary information

University Press of Virginia. The cases include the Imad Elawar case from Lebanon, the Swarnlata Mishra case from India, and the Marta Lorenz case from Brazil, each of which became standard reference cases in the subsequent literature.

transmission could explain the matches. To rule out coincidence, the investigators counted the specific factual statements the child had made and assessed the proportion that corresponded to the prior personality's life. Cases in which the child's statements showed high correspondence to the prior personality's life across many specific points were retained as evidentially strong; cases in which the correspondence was vague, partial, or attributable to general guessing were downgraded or excluded. The protocol has been examined by independent methodological reviewers since its first publication and has been refined in detail but not in its essential structure.[163]

The archive has continued under Tucker's direction since 2007 with the addition of cases from North American and Western European populations, which had been underrepresented in the earlier Asian-focused case work. Tucker's 2013 book Return to Life presents a number of American cases collected under the standard protocol, including the case of James Leininger, the case the present chapter has bookended.[164] The case file is unusual within the archive

---

[163]Methodological synthesis: Stevenson, I. (2001). Children Who Remember Previous Lives: A Question of Reincarnation (rev. ed.). Jefferson, NC: McFarland. Independent methodological review: Edwards, P. (1996). Reincarnation: A Critical Examination. Amherst: Prometheus Books, which contests the cases on philosophical grounds while acknowledging the rigor of the investigative protocol. More sympathetic methodological assessment: Almeder, R. (1992). Death and Personal Survival: The Evidence for Life After Death. Lanham, MD: Rowman & Littlefield.

[164]Tucker, J. B. (2013). Return to Life: Extraordinary Cases of Children Who Remember Past Lives. New York: St. Martin's Press. The James Leininger case is presented in detail in chapters 5 and 6. The case file includes records of statements made by the child before any contact with the prior personality's family, the identification of the specific historical pilot the child claimed to have been (James Huston, killed at Chichi Jima in March 1945), and subsequent verification of details the child provided against military records, including specific aircraft markings and the names of fellow crew

for its extensive documentation, the cooperation of the prior personality's surviving family in the verification process, and the availability of military records that could confirm or refute the specific factual claims. The Leininger case, like the older cases the archive contains, has been examined by skeptical reviewers and has not been overturned. The standard skeptical-explanation candidates — coaching by the parents, exposure to source material the child had encountered through ordinary channels, post-hoc reconstruction of statements to fit known facts — have been examined and have not been demonstrated. The case remains in the published archive under the same evidentiary status as the older Asian cases that Stevenson originally documented.

The Standard Case Pattern

The aggregate features of the over 2,500 cases now in the archive have been the subject of statistical analyses by Stevenson, Tucker, and several independent researchers. The patterns that emerge are consistent across the cultural backgrounds of the subjects and across the historical periods of the prior personalities the cases identify.

The cases typically begin when the child is between two and four years old, often as soon as the child has acquired enough language to make specific factual statements. The child begins, usually spontaneously and without solicitation from the family, to describe events from a life the child claims to have lived prior to the current one. The statements are specific — names of people, names of places, occupational details, family relationships, manner of death — rather than vague or general. The child often expresses an emotional urgency about the prior life, sometimes asking to be taken to the prior personality's family, sometimes expressing grief at being separated from the prior personality's family members, sometimes

---

members the child named before the family had researched the case.

treating the current family as substitutes for the prior family rather than as the child's own kin. The statements continue for a period of months to years, then begin to fade. By the time the child is six or seven, the statements have usually stopped, and by adolescence, in most cases, the child no longer has access to the memories that had been so vivid earlier.

The fading pattern is itself one of the corpus's most consistent features. The cases follow a developmental trajectory that the standard infantile-amnesia framework introduced in Chapter 8 does not predict — the children producing the most detailed verifiable claims are precisely those in the age range when the standard framework holds episodic memory should not be functioning at the level the cases demonstrate, and the fading of the claims correlates with the period when, on the standard framework, episodic memory is normally consolidating into the pattern that will persist through adult life. The pattern is consistent with the receiver model introduced in Chapter 6: the filtering machinery that locally focuses the consciousness signal through the substrate becomes more efficient through childhood development, and the access to pre-arrival content that is briefly available in early childhood is closed off as the substrate matures. The pattern does not require the receiver model to be observed, but it is, on the receiver model, expected.

Behavioral correspondences between the subject children and the prior personalities are a distinct subset of the case material and constitute some of the most striking individual features of specific cases. Children sometimes display phobias of the specific cause of death the prior personality experienced — drowning phobias in children whose prior personalities drowned, fire phobias in children whose prior personalities died in fires, vehicle phobias in children whose prior personalities died in traffic accidents — at ages and under conditions where ordinary conditioning could not have produced the phobia. Children sometimes display unlearned skills consistent with the prior personality's occupation or training: the child of a non-musical family who picks up a musical instrument with unusual facility and prefers the genre the prior personality played, or the child who displays mechanical or linguistic competence the family has not provided. Children sometimes recognize specific individuals from the prior personality's family on first meeting, identifying them by name and recalling details of the relationship that the prior personality had occupied. The behavioral correspondences are documented in the case files with the same standards of contemporaneous recording that apply to the verbal statements.[165]

The Birthmark and Birth-Defect Subset

The most empirically constrained subset of the archive — and the one Stevenson considered the most consequential for the broader claim — is the cases involving birthmarks and birth defects that correspond to wounds the prior personality is independently documented to have sustained. Stevenson devoted two volumes, published in 1997 under the title Reincarnation and Biology: A Contribution to the Etiology of Birthmarks and Birth Defects, to the detailed presentation of this subset.[166] The two volumes present 225 cases in which the subject child carried a birthmark, birthmark cluster, or congenital deformity that corresponded — in location, in shape, in size, and in some cases in specific anatomical detail — to a wound on the prior personality's body. The wounds in question were

[165]Statistical analysis of behavioral correspondences across the corpus: Stevenson, I. (1990). "Phobias in children who claim to remember previous lives." Journal of Scientific Exploration 4(2): 243-254. Synthesis with skill correspondences: Pasricha, S. K. (1990). Claims of Reincarnation: An Empirical Study of Cases in India. New Delhi: Harman Publishing.
[166]Stevenson, I. (1997). Reincarnation and Biology: A Contribution to the Etiology of Birthmarks and Birth Defects (2 vols.). Westport, CT: Praeger. Stevenson, I. (1997). Where Reincarnation and Biology Intersect. Westport, CT: Praeger (the abridged companion volume for general readers).

most frequently the cause of death — bullet entry and exit wounds in cases where the prior personality had been shot, blade wounds in cases of stabbing, burn patterns in cases of fire death, blunt-trauma marks in cases of beating or vehicle accident. In many of the documented cases, autopsy records or photographs of the prior personality's body were available and could be compared directly against the subject child's birthmarks at the same anatomical positions.

The statistical analysis Stevenson performed on the broader corpus indicated that, of the subjects in the case archive whose prior personalities had died of wounds that could in principle leave anatomically specific marks, approximately one-third of the subjects showed birthmarks or birth defects that corresponded to those wounds. This is, on its face, an extraordinary rate. Birthmarks themselves are common in the general population, with most adults carrying several, but the specific correspondence between a birthmark's location and shape and an independently documented prior wound is not predicted by any mainstream developmental mechanism. The standard embryological framework holds that birthmarks form during fetal development in response to vascular, pigmentary, or structural irregularities that are determined by the developmental program, with no available mechanism for the developmental program to encode the specific anatomical pattern of an injury that occurred to a separate individual prior to the subject's conception.

The specific cases Stevenson documented in the two-volume work include several whose photographic documentation is particularly striking. The case of Chanai Choomalaiwong, the Thai boy whose case the chapters of this book have bookended, born with two round birthmarks at the anatomical positions corresponding to the entry and exit wounds of the bullet that had killed the schoolteacher Bua Kai Lawnak five years before Chanai's birth. The case of a Burmese girl born with a long linear birthmark across the throat at the position of the slash that had killed the prior personality. The case of an Indian boy born with multiple defects of the fingers on his right hand, in a pattern corresponding to the loss of fingers the prior personality had sustained in a farming accident before his death. The published case material includes the photographic and medical documentation in each instance, with appropriate redaction of identifying information for the surviving families.[167]

What the birthmark subset contributes to the corpus is the empirical constraint that mere verbal statements do not provide. A child's verbal statements about a prior life can, in principle, be the product of fraud, coaching, or accidental information transmission. A birthmark cannot be coached. The birthmark was present at birth, before the child had any opportunity to produce verbal claims or to receive coaching about a prior personality. The correspondence between the birthmark and an independently documented wound on a deceased individual who was unknown to the subject's family is an objective, photographable, medically documentable feature of the case. The standard skeptical explanations that apply to verbal cases do not apply to the birthmark subset, and the standard developmental mechanisms have no available account for the correspondence. The birthmark subset is, on the available evidence, the strongest single empirical pressure point against the standard model's account of reincarnation cases.

Xenoglossy

A second empirically constrained subset of the broader case material concerns xenoglossy — the demonstrated ability to speak a language the subject has not learned through ordinary means.

[167]On the photographic documentation methodology for the birthmark cases, with particular attention to the Chanai Choomalaiwong case Stevenson photographed in 1979, see Stevenson, I. (1997). Where Reincarnation and Biology Intersect. Westport, CT: Praeger, pp. 1-50 (methodology) and pp. 300-323 (Chanai case).

Stevenson distinguished, in his published work on the phenomenon, between recitative xenoglossy (the production of words or phrases in an unfamiliar language without the ability to use them communicatively) and responsive xenoglossy (the demonstrated ability to carry on a conversation in the unfamiliar language, responding appropriately to questions and statements made by a native speaker). Responsive xenoglossy is the more empirically demanding form, and it is the form that the most carefully documented cases in the literature exhibit.[168]

The most extensively documented case of responsive xenoglossy in the literature is the Sharada case from India, reported by Stevenson and his Indian collaborator Satwant Pasricha in detail across multiple publications. The subject was an Indian woman named Uttara Huddar, born in 1941 in the Marathi-speaking city of Nagpur, who had received her education in Marathi and English and who had had no documented exposure to Bengali in her life. Beginning in 1974, at the age of thirty-three, Uttara began to manifest a secondary personality that called herself Sharada and that spoke fluent Bengali at conversational speed, with native phonology, and with appropriate use of regional vocabulary specific to the Bengal region of the 1820s — the period to which the Sharada personality consistently dated herself.[169] The investigators documented Sharada's communicative competence under controlled conditions, conducted extended conversations in Bengali with the personality through native-speaker investigators, and verified

[168]Stevenson, I. (1974). Xenoglossy: A Review and Report of a Case. Charlottesville: University Press of Virginia. Extended treatment: Stevenson, I. (1984). Unlearned Language: New Studies in Xenoglossy. Charlottesville: University Press of Virginia.

[169]Stevenson, I., and Pasricha, S. (1980). "A preliminary report on an unusual case of the reincarnation type with xenoglossy." Journal of the American Society for Psychical Research 74(3): 331-348. Extended report: Akolkar, V. V. (1992). "Search for Sharada: Report of a case and its investigation." Journal of the American Society for Psychical Research 86(3): 209-247.

specific historical and geographic details Sharada provided about the Bengal region in the period she identified herself with. The case file includes the testimony of multiple Bengali-speaking professors who certified the linguistic authenticity of Sharada's speech under conditions Uttara herself, in her ordinary personality, was unable to reproduce. The case is unusual within the corpus for the linguistic detail it documents, but it is not isolated. Several other responsive-xenoglossy cases in the archive show similar features at lower documentary intensity.

The interpretive significance of responsive xenoglossy is structural. The acquisition of conversational competence in a language requires extensive exposure to the language through ordinary developmental channels — typically thousands of hours of input across years of childhood or adult study. The cases of responsive xenoglossy in the corpus involve subjects who, on extensive documentation, did not have access to the relevant exposure. The standard mainstream explanations for the cases — cryptomnesia, deliberate fraud, dissociative production of language fragments — have been examined and, in the most evidentially constrained cases, have not been supported by the available evidence. The cases remain in the published archive as documented and unexplained-by-standard-mechanism instances of competence that the subject did not acquire through any channel mainstream developmental psychology recognizes.

Cross-Species Cases and the Ascent Thesis

A small but architecturally consequential subset of the archive concerns cases in which the prior personality the subject child reports is not a previous human but a non-human animal, or in which fragments of the subject's reports reference a non-human prior form. Such cases are rare in the corpus — substantially less common than human-to-human cases — and Stevenson treated them with appropriate caution in his published work, classifying them as a separate category and noting that the standards of verification that apply to human cases cannot be straightforwardly extended to non-human prior personalities for whom no documentary record exists.

The rarity of cross-species cases in the archive is structurally consistent with the ascent thesis developed in Chapter 5. On the framing this book has been building, the human substrate is the integration point of the ascent through forms — the configuration at which the accumulated perspectives of prior forms become available to the substrate's integrative capacity. A spirit that has reached the human substrate has, on this framing, traversed the lower-form sections of the substrate hierarchy in its prior ascent. Memory of those traversals would not be expected to be vivid or detailed in the integrated human substrate, both because the substrate-mediated character of the lower-form experiences was less narratively organized than the human form's integration permits and because the ascent thesis predicts that the lower-form perspectives are integrated into the human form's general capacity rather than retained as specific recoverable memories. Cross-species cases in the corpus should therefore be expected to be rare, fragmentary, and more affective than narrative — and that is, by the descriptive features of the cases that have been recorded, what they are.

The Chapter 5 framing also accommodates the post-human re-entry possibility introduced in that chapter, in which the realized spirit can return into any form, including non-human forms, deliberately. Cross-species cases in the archive that involve adult animals subsequently identified with specific deceased humans — a class of cases that is rare in the corpus but is documented at small numbers in several traditional cultures — would be the empirical signature of such re-entries. The archive's case material on these cases is small. The framing is consistent with the cases that exist. The book does not assert that the post-human re-entry hypothesis has been empirically established by this material; it asserts that the framework predicts the rarity of such cases as well as the structural features they exhibit when they occur.

The Falsification Posture

Stevenson, across his career, repeatedly emphasized that his work was offered to the scientific community as an invitation to falsification rather than as a claim of established truth. The published case files were available for examination by any qualified researcher. The investigators' notes, the witness testimony, the documentary records, the photographs and medical records were preserved at the Division and available to scholars under appropriate access protocols. The methodological protocol was published in detail. The cases were presented with their evidential complications, with the alternative-explanation candidates the standard literature would invoke, and with Stevenson's own assessments of the relative evidential strength of each case. The invitation to falsification was explicit and repeated.

The principal published critical examinations of the corpus during Stevenson's lifetime were the 1996 monograph Reincarnation: A Critical Examination by the American philosopher Paul Edwards, and the broader skeptical-movement commentary in the publications of the Committee for Skeptical Inquiry.[170] Edwards's critique is the most substantive of the available counter-positions, and it does not attempt to demonstrate fraud in any of the specific cases the archive contains. The Edwards argument operates at the philosophical level: reincarnation, on Edwards's framing, is metaphysically implausible on grounds independent of the case material, and the case material is therefore likely to admit alternative

[170]Edwards, P. (1996). Reincarnation: A Critical Examination. Amherst, NY: Prometheus Books. The Edwards critique is structured as a philosophical examination of the metaphysical coherence of reincarnation rather than as an empirical demonstration of fraud or error in any specific case Stevenson documented.

explanations even where the alternative explanations have not been specifically identified. The critique is internally consistent but does not engage the empirical question the corpus is designed to address. Of the specific cases in the archive that have been examined under skeptical investigation, none has been demonstrated to be fraudulent in the strict sense, and none has been overturned by the identification of a mainstream explanation that the original investigation had missed. The corpus continues to grow under Tucker's direction. The cases continue to come in. The falsification has not occurred.

The standing of the corpus has been affirmed, from a position of explicit skepticism, by several mainstream scientific figures. The American astronomer and prominent skeptic Carl Sagan, in his 1995 book The Demon-Haunted World: Science as a Candle in the Dark, listed Stevenson's investigation of children who appear to remember previous lives as one of three claims in the broader parapsychological literature that deserved serious empirical investigation, distinguishing the work from the much larger body of unverifiable paranormal claims that the book otherwise dismissed.[171] Sagan was not endorsing reincarnation. He was identifying Stevenson's work as the kind of empirical investigation that, by skeptical standards rather than against them, deserved continued examination. The endorsement, from one of the twentieth century's most prominent skeptical voices, is the kind of hostile-witness assessment that the chapter's broader argument has marked at several points in the book.

---

[171]Sagan, C. (1995). The Demon-Haunted World: Science as a Candle in the Dark. New York: Random House, p. 302. Sagan's specific framing — that the cases should be examined seriously rather than dismissed out of hand — is widely cited in the subsequent Stevenson-Tucker secondary literature as a representative skeptical endorsement of the methodological seriousness of the investigation, distinct from any endorsement of the reincarnation hypothesis itself.

Independent psychological assessment of the subject children has been pursued most systematically by the Icelandic psychology professor Erlendur Haraldsson, whose research career spanned four decades and who collaborated with Stevenson on cross-cultural case investigations beginning in the 1980s. Haraldsson conducted controlled psychological assessments of Stevenson's subject children in Sri Lanka and other South Asian populations, comparing them to matched control children of the same age, sex, socioeconomic background, and educational level.[172] Haraldsson's findings, published in mainstream psychology journals, were that the subject children differed measurably from their matched controls on several psychometric dimensions — vocabulary and verbal ability above age expectations, higher rates of dissociative experiences, more vivid memory in general — but did not show the elevated rates of fantasy proneness, suggestibility, or family dysfunction that the standard skeptical explanations would predict if the cases were the products of confabulation or imaginative play. The subject children were, on the psychological profile, recognizably ordinary children with one unusual cognitive feature: detailed memory of factual content that, on investigation, frequently turned out to correspond to verifiable details of a previously unknown deceased person. The psychological profile rules out several of the standard alternative explanations and is, on Haraldsson's own framing, consistent with but not dispositive for the reincarnation interpretation. The interpretation is left to the reader. The psychological data is part of the case.

Stevenson's own framing of the standing of the corpus, in his final years before his death in 2007, was that he had spent fifty years inviting his colleagues to falsify the cases and that the falsification

[172]Haraldsson, E. (1995). "Personality and abilities of children claiming previous-life memories." Journal of Nervous and Mental Disease 183(7): 445-451. Synthesis: Haraldsson, E., and Matlock, J. G. (2016). I Saw a Light and Came Here: Children's Experiences of Reincarnation. Hove, UK: White Crow Books.

had not been performed. He did not claim the cases proved reincarnation in any final sense — he was clear, throughout his career, that the cases were suggestive rather than demonstrative — but he did claim that the cases existed, that the documentation was rigorous, and that the standard mainstream framework owed an account of the corpus that had not been provided. The framing the present chapter takes is the same framing. The archive is a corpus of over 2,500 cases collected under controlled protocols by trained psychiatrists over six decades, including subsets — the birthmark-correspondence cases and the responsive-xenoglossy cases — that the standard developmental and neurological models have no available account for. The corpus has not been overturned by any of the standard skeptical-explanation candidates. The corpus continues. The interpretation is open to the reader. The corpus is what the reader's interpretation must engage.

What This Establishes

The within-lifetime case for memory in matter, established in Chapter 10, has its across-lifetime extension in the corpus the present chapter has examined. The mechanisms are different. The within-lifetime memory operates through methylation, synaptic remodeling, neuropeptide signaling, and autonomic regulation — substrates the body's matter directly carries. The across-lifetime memory, on the corpus the present chapter has surveyed, operates through whatever substrate carries the consciousness signal between the local substrates that successive lives provide. The receiver model of Chapter 6 holds that the consciousness signal exists in a form the local substrate does not generate. The cross-lifetime memory the UVA archive documents is, on the receiver model, the persistence of the signal's accumulated content across the local substrate transitions that biological death represents. The body's matter disassembles. The signal continues. The next substrate, when it is assembled and configured to receive, locally focuses some portion of what the signal carries. Most of what the signal carries is filtered by the maturing brain's reduction-valve mechanism — the Bergson-Huxley-James framing introduced in Chapter 6 — and is not available to ordinary adult recall. In the early years of the substrate's development, before the filter has fully closed, some fragments occasionally come through. The UVA archive is the body of empirical work that has documented those fragments under controlled investigation.

The corpus does not prove the receiver model. It does not prove reincarnation in any final metaphysical sense. What it does establish is that a substantial body of empirical work, conducted under controlled protocols by trained clinicians for over six decades, has accumulated specific verifiable evidence of memory persistence across substrate transitions that the standard mainstream framework has no available mechanism to account for. The standard skeptical explanations have been examined and, in the most evidentially constrained subsets of the corpus, have not been supported. The framework this book has been developing accommodates the corpus without strain and predicts several of its structural features — the rarity of cross-species cases, the developmental window of recall, the behavioral correspondences, the birthmark subset. The standard framework does not predict these features and does not, on the available examination, explain them after the fact. The corpus, on the chapter's most conservative reading, is the empirical evidence that the across-lifetime extension of the body-remembering thesis the previous chapter established is at least defensible and is, by the cumulative weight of the book's first eleven chapters, the more honest reading of the available data.

This chapter brings the book to within one chapter of the completion of its evidentiary case. Part Two established the blueprint of the pre-selection: universe, address, body, substrate, lineage, conditions, duration. Part Three has examined the evidence: memory in matter within the lifetime in Chapter 10; memory across substrates in the corpus the present chapter has surveyed. The final chapter of Part Three takes up the physical substrate beneath both — the quantum-mechanical framework that the receiver model of Chapter 6 invokes in principle and that the experimental quantum-biology literature of the last twenty years has begun to document in the specific systems where it operates.

✦ ✦ ✦

Anne Huston Barron was eighty-four years old in the spring of 2003 when the telephone rang in her small house in Los Gatos, California, with the call she had been told was coming from the family of the small boy in Louisiana who had been telling his parents, for the previous two and a half years, that he had been her brother. Anne had been twenty-six in March of 1945 when the telegram had arrived at her parents' house in Pennsylvania. She had been preparing, that afternoon, for a family reunion at her home in Los Angeles when she had felt the sudden and unmistakable sense of her brother's presence in the room with her, in the seconds before the doorbell had rung and the Western Union messenger had handed her the envelope with the words she had read three times before they had resolved into what the words were. She had been twenty-six. She had spent the fifty-eight intervening years not speaking of her brother in any sustained way, in the manner of women of her generation who had lost brothers in the war and who had organized their subsequent lives around the work and the marriage and the children and the long quiet decades that followed and that did not, in the social registers of mid-century America, make a great deal of room for sustained discussion of the brothers who had not come home. She had been retired for some years. She lived alone in the house with the small front garden. She had picked up the phone. Andrea Leininger had been on the other end of the line. Andrea had told her that her son James, who would turn five in two weeks, would like to speak with her if she would permit it. Anne had said she would permit it.

The boy came on the line. The voice was the voice of a small boy of nearly five. The boy said, with the casual conversational opening of a person who had been resuming a conversation that had been interrupted rather than initiating a conversation with a stranger he had never met, Hi Annie. Anne did not, in the first half-second after hearing the word, register what she had heard. The half-second after that, she registered what she had heard. No one outside the family had ever called her Annie. No one, in the present configuration of her surviving family, called her Annie anymore. Annie had been what her brother had called her. Annie had been the version of her name she had heard, in the kitchen of the house in Pennsylvania, across the dinner table, in the long letters her brother had sent from the carrier deployments, in the last conversation she had had with him on the telephone the week before he had shipped out for the Pacific in late 1944. Annie was what he had called her. The boy on the line had called her by the name only her brother had called her, in the tone only her brother had used.[173]

She did not interrupt him. She listened. The boy began, in the casual unwinding way a small boy unwinds a series of recollections to a person the small boy has been waiting some time to talk to, to tell her things. He told her she had a sister named Ruth, who had been four years older than her. He told her she had been four years older than him. He told her that their father had drunk too much and had broken things in the house when he had been drunk. He told her that their father had been put in a hospital for six weeks when he had been

[173]On the documented conversation between James Leininger and Anne Huston Barron in spring 2003, including James's use of the nickname "Annie" that only Huston had used, and the specific corroborated family details (the older sister Ruth, the father's alcoholism, the six-week hospitalization, the mother working as a maid during that period, Anne moving in with her grandparents afterward), see Leininger, Leininger, and Gross (2009), Ch. 11; Tucker (2013), pp. 81-83; and Tucker, J. B. (2016). "The case of James Leininger: An American case of the reincarnation type." Explore: The Journal of Science and Healing 12(3): 200-207.

thirteen, because their father had broken some glass in the house and the family had not known what else to do. He told her that while their father had been in the hospital their mother had had to work as a maid. He told her that after their father had come out of the hospital, Anne had moved away to live with the grandparents in another state. He told her each of these things, in the tone of a person reciting from a list of things he had been wanting to say for a long time, while she sat in the chair in the front room of the house in Los Gatos and her hand on the receiver was the hand of a woman who had been told, in the previous ninety seconds of conversation with a small boy in Louisiana she had never met, every detail of a family history she had not shared in any meaningful way with anyone outside her immediate family for fifty-eight years.

He thanked her for the painting. She had, in the previous month, sent the Leininger family a small package that had included a childhood portrait of her brother that their mother had painted of him when he had been six or seven. The portrait had hung in the family's living room in Pennsylvania for the years of her brother's life and had been put away after the telegram. She had retrieved it from the attic when Andrea had asked, and she had sent it to Louisiana. The boy thanked her for it. He thanked her in the tone of a person thanking his sister for having retrieved a thing that had belonged to him and that he had been wondering whether anyone had remembered to keep. He paused. He asked her, then, where the one was that their mother had painted of her. Anne had not, in the previous five decades of her life, ever spoken to anyone outside the family about the matching portrait. The matching portrait had been painted by their mother in the same year that the brother's portrait had been painted, and it had hung beside the brother's portrait in the family's living room across the years of their childhood. The matching portrait had been removed from the wall when the brother's had been removed, and had been put away in the same attic. Anne had not looked at the matching portrait, by her own subsequent recollection, since perhaps the mid-1970s. She had not mentioned it in the package she had sent to Louisiana. No one in the boy's family, on every available evidence, had ever heard of it. The boy was asking her where it was.[174]

She told him she would have to look for it. She told him she thought it might still be in the attic. He told her, in the way a small boy tells an older sister a thing the small boy expects the older sister to be able to manage on her own, that she should look. He told her he would like to see it again. The phone call lasted approximately twenty minutes. When it had ended, Anne sat in the chair in the front room for a long time. She got up. She went up the stairs to the attic. She opened the trunk in which the photographs and the family-album pages and the small wooden box of her mother's correspondence had been kept since she had moved from the house in Pennsylvania to California in 1947. She moved aside the photographs and the album pages and the small wooden box. The matching portrait was at the bottom of the trunk, wrapped in a piece of the same canvas the larger portrait had been wrapped in. She unwrapped it. She looked at the portrait of the seven-year-old girl her mother had painted in the spring of 1928, in the dress the seven-year-old girl had been wearing on the afternoon the painting had been made, with the hair the seven-year-old girl had worn at the length the seven-year-old girl had worn it then, with the slight inclination of the head that her mother had said had been the most difficult part of the composition to get right. She carried the portrait downstairs. She sat with it in the chair in the front

[174]On the specific portrait detail — James asking Anne about "the one she painted of you" referring to the matching mother-painted childhood portrait of Anne herself that no one outside the immediate family knew existed — see the corroborated journalistic account in the Lafayette Daily Advertiser interview with Bruce Leininger, November 18, 2025; and Leininger, Leininger, and Gross (2009), pp. 215-220. Anne subsequently retrieved the matching portrait from her attic and sent a copy of it to the Leininger family in Lafayette, where it remains in Bruce Leininger's collection of case-related artifacts.

room until the late afternoon. The boy was on the line. The boy was not on the line. The brother had been on the line. The brother had been the boy. The boy had been the brother. The portrait was the portrait. She wrapped it again and put it on the table by the door. She would send it to Louisiana in the morning. The chapter that follows takes up the substrate by which the brother and the boy had remained, across the fifty-eight years that had separated them and the substrate transition the telegram in March of 1945 had marked, the same.

Where the Argument Goes Next

The next chapter takes up the quantum-biological literature that grounds the receiver model in measured physics rather than speculation. The European avian-magnetoreception literature has established that migratory birds navigate by detecting Earth's magnetic field through a quantum-coherent radical-pair mechanism in cryptochrome proteins in the retina. The photosynthesis literature has established that the energy-transfer efficiency of the Fenna-Matthews-Olson complex in green sulfur bacteria depends on quantum coherence sustained at physiological temperatures. The olfaction literature has investigated the vibrational theory of smell, in which the perception of specific odorants depends on quantum-tunneling effects in the olfactory receptors. The microtubule literature of Bandyopadhyay, introduced in Chapter 6, has documented quantum-coherent vibrational modes in the cytoskeletal structures of neurons. The convergence of these findings establishes that quantum effects, long thought to operate only at scales below biological relevance, are in fact central to several of the most basic biological functions. The implications for the receiver model are direct. The substrate the brain receives the consciousness signal through is, on the measured experimental physics, the kind of substrate quantum signals can in principle couple to. The next chapter examines what that coupling implies for the framework this book has been developing, and for the time and identity questions the framework has so far deferred. Part Three closes with that examination.

CHAPTER 12

The Quantum Substrate

The monastery at Bodh Gaya that Yongey Mingyur Rinpoche had completed in 2007 sat on a small road approximately a kilometer from the Mahabodhi Temple complex, which housed the descendant Bodhi tree under which the historical Buddha was understood, in the framework Mingyur had been raised inside and had subsequently devoted his adult life to, to have attained the awakening that the entire Buddhist tradition had been organized around since approximately the fifth century before the common era. Tergar Monastery — the name derived from the Tibetan terga, meaning revealed treasure — had been Mingyur's principal residence since its completion. It housed several hundred monks. It served as the international headquarters of the meditation network Mingyur had been building since the publication of his first English-language book in 2007. The monastery had four floors. The third floor housed the abbot's residence. The abbot, at the beginning of June 2011, was Mingyur, who was thirty-six years old and had spent the previous nine years, since the Madison airport sequence in September 2002 that had produced the gamma-coherence measurements the first chapter of this book opened with, as the most internationally visible Tibetan Buddhist teacher of his generation.

He had announced the retreat the year before. The announcement had specified a three-year duration and had described the retreat as wandering rather than confined, in the older tradition of Milarepa and the wandering yogis of the Tibetan and Himalayan past whose example had structured his own contemplative formation since his three-year confined retreat at Sherab Ling between the ages of thirteen and sixteen. The community at Tergar had been preparing for his departure for some months. The arrangements for the continuation of the monastery's operations during his absence had been completed. The senior monks who would serve as administrators during the retreat had been appointed. The international teaching schedule had been wound down to its closing engagements. The plan, on every outward sign the community had access to, was for a formal ceremonial departure at a date Mingyur would announce when the preparations had been completed and the community had been gathered to receive the parting transmissions. The community was waiting for the announcement.[175]

The announcement had not been Mingyur's plan. The announcement Mingyur had been planning was the one his teacher Nyoshul Khen Rinpoche had given him the framework for in the long instructions he had received in the years before Nyoshul Khen's death in 1999 — the framework of the silent departure, the unannounced disappearance into the road, the renunciation of the privilege of monastery and recognition that the Tibetan-Buddhist tulku system had organized around him since the recognition of his prior incarnation in his second year of life. The framework was not new. Milarepa, the eleventh-century yogi whose biography Mingyur had been reading and meditating on since boyhood, had organized his entire post-recognition life around the silent disappearance. Nyoshul Khen had done the same, in the version that the late twentieth century had permitted, between his formal teaching

[175]On the lead-up to Mingyur Rinpoche's June 2011 departure from Tergar Monastery in Bodh Gaya, the public announcement made the previous year, and the community's expectation of a formal ceremonial departure that was instead pre-empted by Mingyur's mid-night escape, see Mingyur Rinpoche, Y., and Tworkov, H. (2019). In Love with the World: A Monk's Journey through the Bardos of Living and Dying. New York: Spiegel & Grau, Prologue and Ch. 1. Mingyur's two parting letters — one to the Tergar community, one to the Tibetan sangha — were published on the Tergar Meditation Community website on June 9 and 20, 2011, respectively, after Lama Soto found them in Mingyur's room the morning after his departure. The English translations of both letters are archived at the Tergar Learning Community at https://learning.tergar.org/.

periods. The framework was the framework. Mingyur had decided he would do the departure as the framework had been done. The community's expectation of the ceremonial event was the community's expectation; the framework's expectation was different.

He left at approximately three in the morning, in the first week of June, after the night-watch monk had completed the rotation and the corridor outside his quarters had been empty for the better part of an hour. He had on his person a small knapsack of the kind sold at the Indian railway-station vendors, containing two saffron shawls of the type the Hindu sadhus of the region wore, a thin cotton dhoti, a small ceramic alms bowl, a few thousand rupees in small bills — the amount he had been able to assemble without arousing suspicion in the previous weeks — and a single change of underclothing. He had on his body the maroon-and-yellow robes of a Tibetan monk, which he was wearing for the first stage of the journey on the practical grounds that the railway-station personnel and the train conductors would not, in the maroon robes, find his presence on the third-class car remarkable. He had under the robes, layered against his skin, the inner garment he had been given by his three-year-retreat teacher Saljey Rinpoche at the conclusion of the retreat in 1992. He did not have a passport. He did not have a mobile phone. He did not have a toothbrush. He did not have an attendant. He had not, in the way of a tulku who had been accompanied by attendants since the age of two, traveled anywhere without an attendant in approximately the previous thirty-three years.[176]

[176]The specific belongings Mingyur Rinpoche carried on his departure — the saffron sadhu shawls and dhoti, the alms bowl, the small amount of rupees, the absence of passport, mobile phone, and toothbrush — are documented in Mingyur Rinpoche and Tworkov (2019), Ch. 1, and in the recollections of his older brother Tsoknyi Rinpoche reported in Vaidyanathan, A., "Mingyur Rinpoche, the millionaire monk who renounced it all," The Guardian, December 22, 2015. The shawls were intended for the later phase of the wandering, when Mingyur planned to

He left the letters on the desk in his quarters. The first was addressed to the Tergar community in English, in his own handwriting, on the stationery the monastery had been using since its founding. The second was addressed to the broader Tibetan sangha in Tibetan, in the more formal register the Tibetan religious correspondence of the late twentieth and early twenty-first centuries had inherited from the older monastic literature. Both letters explained, in the careful phrasing of a man who had been writing parting letters of one kind or another since adolescence, that he had departed for the wandering retreat the framework had been organizing around. Both letters asked that his absence not be interpreted as abandonment of the community, and both asked that the practice the community had been organized around be continued in his absence. He folded the letters. He placed them on the desk in the position where the morning attendant would find them. He closed the door of his quarters behind him. He walked along the third-floor corridor to the staircase at the east end of the floor. He descended the staircase to the ground floor. He walked along the ground-floor corridor to the side door that opened onto the small lane that ran along the eastern boundary of the monastery compound. He opened the door. He stepped out into the lane.

It was approximately three-fifteen in the morning. The lane was empty. The night sounds were the night sounds of any Bihar village in the early summer — the distant call of a dog, the cycle of insect noise that the small irrigation canals along the road sustained, the occasional motor of a rickshaw on the larger road two hundred meters away. He stood in the lane for a moment with the knapsack on his shoulder and the robes settled against his body and the door of the monastery closed behind him. He had been the abbot of the monastery for the previous nine years. He had been a recognized tulku, in the public framework of the Tibetan tradition, since he had

---

assume the saffron-clad persona of a Hindu sadhu to avoid recognition as a Tibetan monk.

been two. He had been Yongey Mingyur Rinpoche, the seventh recognized incarnation of a line of teachers that the Tibetan calendars dated to the seventeenth century, since the recognition by Tai Situ Rinpoche when he had been twelve. He had been the public figure that the international meditation network and the popular books and the gamma-coherence studies and the Madison airport welcome and the dozens of teaching tours had organized around since 2002. He was standing in the lane at three-fifteen in the morning in early June of 2011 with a knapsack containing two saffron shawls and a thin cotton dhoti, in maroon robes that he would shed at the second train transfer, with the intention of walking to the railway station that was approximately a kilometer down the larger road and boarding the early local to Varanasi as a passenger of no fixed identity in the third-class car.[177]

He walked. The lane opened onto the road. The road was empty at that hour. He walked toward the station, which he had been to several times in the previous decade in the company of attendants and never alone, and which he was now approaching alone. The dog that had been calling from the direction of the temple complex continued to call. The cycle of insect noise continued. He walked. The man walking in the maroon robes at three-thirty in the morning along the road toward the Bodh Gaya railway station with a small knapsack on his shoulder was the same man who had stood in the Tibetan delegation at Bodh Gaya in 1991 at the Kalachakra empowerment given by the Dalai Lama, the same man who had spent fifty-eight hours in the Madison electrode cap in September

[177]On the Bodh Gaya railway station and the early-morning local train service to Varanasi in the configuration of the Indian Railways network in 2011, see the published timetables of the East Central Railway zone for the relevant period, available in the Indian Railways archive at https://indianrailways.gov.in. The Bodh Gaya station, which serves the pilgrimage traffic to the Mahabodhi Temple complex, is approximately one kilometer from the location of Tergar Monastery on the road to the temple complex.

2002 producing the gamma-coherence measurements Antoine Lutz had eventually published, the same man who had written three best-selling books on the convergence of contemplative practice and contemplative neuroscience, the same man who would, in the four and a half years that followed the walk to the railway station, contract food poisoning at Kushinagar to the point of near-death, recover, complete a circuit of the major Indian pilgrimage sites in the saffron robes of an anonymous sadhu, retreat into the remote Himalayan caves of Yolmo and Dolpo for the deep silent practice the wandering had been the preparation for, and re-emerge at Tergar in November of 2015 with the brain-age measurements that the Madison lab would, in the follow-up scanning conducted in May of 2016 that subsequently produced the Adluru et al. publication in Neurocase in 2020, find to be approximately eight years younger than the chronological age the body was carrying. He did not know any of that. He was, in the three-thirty-five of the early morning in the first week of June of 2011, a thirty-six-year-old man walking toward a railway station alone. He had begun the practice. The substrate the practice would operate on was the substrate the next four and a half years would document. The substrate the next four and a half years would document is the substrate the chapter that follows takes up.

✦ ✦ ✦

The previous chapters of Part Three have assembled the empirical case for memory across substrates. Chapter 10 established that the body, within a single lifetime, holds the durable trace of what it has lived through at the level of methylation marks, synaptic conformational states, neuropeptide signaling, and autonomic regulation. The body remembers. Chapter 11 established that the corpus collected over six decades at the University of Virginia Division of Perceptual Studies documents memory persistence across the substrate transition that biological death represents — a persistence the standard mainstream framework has no available mechanism to account for and that the receiver model of Chapter 6 directly predicts. The body's archive does not, on the convergence of these two literatures, end with the body's matter.

The man walking to the Bodh Gaya railway station in the early morning of June 2011 was, on every measurable indicator the contemplative-neuroscience laboratories had been able to extract from his nervous system in the previous decade, an exemplar of the substrate-coupling the receiver model requires consciousness to be doing through the substrate it inhabits. The gamma-band coherence Antoine Lutz had recorded in September 2002. The empathy-circuit activation that had run at seven to eight times control levels under the compassion meditation conditions. The brain-age measurements that would, four years and five months later, find his body's neural substrate to be running approximately eight years younger than his chronological age. The substrate Mingyur was carrying along the road was, in measurable physiological terms, a substrate whose coupling to whatever signal the substrate was tuned to was operating at the edge of what the contemplative-neuroscience instruments had so far been able to register. The substrate was doing what the framework needed the substrate to be doing. The question the present chapter takes up is what kind of physical substrate the doing operates in.

If consciousness is a signal the brain is configured to receive rather than produce, what is the substrate the signal operates in? If memory persists across the disassembly of the body that locally held it, what kind of substrate is the persistence implemented in? The receiver model has been argued from the empirical side — the clinical reports, the case archive, the convergent literatures. The chapter now examines whether the receiver model has, in the deepest physics currently available, a substrate consistent with the framework rather than merely compatible with the framework's silence on the question. The answer the chapter develops is that it does. Quantum biology, established across the last two decades as a productive experimental field, has documented several biological systems whose operation depends on quantum coherence sustained at physiological temperatures and timescales. The measurement problem at the foundations of quantum mechanics has remained, after a century of investigation, an unresolved structural question about the role of observation in physical reality. Bell's theorem and the experimental confirmations that earned the 2022 Nobel Prize in Physics have established that the world is not locally real in the classical sense — at least one of locality or definite-prior-to-measurement properties must be abandoned. And the deepest current quantum-gravity frameworks, including the Wheeler-DeWitt equation that has anchored the field for over fifty years, do not contain time as a fundamental variable. The receiver model the book has developed sits inside a physical picture in which each of these features is already present. The chapter examines that picture in detail.

A note on register before the chapter proceeds. The physics this chapter engages is technical, and the chapter does not attempt to pretend otherwise. The chapter aims for the level of detail at which a careful reader without specialist physics training can follow the argument's substance, while preserving the actual scientific content the argument depends on. The reader who wants a deeper technical treatment is directed, in the footnotes, to the primary literature. The chapter's argument does not require the reader to accept any specific interpretation of quantum mechanics; it requires the reader to accept the experimental findings that all major interpretations of quantum mechanics agree on. Those findings are what the chapter examines.

Quantum Biology as an Established Field

For most of the twentieth century, the assumption among working biologists and most physicists was that quantum mechanical effects were irrelevant to biological function. The reasoning was based on a calculation of decoherence times — the rate at which a quantum superposition collapses when the system interacts with a warm, wet environment of the kind that biological systems consist of. The calculated decoherence times for biological systems at body temperature were, by orders of magnitude, shorter than the timescales of any functionally relevant biological process. The conclusion was that whatever consciousness was, and whatever life was, both must operate at the classical level, with quantum effects averaged out into the bulk thermodynamic behavior of large molecular ensembles. This conclusion was widely held and rarely challenged.

The conclusion has not survived the last twenty years of experimental work. Several biological systems are now known to operate through quantum mechanical effects that the older calculations had ruled out. The systems are diverse, the mechanisms are different in each case, and the empirical demonstrations have been published in mainstream peer-reviewed journals including Nature, Science, and the Proceedings of the National Academy of Sciences. Quantum biology is, as of the time of this writing, an established subfield with active research programs at major universities in Europe, North America, and Asia. The chapter takes the four most consequential systems in sequence — avian magnetoreception, photosynthetic energy transfer, olfactory reception, and the microtubule findings already introduced in Chapter 6 — before turning to what the convergence implies for the broader argument of the book.

Avian Magnetoreception and the Radical Pair

Migratory birds navigate over distances of thousands of kilometers using a sensory capacity that ornithologists have studied for over a century: the perception of Earth's magnetic field. The behavioral capacity is well-documented across many species, but the physical mechanism by which the perception operates remained a mystery until the late 1970s, when the German biophysicist Klaus Schulten proposed that the relevant mechanism was a quantum-mechanical effect occurring in molecules called cryptochromes in the birds' retinas.[178] Schulten's proposal was that absorption of blue light by a cryptochrome molecule produces a pair of unpaired electrons — a so-called radical pair — whose spin states are quantum mechanically entangled. The subsequent recombination of the radical pair depends on the relative orientations of the electron spins, and those orientations are influenced by external magnetic fields. The bird's sensitivity to magnetic field direction, on Schulten's proposal, depends on the bird's brain reading out the chemical products of cryptochrome activation, which in turn depend on the quantum-coherent evolution of the entangled electron spin states between the moment of light absorption and the moment of recombination.

The proposal was confirmed experimentally over the subsequent decades. The behavioral demonstration that migratory birds use a light-dependent magnetic compass was established in the 1990s. The molecular identification of cryptochrome as the active sensory molecule was established in the early 2000s. The most recent and consequential confirmation was published in Nature in 2021 by a collaboration led by Henrik Mouritsen at the University of Oldenburg, who demonstrated directly that the cryptochrome from a migratory songbird, the European robin, exhibits the magnetic field sensitivity Schulten's proposal had predicted.[179] The 2021 paper

[178]Schulten, K., Swenberg, C. E., and Weller, A. (1978). "A biomagnetic sensory mechanism based on magnetic field modulated coherent electron spin motion." Zeitschrift für Physikalische Chemie 111: 1-5. The proposal was theoretical at the time and was treated with skepticism by mainstream avian biology for two decades.

[179]Xu, J., Jarocha, L. E., Zollitsch, T., Konowalczyk, M., Henbest, K. B., Richert, S., Golesworthy, M. J., Schmidt, J., Déjean, V., Sowood, D. J. C., Bassetto, M., Luo, J., Walton, J. R., Fleming, J., Wei, Y., Pitcher, T. L., Moise, G., Herrmann, M., Yin, H., Wu, H., Bartölke, R., Käsehagen, S. J., Horst, S., Dautaj,

measured the quantum-coherence properties of the relevant cryptochrome under laboratory conditions and demonstrated that the molecule's spin-state evolution exhibits sensitivity to magnetic fields at the magnitudes Earth's field provides. The avian magnetic compass is, on the current published consensus, a quantum-coherent biological sensor whose function depends on the maintenance of electron spin entanglement at physiological temperature and timescale. The earlier decoherence calculations were wrong. The system works.

Photosynthetic Quantum Coherence

The second major demonstration of quantum coherence in biological function concerns the photosynthetic apparatus of green sulfur bacteria, and by extension the broader photosynthetic machinery of plants and other photosynthetic organisms. The relevant published work began with a 2007 paper in Nature by a team led by Graham Fleming at the University of California at Berkeley, examining the Fenna-Matthews-Olson complex — a protein-pigment complex that serves as an energy-transfer antenna in the photosynthetic apparatus of certain bacteria.[180] The team used two-dimensional electronic spectroscopy, a technique that resolves the dynamics of quantum-mechanical energy transfer at femtosecond timescales, to demonstrate that the energy absorbed by the antenna pigments did not transfer to the photosynthetic reaction center by classical hopping among the pigments but rather propagated through

G., Murton, P. D. F., Gehrckens, A. S., Chelliah, Y., Takahashi, J. S., Koch, K.-W., Weber, S., Solov'yov, I. A., Xie, C., Mackenzie, S. R., Timmel, C. R., Mouritsen, H., and Hore, P. J. (2021). "Magnetic sensitivity of cryptochrome 4 from a migratory songbird." Nature 594: 535-540.

[180]Engel, G. S., Calhoun, T. R., Read, E. L., Ahn, T.-K., Mančal, T., Cheng, Y.-C., Blankenship, R. E., and Fleming, G. R. (2007). "Evidence for wavelike energy transfer through quantum coherence in photosynthetic systems." Nature 446(7137): 782-786.

the complex as a coherent quantum wavefunction that simultaneously explored multiple paths and selected the path of optimal energy transfer.

The 2007 finding was unexpected and was initially contested on the grounds that the quantum coherence the experiment had measured could not, on standard decoherence calculations, persist long enough at the physiological temperatures the bacteria actually live at to be functionally relevant. The objection was tested experimentally in 2010 by a team that performed the same measurements at the room-temperature range, with the result that the quantum coherence persisted at functionally relevant timescales even at the temperatures of warm aqueous bacterial cells.[181] The 2010 paper established that the photosynthetic quantum coherence is not an artifact of cold-laboratory conditions; it is the operational mechanism of the energy transfer in vivo. The efficiency of the FMO complex's energy capture — approaching the theoretical maximum allowed by physical law — depends on the simultaneous quantum exploration of all available pathways from the antenna to the reaction center. Classical hopping would yield substantially lower efficiency. The quantum mechanism is what makes photosynthesis as efficient as it is. The matter the photosynthetic apparatus consists of operates, in its principal function, by quantum-mechanical principles that the older biological framework had ruled out as impossible.

Olfaction and Quantum Tunneling

The third candidate quantum mechanism in biology is more contested than the first two but has accumulated enough experimental support to warrant inclusion in the chapter's case. The

---

[181]Panitchayangkoon, G., Hayes, D., Fransted, K. A., Caram, J. R., Harel, E., Wen, J., Blankenship, R. E., and Engel, G. S. (2010). "Long-lived quantum coherence in photosynthetic complexes at physiological temperature." Proceedings of the National Academy of Sciences 107(29): 12766-12770.

vibrational theory of olfaction, developed in its modern form by the American-British biophysicist Luca Turin in a 1996 paper in Chemical Senses, proposes that the mammalian sense of smell does not operate solely by the lock-and-key mechanism the standard receptor-biology framework had assumed.[182] On the standard view, the perception of a specific odorant depends on the molecule's three-dimensional shape, which determines which olfactory receptors the molecule can bind to. On Turin's proposal, the olfactory receptors do not detect shape alone; they detect the vibrational frequencies of the odorant molecule through a quantum-tunneling mechanism in which an electron in the receptor protein tunnels through the bound odorant and, in doing so, reads out the odorant's specific pattern of molecular vibrations. The perceived smell, on this account, depends on the vibrational frequencies the receptors detect, not only on the molecular shape.

The theory has been controversial in the olfactory-science community for the better part of three decades, and the standard textbook account of olfaction remains the shape-based mechanism. The theory has, however, accumulated experimental support that the shape-based mechanism alone cannot account for. The strongest single demonstration was published in 2011 by a collaboration led by Maria Isabel Franco and Efthimios Skoulakis at the Alexander Fleming Biomedical Sciences Research Center in Athens.[183] The team demonstrated that fruit flies, Drosophila melanogaster, can distinguish between two molecules — acetophenone and its deuterated isotopologue, in which ordinary hydrogen atoms are

---

[182]Turin, L. (1996). "A spectroscopic mechanism for primary olfactory reception." Chemical Senses 21(6): 773-791. Extended treatment: Turin, L. (2006). The Secret of Scent: Adventures in Perfume and the Science of Smell. New York: Ecco.

[183]Franco, M. I., Turin, L., Mershin, A., and Skoulakis, E. M. C. (2011). "Molecular vibration-sensing component in Drosophila melanogaster olfaction." Proceedings of the National Academy of Sciences 108(9): 3797-3802.

replaced by their heavier deuterium isotope — that have essentially identical three-dimensional shapes but substantially different vibrational frequencies. The standard shape-based theory of olfaction predicts that the two molecules should be indistinguishable to the olfactory system because they have the same shape. The flies, however, distinguish them reliably, suggesting that the olfactory receptors are detecting something other than shape alone. The vibrational frequencies are the most plausible candidate for what the receptors are reading. The mechanism Turin proposed is, on the current published experimental record, supported but not yet universally accepted within the olfactory-science community. The chapter notes the contested status and proceeds without depending on the final resolution of the question. What the olfaction case contributes to the broader argument is the demonstration that quantum tunneling effects in biological receptors are at least empirically plausible at the molecular level. The case strengthens the broader convergence even if its specific mechanism is not yet textbook physiology.

The Microtubule Substrate, Revisited

The fourth quantum-biological system relevant to the chapter's argument was introduced in Chapter 6 and is worth reprising in the present context. The experimental findings of the Indian biophysicist Anirban Bandyopadhyay and his collaborators in Japan demonstrated, in a series of papers published between 2013 and 2015, that single brain microtubules exhibit organized vibrational coherence at multiple frequency bands extending from the kilohertz range to the gigahertz range, with quantum-coherent properties that persist at physiological temperatures for durations far longer than the older theoretical calculations had predicted. The microtubules are present in every neuron, in particularly dense populations in the dendritic and axonal processes of the cortex, and constitute a candidate substrate for the kind of quantum-coherent state that the Penrose-Hameroff Orch-OR theory proposed as the substrate of conscious moments.

In the broader context of the present chapter, the microtubule findings are not isolated. They are the fourth in a series of established biological systems whose function depends on quantum-coherent mechanisms operating at physiological temperatures and timescales — a series whose first three members are the avian magnetic compass, the photosynthetic energy transfer apparatus, and the olfactory receptor system. The microtubule findings differ from the other three in that they are more directly relevant to the question of consciousness, because they concern the neural cytoskeleton rather than peripheral sensory or metabolic systems. The directness is what makes the microtubule findings particularly significant for the framework this book has been developing. The neurons of the cortex contain, on the experimental evidence, structures whose physical properties are of the kind that the receiver model requires for consciousness to operate as a substrate-coupled quantum signal. The substrate is in the brain. The substrate behaves the way the framework needs it to behave. The framework is not without a physical mechanism. It has one.

The Measurement Problem

The empirical case for quantum effects in biological systems is the chapter's first major thread. The second major thread takes up the unresolved foundational question at the heart of quantum mechanics itself — the question of what the role of observation is in the physical processes the theory describes. The question is not new and has not been resolved across the century since quantum mechanics was formulated. It is, however, more directly relevant to the framework this book has been developing than any other open question in fundamental physics.

The standard formulation of quantum mechanics, developed in the 1920s and codified in the textbooks of the subsequent decades, describes physical systems by a mathematical object called the wavefunction. The wavefunction evolves deterministically according to the Schrödinger equation when the system is not being measured. When the system is measured, the wavefunction is said to collapse to a definite outcome with probabilities determined by the wavefunction at the moment of measurement. The collapse is not described by the Schrödinger equation. The collapse is a separate postulate of the theory, introduced to account for the fact that measurements yield definite results rather than superpositions. What exactly counts as a measurement, and what physical process the collapse refers to, has been the subject of continuous debate within the foundations-of-physics community since the theory was first formulated. The Hungarian-American mathematician John von Neumann, in his 1932 textbook on quantum mechanics, identified the problem in its modern form and showed that any line drawn between the quantum system being measured and the classical apparatus doing the measuring could be drawn at any point in the chain from the system to the observer's brain without changing the predictions of the theory.[184]

The implication von Neumann drew, and that the Hungarian-American physicist Eugene Wigner extended in a 1961 paper, was that the line between the quantum and the classical realms might in principle lie at the level of the conscious observer rather than at any earlier point in the measurement chain.[185] The von Neumann-Wigner

---

[184]von Neumann, J. (1932/1955). Mathematical Foundations of Quantum Mechanics (R. T. Beyer, trans.). Princeton: Princeton University Press. The English translation by Beyer remains standard.

[185]Wigner, E. P. (1961). "Remarks on the mind-body question." In I. J. Good (ed.), The Scientist Speculates (pp. 284-302). London: Heinemann. Reprinted in Wigner, E. P. (1967). Symmetries and Reflections. Bloomington: Indiana University Press.

interpretation of quantum mechanics held that consciousness itself is what causes wavefunction collapse, with the quantum-mechanical formalism describing all of physical reality except the conscious observer's encounter with the experimental result. The interpretation has been a minority position throughout its history, but it has never been ruled out by experiment, and it remains one of the live interpretations of the theory's foundations. Other interpretations — the Copenhagen interpretation, the many-worlds interpretation, decoherence-based interpretations, Bohmian mechanics, QBism — propose different accounts of what the collapse is or whether collapse occurs at all. What they all share is the recognition that quantum mechanics, as a complete physical theory, cannot dispense with some account of how the probabilities the theory assigns become the definite outcomes that experimenters in fact observe. The role of observation is not, in quantum mechanics, eliminable from the theory's foundations.

The American physicist John Wheeler, who supervised both Hugh Everett's development of the many-worlds interpretation and Richard Feynman's early work in quantum electrodynamics, proposed a series of thought experiments and theoretical framings collectively known as participatory quantum mechanics, in which the act of observation is treated as constitutive of the physical outcomes the act records.[186] Wheeler's famous formulation was that we live in a participatory universe — a universe in which the observer is not separate from the observed but is, at the quantum level, a participant in producing the outcomes the observation registers. The Wheeler framing is not the consensus view. It is, however, one of the most thoughtful articulations of the structural

[186]Wheeler, J. A. (1983). "Law without law." In J. A. Wheeler and W. H. Zurek (eds.), Quantum Theory and Measurement (pp. 182-213). Princeton: Princeton University Press. Synthesis: Wheeler, J. A. (1990). "Information, physics, quantum: The search for links." In W. H. Zurek (ed.), Complexity, Entropy, and the Physics of Information. Redwood City, CA: Addison-Wesley.

feature of quantum mechanics that all working interpretations of the theory share: observation is not, in the quantum realm, a passive registration of pre-existing facts. The act of observation is part of what produces the facts that are observed.

Bell's Theorem and the End of Local Realism

The deepest empirical finding bearing on the foundations of quantum mechanics in the last seventy years is the experimental confirmation of the violation of Bell's inequalities. The Irish physicist John Bell, in a 1964 paper, derived a mathematical inequality that any theory satisfying two intuitively reasonable conditions — locality (no influences propagate faster than light) and realism (physical properties exist independently of measurement) — must satisfy in the statistical correlations of certain experimental setups involving entangled particles.[187] Bell showed that quantum mechanics predicts violations of his inequality in specific experimental conditions, while any theory satisfying both locality and realism cannot produce such violations. The prediction was sharp and testable. The question was whether the actual experiments would confirm quantum mechanics or one of the local-realist alternatives.

The first conclusive experimental test was performed by the French physicist Alain Aspect and his collaborators in 1982, using entangled photon pairs and measurements with rapidly switching polarization analyzers that closed the locality loophole — the possibility that the measurement results were correlated by signals propagating between the analyzers at or below the speed of light.[188]

---

[187]Bell, J. S. (1964). "On the Einstein-Podolsky-Rosen paradox." Physics Physique Физика 1(3): 195-200. Synthesis: Bell, J. S. (1987). Speakable and Unspeakable in Quantum Mechanics. Cambridge: Cambridge University Press.
[188]Aspect, A., Dalibard, J., and Roger, G. (1982). "Experimental test of Bell's inequalities using time-varying analyzers." Physical Review Letters 49(25): 1804-1807. Loophole-free

The Aspect experiments and the subsequent generations of refinements through the 2010s — including the loophole-free experiments of 2015 that closed the remaining theoretical objections — have all confirmed the quantum-mechanical predictions and violated Bell's inequalities to the precision the experimental apparatus permits. The 2022 Nobel Prize in Physics was awarded to Aspect, the American physicist John Clauser, and the Austrian physicist Anton Zeilinger for the experimental work establishing this result. The implication, accepted by every working interpretation of quantum mechanics, is that the world is not locally real in the classical sense. Either physical influences propagate faster than light, or physical properties do not exist independently of measurement, or both. The reader is permitted to choose which of these features to surrender; the reader is not permitted to keep both.

What the Bell-Aspect result means for the framework of this book is structural rather than mechanistic. The book has been arguing for a model in which consciousness is a signal that the brain receives rather than produces, in which memory persists across substrate transitions, in which arrangement of arrival conditions occurs before the substrate is assembled, in which the visit selected the universe before the universe contained the substrate the visit would inhabit. Each of these claims is, in the classical local-realist picture of physics, structurally impossible — the visitor would have to influence physical events without being a physical event, the memory would have to operate non-locally across the boundary that biological matter establishes, the selection would have to occur before there was anywhere local for the selection to occur from. The Bell-Aspect result removes the classical local-realist picture from

---

confirmation: Hensen, B., Bernien, H., Dréau, A. E., Reiserer, A., Kalb, N., Blok, M. S., Ruitenberg, J., Vermeulen, R. F. L., Schouten, R. N., Abellán, C., Amaya, W., Pruneri, V., Mitchell, M. W., Markham, M., Twitchen, D. J., Elkouss, D., Wehner, S., Taminiau, T. H., and Hanson, R. (2015). "Loophole-free Bell inequality violation using electron spins separated by 1.3 kilometres." Nature 526: 682-686.

the table. The physics that has replaced it is one in which non-local correlations are documented experimental fact, in which physical properties are not all definite prior to the act of observation, and in which the role of measurement and observation is, at the foundational level, an open structural question. The framework this book has been developing is not, in this physical picture, in conflict with established science. It is consistent with the deepest established science currently available.

Time as Construct

The deepest version of the chapter's argument concerns time itself. The framework the book has been developing requires that the visitor exist before the substrate, that pre-arrival arrangement occur before the substrate is assembled to be arrived into, and that memory persist after the substrate disassembles. Each of these requirements appears, in the ordinary common-sense picture of time, to require a temporal sequence in which non-physical events precede or follow physical ones in a linear chronology. The picture is contested at the deepest levels of contemporary physics, and the contestation has direct consequences for the framework's coherence.

Einstein's special relativity, formulated in 1905, established that time is not absolute — that the duration between two events depends on the reference frame in which the duration is measured, and that two observers in relative motion will disagree about the timing of events in ways that the theory specifies. Einstein's general relativity, formulated in 1915, extended the framework: time is not only relative to motion but also to gravitational potential, and the rate at which time passes depends on the local curvature of spacetime. The implications were verified experimentally many times — most precisely by the Gravity Probe B mission in 2004 — and constitute the foundations of the global positioning system, whose satellites must correct for both special-relativistic and general-relativistic time-dilation effects to maintain the accuracy the system requires.[189] Time, on Einstein's framework, is not a universal flowing parameter that all observers share. Time is a local feature of the spacetime geometry, varying with motion and gravity, with no single global present moment that all observers can agree on. The intuitive picture of time as a universal river flowing equally for everyone is, on the experimental physics, false.

The deeper revision of the picture comes from the attempt to combine quantum mechanics with general relativity, in the project that physicists call quantum gravity. The most famous equation in the field, the Wheeler-DeWitt equation derived in 1967 by Bryce DeWitt building on earlier work by John Wheeler, describes the quantum state of the universe as a whole — and the equation contains no time variable at all.[190] The absence of time from the fundamental equation of quantum gravity is known in the field as the problem of time and has been the subject of continuous research since the equation was first written down. Several proposed resolutions exist. One, advanced by the British physicist Julian Barbour, holds that time does not in fact exist at the fundamental level — that what we experience as the flow of time is an emergent property of the relationships among physical configurations, with no

[189]Everitt, C. W. F., DeBra, D. B., Parkinson, B. W., Turneaure, J. P., Conklin, J. W., Heifetz, M. I., Keiser, G. M., Silbergleit, A. S., Holmes, T., Kolodziejczak, J., Al-Meshari, M., Mester, J. C., Muhlfelder, B., Solomonik, V. G., Stahl, K., Worden, P. W., Bencze, W., Buchman, S., Clarke, B., et al. (2011). "Gravity Probe B: Final results of a space experiment to test general relativity." Physical Review Letters 106(22): 221101.
[190]DeWitt, B. S. (1967). "Quantum theory of gravity. I. The canonical theory." Physical Review 160(5): 1113-1148. Synthesis of the problem of time: Isham, C. J. (1993). "Canonical quantum gravity and the problem of time." In L. A. Ibort and M. A. Rodríguez (eds.), Integrable Systems, Quantum Groups, and Quantum Field Theories. Dordrecht: Kluwer Academic.

underlying temporal substrate beneath the configurations themselves.

Barbour's framing was given a more precise technical treatment in the Page-Wootters mechanism, developed by the American physicists Don Page and William Wootters in 1983 and substantially elaborated in subsequent decades.[191] The Page-Wootters mechanism proposes that time emerges from quantum entanglement between subsystems of the universe. There is no time at the level of the universe as a whole. There is only entanglement among subsystems. From the perspective of any one subsystem, the rest of the universe appears to be evolving in time — but the appearance is a feature of how the subsystem is correlated with the rest, not a feature of any underlying temporal substrate. The Italian theoretical physicist Carlo Rovelli has been the most prominent contemporary synthesizer of this framing for general audiences, and his 2018 book The Order of Time presents the technical foundations in accessible form. The book's central claim, controversial within the physics community but well within the mainstream conversation, is that time is not fundamental. Time is emergent, local, relational, and observer-dependent. The picture of time as a flowing universal medium is, on this framing, an artifact of how the human substrate is configured to integrate the information from its sensory systems and its memory architecture, not a feature of the underlying physics.

The implications for the framework of this book are direct. The framework requires that the visitor exist before the substrate, that arrangement occur before arrival, that memory persist after

[191]Page, D. N., and Wootters, W. K. (1983). "Evolution without evolution: Dynamics described by stationary observables." Physical Review D 27(12): 2885-2892. Experimental implementation of the Page-Wootters mechanism: Moreva, E., Brida, G., Gramegna, M., Giovannetti, V., Maccone, L., and Genovese, M. (2014). "Time from quantum entanglement: An experimental illustration." Physical Review A 89(5): 052122. Synthesis for general readers: Rovelli, C. (2018). The Order of Time. New York: Riverhead Books.

disassembly. These requirements appear paradoxical only if time is what the common-sense picture takes it to be — a universal flowing medium in which before and after are absolute relations. If time is what contemporary physics increasingly suggests it is — an emergent local property of the entanglement structure of the physical world, with no underlying universal flow — then the framework's temporal claims are not paradoxical. The visitor existing before the substrate is a statement about the entanglement structure of the relevant subsystems, not about a flowing universal time the visitor temporarily steps outside of. The arrangement that produces the matching of substrate to lineage is an entanglement feature of the conditions, not a temporally prior cause in the classical sense. The memory that persists is a feature of the substrate-independent informational structure of what the visit has accumulated, not of a temporal record stored in matter that has now disassembled. The framework's metaphysical commitments turn out to be statements about the structure of physical reality at the level the deepest contemporary physics is examining, not statements opposed to physical reality.

The Physics of the Receiver Model

Pulling the threads together. The chapter has examined the empirical case for quantum coherence in biological systems — established in avian magnetoreception, in photosynthesis, in olfaction at the contested but experimentally supported level, and in the microtubule findings that are most directly relevant to the receiver model of consciousness. The chapter has examined the measurement problem in quantum mechanics and the foundational role of observation in the theory's structure. The chapter has examined the Bell-Aspect experimental result and the end of classical local realism. The chapter has examined the time-emergence framing of contemporary quantum gravity and the experimental work on time-relativity that constitutes its foundation.

Each of these threads is mainstream science. Each has been the subject of substantial published work, peer-reviewed and replicated. Each is consistent with the framework this book has been developing in a way the older classical picture of physics was not.

The receiver model proposes that consciousness is a signal that the brain is configured to receive rather than produce, that the substrate the signal operates in is not exhausted by the local matter of the substrate the brain inhabits, and that the signal carries content across the substrate transitions that biological death represents. On the classical picture of physics, each of these claims would be structurally inadmissible. The signal would require a non-physical substrate; the non-locality would require faster-than-light influence; the persistence would require a kind of identity that the local-realist framework does not accommodate. On the contemporary picture of physics that this chapter has examined, none of these claims is inadmissible. The non-physical substrate is not required: the substrate of the signal could be the quantum-mechanical entanglement structure of the underlying physical reality, with the brain coupling to it through quantum-coherent mechanisms of the kind the four biological systems the chapter examined have already demonstrated. The non-locality is not problematic: non-local correlations are experimental fact, established by the Bell-Aspect result, and at the foundation of how quantum mechanics actually works. The persistence is not paradoxical: if time is emergent rather than fundamental, the identity of the signal does not require a temporal substrate beneath the substrates the signal locally inhabits.

The receiver model is, on the chapter's argument, a model that current physics could accommodate. The chapter has not proven that the receiver model is correct. It has shown that the receiver model is not in conflict with the deepest physics currently available, and that the alternative classical picture under which the receiver model would be inadmissible has, by the cumulative weight of the last hundred years of experimental physics, been displaced from the foundations. The framework this book has been developing is not metaphysics opposed to science. It is metaphysics consistent with the science that has replaced the older metaphysics the standard view inherited from a physics that experiments have already moved beyond.

What This Establishes

Part Three of the book closes here. The within-lifetime memory of Chapter 10, the across-substrate memory of Chapter 11, and the quantum-substrate physics of the present chapter constitute the empirical case for the framework the book has been developing. The blueprint of Part Two specified what the framework requires. Part Three has assembled the evidence that the requirements are met. The body holds the visit's accumulating memory in its molecular and cellular substrates within the lifetime; the case archive documents the persistence of memory across substrates at corpus sizes the random-coincidence model cannot accommodate; and the quantum substrate that the receiver model invokes is, on the experimental physics, the kind of substrate the empirical findings describe.

The framework is now earned. The remaining chapters of the book turn from the evidentiary case to its implications. If the framework is correct, the visitor is not assigned to the life the visitor inhabits; the visitor chose it. If the visitor chose it, the grievance posture that treats one's life as something that happened to one is incoherent. If the grievance posture is incoherent, the practical orientation that follows is one of active stewardship of the visit rather than passive endurance of its conditions. The framework has consequences. Part Four examines them.

The man walking toward the Bodh Gaya railway station in the early morning of June 2011 had, in the framework his contemplative tradition had been organized around for over a thousand years and in the framework the four chapters of Part Three of this book have assembled in mainstream scientific terms, undertaken the practice the substrate the framework describes was built for. The four and a half years that followed his departure from the lane outside Tergar Monastery would produce, on the brain-imaging measurements the Madison lab would take when he returned, a substrate running approximately eight years younger than his chronological age. The substrate had been doing what the framework had been describing. The doing had left its measurable trace in the substrate the doing had operated through. The next chapter takes up the implications of the framework for the substrate any visit's contribution to itself is conducted through.

Where the Argument Goes Next

Part Four opens with the implication that has been implicit in the framework throughout the book and that the next chapter takes up directly. The life the reader is in was not assigned. It was chosen. The choice was made before arrival, in the conditions that the previous chapters have characterized, by the same consciousness that is now reading this sentence. The implications for how the reader holds the difficulties, the failures, the disappointments, and the unchosen-seeming conditions of the present life are substantial. The next chapter develops them. The chapter that follows it examines coherence as the operational technology by which the visit's accumulated content becomes accessible to the substrate while the substrate is still functional. The chapter after that examines the Gaia Method as the system the broader work has been organized around. The closing chapter examines the symbiont obligation that the ascent thesis of Chapter 5 has been pointing toward since the book's structural framing was first laid out. Part Four is the operational consequence of the eleven chapters of evidence that have preceded it. The case has been made. The application begins now.

CHAPTER 13

This Life Was Chosen, Not Assigned

The ridge above Futami Ko harbor on the island of Chichi Jima, in the Ogasawara archipelago approximately one thousand kilometers south of Tokyo and two hundred kilometers north of Iwo Jima, rises from the inland slope of the harbor's eastern arm to an elevation of roughly two hundred meters and provides, from its southern face, a view across the sheltered water that the United States Navy carrier strikes of late February and early March of 1945 had pressed against in support of the larger Iwo Jima invasion. The harbor on the morning in the spring of 2010 was the same harbor it had been in March of 1945. The eastern arm was the same eastern arm. The wooded slope down to the small fishing village at the head of the bay was the same wooded slope. The Japanese fishing boat at the dock at the foot of the slope, which had brought the family up from the larger Ogasawara-Maru ferry that morning from the main pier at the southern end of the harbor, was tied to the same wooden dock the Japanese fishing boats of the 1940s had used. The boy standing on the ridge in the spring of 2010 was twelve years old. He had been born in San Francisco in April of 1998. He had spent the previous decade and a half of his life in Texas and Louisiana. He had never, in the life the body had been carrying, been to Japan.

The Japanese television crew that had paid for the family's passage from Tokyo to Chichi Jima had been organizing the trip for the better part of a year, since the original invitation had been extended to the family in 2006 and the schedule had been repeatedly postponed for the various practical reasons international television production tends to be postponed for. The crew was small — a director, a cameraman, a sound engineer, a producer who handled the local arrangements, and a translator who moved between the family and the village contacts the production had been working with. The footage they had collected over the previous three days included the family's arrival on the inter-island ferry from Tokyo, the boy's reactions at the various sites on the larger Iwo Jima itself where the family had stopped before continuing south to Chichi Jima, the interviews with the elderly Japanese veterans of the local garrison the production had located in the small towns along the coast, and the long sequence at the wartime cemetery on the south side of the island where the boy had stood for the better part of an hour in front of the small stone memorial without speaking.[192]

The morning of the third day was the morning the production had organized around the harbor. The boy had been quiet at breakfast. The Japanese family who ran the small inn at the village had given the family a steamed rice breakfast at the low table in the inn's front room, and the boy had eaten without commenting on the food in a way the parents had not, in the boy's twelve years of life so far, seen the boy refuse to comment on food. The walk from the inn to the ridge had taken about twenty minutes. The path went up through the wooded slope on the east side of the village, around the base of the hill that separated the village from the eastern arm of the harbor, and onto the ridge at the point where the slope flattened into the small clearing the local council had maintained as a viewing

[192]On the 2010 Chichi Jima trip arranged by Fuji TV at the family's age-twelve point in the case timeline, including the flower-drop ceremony over the estimated wreck site, the ridge above Futami Ko where James broke down, and the documentary that resulted, see Leininger, B., Leininger, A., and Gross, K. (2009). Soul Survivor: The Reincarnation of a World War II Fighter Pilot. New York: Grand Central Publishing, Ch. 17 and Epilogue (the Japan trip was included in later editions); Tucker, J. B. (2013). Return to Life: Extraordinary Cases of Children Who Remember Past Lives. New York: St. Martin's Press, pp. 87-89; and the corroborating Bruce Leininger account in Leininger, B. (2022). "Consciousness Survives Physical Death: The James Leininger Case." Bigelow Institute for Consciousness Studies essay contest entry, pp. 23-25, available at https://www.bigelowinstitute.org.

point for the harbor. The clearing had a small wooden bench, a wooden post with a brass plaque that the Japanese had set there in the 1990s to mark the war dead of both sides at the harbor, and an unobstructed view across the water to the western arm of the harbor where the cargo ships had been moored in March of 1945 and where the formation of FM2 Wildcats from VC-81 squadron of the USS Natoma Bay had pressed their strafing run in the early morning of the third of that month.

The boy walked out onto the clearing. He stood at the edge of the slope. He looked at the water. The Japanese cameraman, who had been instructed by the director to stay back from the family at this part of the morning, was approximately twenty meters behind the boy with the camera framing the boy's small figure against the water below. The sound engineer was behind the cameraman with the long boom microphone extended toward the boy's position. The parents were standing approximately five meters behind the boy and slightly to the side, in the position the director had asked them to take, so that the boy's face could be filmed in profile against the harbor without the parents being in the frame. The wind on the ridge that morning was coming from the northwest at perhaps fifteen kilometers per hour. The water below was lightly chopped. A small fishing boat was crossing the bay from the western arm to the village dock. The cycle of insect noise the wooded slope produced was the cycle of insect noise of any subtropical Pacific island in early spring.

The boy stood for perhaps thirty seconds without speaking. He raised his right arm and pointed, in the direction of a position perhaps two hundred meters off the southern end of the western arm of the harbor, at a spot where the water was a slightly darker shade than the water around it because of the depth of the harbor floor at that position. The father, who had been watching the boy's face, walked the five meters that separated them and stood next to the boy at the edge of the slope. The boy did not look at the father. The boy continued to look at the water. The boy's right arm continued to point. The boy spoke, in the small voice of a twelve-year-old who had stopped trying to manage the moment a few seconds earlier, the sentence that the father had been carrying in the case file since the boy had been three. The boy said that the spot the boy was pointing at was where the boy's plane had been when the boy's plane had been shot down. The boy then began to cry, in a way that the father had not seen the boy cry since the boy had been four and the nightmares had been at their peak, and the boy turned and put his arms around the father's waist and pressed his face against the father's shirt and continued to cry for the better part of the next ten minutes while the father stood on the ridge with his hand on the back of the boy's head and the Japanese camera continued to film from twenty meters back and the wind continued to come from the northwest and the water below continued to be lightly chopped.[193]

After the crying had subsided, the family walked down from the ridge to the village dock. The Japanese fishing boat that the production had chartered for the morning was waiting at the dock with its small diesel engine running. The boat's captain, an elderly Japanese man from the village who had been a child at the village during the war and who had agreed to participate in the production after the producer had explained the family's situation to him over several weeks of preparation, was at the wheel in the small cabin at the stern. The director and the cameraman and the sound engineer boarded the boat. The family boarded the boat. The boy was carrying the small bouquet of yellow chrysanthemums and white lilies the family had brought from Tokyo for the ceremony. The boat pulled

[193]Bruce Leininger's contemporaneous recollection of the moment on the ridge — "James breaks down while we were standing on a ridge overlooking the harbor and tugs at my arm and said, 'This is where I flew in the day my plane was shot down'" — is preserved in Leininger (2022), pp. 24-25, and was subsequently verified by Bruce against the position of the wreck site documented in VC-81 Aircraft Action Report No. 80 of March 3, 1945. The AAR-80 wreck position is consistent with the spot James pointed to from the ridge.

away from the dock and made its way across the bay to the position the boy had pointed to from the ridge, with the captain navigating by the chart the production had prepared from the wreck-site coordinates the United States Naval Historical Center had provided. The position the chart showed and the position the boy had pointed to from the ridge were the same position, within the precision the chart's coordinate system permitted.

When the boat had reached the position, the captain cut the engine. The boat drifted on the harbor swell. The cameraman and the sound engineer set up at the bow. The boy walked from the cabin to the bow with the bouquet in his right hand and his left hand on the rail. The father followed the boy to the bow and stood a step behind him. The boy looked at the water for perhaps a minute. He then held the bouquet out over the rail, brought his right hand to his forehead in the salute his father had taught him three years earlier when they had attended the September dedication ceremony at the Natoma Bay memorial at the Nimitz Museum in Fredericksburg, Texas, and said, in the voice of a twelve-year-old who had been carrying the moment for the previous ten years of his substrate's existence, the sentence the family would subsequently include in the documentary's closing minute. The boy said that he would never forget James M. Huston Jr. He held the salute for perhaps three seconds. He dropped the salute. He released the bouquet over the rail. The bouquet fell the meter and a half from the rail to the water. It floated on the surface. The harbor swell carried it gently away from the boat. The boy stood at the rail and watched it. He did not cry again. He stood there until the bouquet had drifted approximately twenty meters from the boat's position, and then he turned and walked back to the cabin and sat down on the bench at the rear and did not, for the remainder of the boat ride back to the village, say anything.[194]

[194]The flower-drop ceremony over the AAR-80 wreck site and James's salute and the sentence "James M. Huston, I'll never forget you" are documented in Bruce Leininger's recollection in the Lafayette Daily Advertiser interview of November 18, 2025

The drawings the boy had been producing since the age of two and a half — the violent dog-fight drawings with the planes signed James 3 in the corner, the burning cockpits, the falling figures, the cycle of war imagery that had filled the family's refrigerator and the walls of the boy's bedroom and the back of every piece of paper the family had owned for nine and a half years — stopped, in the months that followed the trip, in the way the case file would subsequently document and the trauma-medicine literature would subsequently confirm as the closure pattern that the more developed Stevenson-Tucker cases routinely produced at the conclusion of the access window. The new drawings the boy began to produce in the late spring and summer of 2010 were of mountains, of the small Japanese fishing village they had stayed in on Chichi Jima, of the harbor in the morning light, of the family in the back of the boat with the captain at the wheel, of the small bouquet on the water with the morning light on the petals. The boy's drawings had shifted from war to peace. The substrate's filter had completed the closing the developmental literature had predicted. The visit, on the framework, had completed the integration the visit had been organized around.[195]

✦ ✦ ✦

(Acadiana News); see also the corroborating account in Leininger (2022), p. 25. The ceremony was filmed by the Fuji TV crew and was included in the documentary that aired in Japan in 2010.

[195]On the closure pattern documented across the Stevenson-Tucker archive — the marked decrease in past-life statements, drawings, nightmares, and behavioral correspondences that typically follows the subject's visit to the prior personality's relevant locations or family members — see Tucker (2013), pp. 197-203; and the broader synthesis in Tucker, J. B. (2008). Life Before Life: Children's Memories of Previous Lives. New York: St. Martin's Press, pp. 161-172. The Leininger case is unusual in the archive for the documentary intensity of the closure sequence, including the Chichi Jima trip and the subsequent stabilization of the boy's drawings and dreams across the years that followed.

The case is now made. Twelve chapters of evidence — from the cosmological fine-tuning of Chapter 1 through the quantum substrate of Chapter 12 — have built a framework in which the life the reader is in was not assigned by random distribution and is not endured against a backdrop of accident. The universe was configured. The address was specific. The body was curated. The substrate was tuned. The signal was real. The lineage was matched. The arrival was chosen. The duration was arranged. The body is the archive. The signal carries across substrates. The physics permits the framework. The framework, on the cumulative weight of the evidence the book has surveyed, is more demanding to dismiss than to provisionally accept.

Part Four examines what follows. The framework has practical consequences, and the consequences are operational rather than merely interpretive. If the life the reader is in was chosen, the reader's relationship to the life cannot coherently be the relationship that the lottery model invites. The lottery model invites the reader to treat the conditions of life as a hand that was dealt — as circumstances that happened to one rather than circumstances that one selected. The framework the book has developed does not permit that posture. The posture is internally inconsistent with the framework's documented evidence. Holding the framework while maintaining the grievance posture would require the reader to believe simultaneously that the conditions of arrival are pre-selected and that the conditions are unfair impositions on a self who did not choose them. The two beliefs cannot both be true. The chapter develops the implication.

The chapter proceeds in five movements. The first examines the grievance posture in its standard form — the cluster of attitudes that the lottery model authorizes and that the framework the book has developed renders incoherent. The second examines what the framework's pre-selection thesis specifically implies for the difficult conditions of any individual life. The third examines the trap of the spiritual-bypass reading — the misuse of the pre-selection thesis to dismiss real suffering, blame the suffering on the sufferer, or substitute a flattened pseudo-acceptance for actual engagement with the conditions of arrival. The fourth examines what the pre-selection thesis actually permits and requires by way of practical orientation. The fifth bridges to the operational chapters that follow — coherence as the technology of remembering, the integrated method as the system the technology is organized into, and the closing chapter on the symbiont obligation that the framework's structural commitments imply.

A note on register before the chapter proceeds. The earlier chapters of the book operated in the institutional voice of the scientific case. The present chapter is the first to operate in the voice of consequence. The shift is deliberate. The chapter speaks to the reader directly, in the second person where it lands, because the framework's implications are personal in a way that the evidence the book has assembled has not been. The reader is not, in what follows, a generic audience reading an argument about other people. The reader is the visitor whose visit the prior chapters have characterized. The chapter is about your visit. The chapter writes accordingly.

The Grievance Posture

The grievance posture is the cluster of orientations that the lottery model authorizes by default. It holds, in its purest form, that the conditions of one's life are impositions on a self who did not choose them. The family was the family one happened to be born into; the body was the body one happened to inhabit; the era was the era one happened to arrive in; the constraints were the constraints one happened to encounter. None of it was chosen. All of it was assigned. The self who is now living the consequences is, on the grievance posture, the victim of an arrangement the self had no part in making and no responsibility for accepting.

The grievance posture is not a marginal cultural attitude. It is, in the contemporary register, one of the dominant frameworks through which difficult experience is held. The contemporary therapeutic vocabulary, the contemporary political vocabulary, and substantial portions of the contemporary spiritual vocabulary all operate, at varying levels of explicitness, within its terms. The conditions of one's life are framed as the impositions of a system, of a family, of a body, of a culture, of an era — all of them external to the self that is experiencing them, all of them objects of legitimate grievance, all of them properly held at a distance from the self that is the locus of one's identity. The self is innocent. The conditions are the violation. The work of the life, on this framing, is to resist or escape or recover from the violation that the conditions constitute.

The grievance posture has, within the lottery model that authorizes it, a certain internal coherence. If the conditions of arrival are random, then they have no meaning. If they have no meaning, then the self who is enduring them has no responsibility for them. If the self has no responsibility, then the grievance is appropriate. The grievance posture is, on the lottery model, the honest and accurate framing of one's relationship to one's conditions. Anything else would be a kind of self-deception, the substitution of a comforting story for the unflinching recognition that one is the victim of an arrangement one did not author.

The grievance posture's internal coherence ends when the lottery model ends. The framework the book has developed has, through twelve chapters of scientific and clinical evidence, replaced the lottery model with a framework in which the conditions of arrival are matched at the molecular level, the arrival is reported as chosen by the visitor whose recovered memory is most direct, the duration is arranged with measurable empirical signatures, and the signal that constitutes the visitor exists in a substrate that current physics can accommodate. The framework is not, on the available evidence, more speculative than the lottery model. It is, on the cumulative weight of the case, less so. And the framework does not authorize the grievance posture, because the framework does not contain the random imposition that the grievance posture requires as its premise. The conditions of life, on the framework, were not impositions. They were selections. The visitor selected them. The visitor was, on the framework, the agent of the arrangement the visitor is now experiencing as the visit.

The grievance posture has been studied, in mainstream psychology, under a different name. The American psychologist Martin Seligman, working at the University of Pennsylvania, established in the late 1960s and early 1970s that animals subjected to inescapable aversive conditions develop a behavioral pattern that Seligman called learned helplessness — a generalized expectation that one's actions have no effect on one's conditions, which transfers to subsequent situations where escape is in fact possible.[196] The animals that developed learned helplessness in the original experiments showed cognitive, emotional, and motivational deficits that mirrored the symptoms of clinical depression in humans, and the framework Seligman developed became one of the foundational accounts of how persistent depression operates at the level of the organism's expectations about its own efficacy. The grievance posture is the cognitive and affective signature of learned helplessness applied to the conditions of one's life as a whole. The visitor who treats the conditions as impositions has, on Seligman's framework, generalized the expectation that one's actions cannot affect one's conditions into a stable orientation toward existence itself. The clinical consequences are documented at the level of depressive symptoms, reduced life satisfaction, and impaired

[196]Seligman, M. E. P. (1972). "Learned helplessness." Annual Review of Medicine 23: 407-412. Book-length treatment: Seligman, M. E. P. (1975). Helplessness: On Depression, Development, and Death. San Francisco: W. H. Freeman. Updated framework: Seligman, M. E. P., and Maier, S. F. (2016). "Learned helplessness at fifty: Insights from neuroscience." Psychological Review 123(4): 349-367.

capacity for the kind of agency the next sections of the chapter will return to.

The contrasting orientation — what the American-Canadian psychologist Albert Bandura, working at Stanford from the 1960s through the 2000s, called self-efficacy — has been studied with similar rigor and is the empirical signature of the agency the framework's working orientation produces.[197] Self-efficacy is the belief that one's actions are consequential to one's conditions — that the actions one takes can in fact alter the situations one is in. The construct has been measured across thousands of studies and is among the most robustly predictive variables in the entire psychological literature for health behaviors, academic and occupational performance, recovery from illness and trauma, and overall life satisfaction. The conditions that produce self-efficacy include the experience of successful action under difficulty, observation of others succeeding under difficulty, encouragement from those whose judgment one trusts, and the framing that one's situation is responsive to effort. The grievance posture systematically undermines self-efficacy by foreclosing the framing on which efficacy depends. The framework's working orientation systematically supports self-efficacy by making the framing available. The two orientations are not, in their clinical signatures, neutral alternatives. The clinical literature documents which one produces the functioning the visitor needs to do the work the visit was assembled around.

What Pre-Selection Implies for the Difficult Conditions

The framework's pre-selection thesis has a specific implication for the difficult conditions of any individual life. The

[197]Bandura, A. (1977). "Self-efficacy: Toward a unifying theory of behavioral change." Psychological Review 84(2): 191-215. Synthesis: Bandura, A. (1997). Self-Efficacy: The Exercise of Control. New York: W. H. Freeman.

implication is not that the difficulties are easy, are deserved, are minor, or are to be dismissed. The implication is that the difficulties were known to the visitor at the time of selection, were factored into the arrangement, and constitute part of what the visit was assembled to encounter rather than incidental obstacles to a visit that was about something else. The difficulties are not the friction the visit had to overcome to be what it was supposed to be. The difficulties are, on the framework, part of what the visit was supposed to be.

This is a hard claim, and the chapter does not soften it. The visitor who selected the family that produced the abuse, the body that carries the illness, the lineage that carries the trauma, the era that contained the war — that visitor knew, on the framework, that the conditions of arrival would include those features. The selection was not made in ignorance. The selection was made with the conditions specifically in view, for reasons that the visitor's recovered memory in the few cases where recovery is most direct describes as related to specific work the visitor wanted to do, specific relationships the visitor wanted to be in, specific perspectives the visitor wanted to integrate. The Wambach dataset surveyed in Chapter 8 documented that 81 percent of the subjects who recovered pre-birth memories reported having actively chosen to be born into their current life, with the choice often expressed as reluctance, hesitation, and reflection on the difficulty of the conditions rather than as eager anticipation. The visitor knew what was coming. The visitor came anyway.

What follows from this, on the framework's working logic, is not that the difficulties cease to be difficult or that they were not really painful. The difficulties are real. The pain is real. The framework does not require the visitor to deny either, and the chapter does not invite the reader to do so. What the framework does require, in a way the grievance posture cannot accommodate, is that the visitor recognize the difficulties as part of the visit's arranged content rather than as impositions on a visit the visitor would otherwise have chosen differently. The difficulties are not deviations from the visit's plan. The difficulties are part of the plan. They were assembled into the conditions of arrival, with the visitor's prior knowledge, for reasons the visitor at the time of selection considered sufficient. The reasons are not, in most cases, accessible to the visitor's current substrate-bound recall. The reasons existed. They were the visitor's own.

A consequence of the framework that the chapter will mark and then return to is that the standard categories of complaint do not apply to the conditions of one's own life in the way they continue to apply to the conditions of others' lives. The framework's implications for the visitor's relationship to the visitor's own conditions are different from the framework's implications for the visitor's relationship to anyone else's conditions. The chapter is not arguing that the abused child deserved the abuse, that the dying friend chose the illness, that the structurally disadvantaged group selected the disadvantage. The framework does not authorize any such reading. The framework's pre-selection thesis is a first-person claim about one's own conditions, not a third-person claim about anyone else's. The next section addresses this distinction directly, because it is the point at which the framework is most often misused.

The Empirical Signature: Meaning-Making and Post-Traumatic Growth

The clinical literature contains a substantial body of work that bears directly on the framework's working orientation, though it has been developed within mainstream psychology without reference to the framework's metaphysical commitments. The principal originator of the relevant tradition is the Austrian psychiatrist and Holocaust survivor Viktor Frankl, whose 1946 book Man's Search for Meaning developed the framework Frankl called logotherapy from his clinical practice in Vienna and from his observations of the Nazi concentration camps.[198] Frankl's central observation, drawn from the camps but extended into ordinary clinical practice afterward, was that the human capacity to endure and even to find significance in extreme suffering depended on the sufferer's ability to locate meaning in the suffering — to understand the experience as serving some purpose the sufferer could affirm, even when the conditions of the suffering were unchosen and the suffering itself was severe. The clinical framework Frankl developed has been substantially elaborated by subsequent therapists and researchers and remains one of the major branches of mainstream psychotherapy. The empirical observation underlying the framework — that meaning-making capacity is among the strongest single predictors of psychological survival under conditions of severe difficulty — has been validated across many subsequent studies of trauma, illness, bereavement, and other major adversities.

The contemporary research program that has formalized Frankl's observation is the post-traumatic growth literature developed by the American psychologists Richard Tedeschi and Lawrence Calhoun at the University of North Carolina at Charlotte beginning in the 1990s.[199] Tedeschi and Calhoun documented, across a substantial body of clinical research, that a significant fraction of people who experience major adversity — bereavement, serious

[198]Frankl, V. E. (1946/2006). Man's Search for Meaning (rev. ed., I. Lasch, trans.). Boston: Beacon Press. Therapeutic framework: Frankl, V. E. (1969). The Will to Meaning: Foundations and Applications of Logotherapy. New York: New American Library.

[199]Tedeschi, R. G., and Calhoun, L. G. (2004). "Posttraumatic growth: Conceptual foundations and empirical evidence." Psychological Inquiry 15(1): 1-18. Comprehensive treatment: Calhoun, L. G., and Tedeschi, R. G. (Eds.) (2006). Handbook of Posttraumatic Growth: Research and Practice. Mahwah, NJ: Lawrence Erlbaum. Integrative review of the broader meaning-making literature: Park, C. L. (2010). "Making sense of the meaning literature: An integrative review of meaning making and its effects on adjustment to stressful life events." Psychological Bulletin 136(2): 257-301.

illness, accident, assault, combat trauma, natural disaster — report not only the expected negative consequences but also specific positive changes in their lives that they themselves attribute to the experience of the adversity. The positive changes include increased appreciation for life, deepening of important relationships, recognition of personal strength, identification of new possibilities for the life, and changes in spiritual or existential orientation. The phenomenon is not universal — many trauma survivors do not experience post-traumatic growth, and the literature is careful not to expect or impose it — but the phenomenon is documented at significant rates, and the factors that predict it have been studied extensively.

The factor that emerges most consistently in the post-traumatic growth literature as a predictor of the phenomenon is the survivor's capacity to construct a meaning-frame in which the difficult experience is integrated into the larger arc of the life rather than held as an isolated imposition on the life. The survivors who develop the meaning-frame show measurable improvements in well-being relative to comparable survivors who do not. The survivors who do not develop the meaning-frame, or who actively resist developing one, show poorer long-term outcomes on the standard psychological measures. The clinical literature does not require any specific metaphysical commitment for the meaning-frame to operate; the meaning-frame can be religious, secular, philosophical, relational, or constructed from whatever resources the survivor's situation makes available. What matters is that the frame integrates the difficulty into a structure the survivor can affirm rather than holding the difficulty as an arbitrary external imposition the survivor merely endures.

The framework this book has developed is not the only meaning-frame that produces the post-traumatic growth signature, and the chapter is not arguing that the framework is required for the signature to occur. The chapter is observing, however, that the framework produces precisely the kind of meaning-integration the clinical literature has identified as the operational mechanism by which difficult experience contributes to growth rather than to ongoing deterioration. The pre-selection thesis is, in operational terms, a meaning-frame in the technical sense the post-traumatic growth literature uses the term. The frame integrates the difficulties of the visit into a structure the visitor can affirm. The visitor's affirmation is not pseudo-acceptance imposed by the framework's vocabulary; the visitor's affirmation is the recognition that the visit was the visitor's own selection. The recognition produces, on the available clinical evidence, the orientation the post-traumatic growth literature has documented as the empirical signature of psychological resilience under difficulty. The framework is, in this sense, empirically congruent with the most rigorously studied protective factor in the contemporary trauma literature.

The Spiritual-Bypass Trap

The American psychologist John Welwood introduced the term spiritual bypassing in 1984 to name a pattern that has been the principal pathology of contemporary spiritual practice and that the framework of this book is particularly vulnerable to enabling if its implications are not handled carefully.[200] Welwood's definition was the use of spiritual ideas and practices to avoid facing unresolved emotional issues, psychological wounds, and unfinished developmental tasks. The pattern is recognizable in many contemporary spiritual frameworks: the practitioner uses the framework's vocabulary to dismiss feelings the framework had nothing to say about, to substitute conceptual acceptance for actual

[200]Welwood, J. (1984). "Principles of inner work: Psychological and spiritual." Journal of Transpersonal Psychology 16(1): 63-73. Extended treatment: Welwood, J. (2000). Toward a Psychology of Awakening: Buddhism, Psychotherapy, and the Path of Personal and Spiritual Transformation. Boston: Shambhala.

processing of difficult experience, and to maintain the appearance of spiritual progress while leaving the underlying difficulties unaddressed.

The framework of this book is vulnerable to spiritual-bypass misuse in a specific way. The framework's pre-selection thesis can be deployed, by a reader who is using the framework to avoid rather than to engage, as a universal solvent for difficult emotions. The grief at a loss can be dissolved with you chose this. The anger at an injustice can be dissolved with you chose this. The despair in the face of intractable conditions can be dissolved with you chose this. The deployment is, on the framework's terms, technically consistent — the visitor did, on the framework, choose the conditions that produced the loss, the injustice, the intractability. But the deployment is, in its effect, the spiritual-bypass pattern Welwood identified. The framework's vocabulary is being used to short-circuit the processing the conditions actually call for, with the framework's authority lending the bypass a spiritual legitimacy it would not otherwise have. The result is not the framework's working integration. The result is the framework's misuse to avoid the visit's actual content.

The American psychotherapist Robert Augustus Masters extended Welwood's framework in a 2010 monograph specifically focused on the clinical patterns spiritual bypassing produces and the practical interventions that distinguish working integration from evasive deployment.[201] Masters identified several specific signatures of the bypass pattern: a chronically positive presentation that does not match the underlying affective state; a tendency to interpret personal interpersonal conflict as belonging to the conflict's other party rather than examining one's own contribution; a flattening of

[201]Masters, R. A. (2010). Spiritual Bypassing: When Spirituality Disconnects Us from What Really Matters. Berkeley: North Atlantic Books. Methodological clinical assessment: Caplan, M. (2009). Eyes Wide Open: Cultivating Discernment on the Spiritual Path. Boulder: Sounds True.

emotional range with particular suppression of anger and grief; and a habitual deployment of spiritual vocabulary in moments when the situation actually called for ordinary emotional response. The signatures are diagnostic and have clinical correlates. Bypass practitioners present, in clinical settings, with characteristic difficulties — relational instability, somatic complaints whose physical correlates are minimal, episodes of breakthrough rage or depression that surprise the practitioner because the rage or depression had been bypassed rather than processed for years. The patterns are recognizable and the interventions Masters and his clinical successors have developed are specific. The framework's working integration requires that the reader be able to recognize the patterns in oneself and to interrupt them when they appear.

The chapter marks this trap explicitly because the framework's working integration requires not falling into it. The visitor who chose the life did not choose the life in order to bypass the life's emotional content. The visitor chose the life precisely to engage that content — to experience the grief, to feel the anger, to know the despair, to live through the relationships, the difficulties, the failures, and the redemptions that the conditions of arrival made possible. The pre-selection thesis is not, on the framework's working logic, a license to skip the visit. It is the recognition that the visit's content is part of what the visitor came for. Skipping the content by deploying the framework's vocabulary to dissolve the content is not the framework's application; it is the framework's evasion. The chapter is not interested in the evasion. The chapter is interested in the application.

A practical test of whether one is applying the framework or evading the visit is available. The application increases the visitor's capacity to be present to the difficulty. The evasion decreases that capacity. The application produces grief that is met, anger that is felt, despair that is sat with — and through the meeting, the feeling, the sitting with, the visitor's relationship to the conditions changes in ways the evading deployment cannot produce. The evading deployment produces a thin acceptance that has not actually been earned, a vocabulary that has not been integrated into the body the visitor inhabits, a posture of pseudo-equanimity that any sufficient stress will collapse. The application produces an actual equanimity. The evasion produces only the appearance of one. The framework, used correctly, makes the visit more present. The framework, used as bypass, makes the visit more distant. The reader will know the difference by checking the visitor's actual experience, not the framework's vocabulary applied to it.

What the Framework Permits and Requires

With the spiritual-bypass trap marked, the chapter can now describe what the framework permits and requires by way of practical orientation. The framework permits and indeed requires that the visitor experience the visit's content fully. Grief is grief. Anger is anger. Despair is despair. None of these is dismissed by the framework; all of these are part of what the visit was assembled to include. The visitor's permission to feel what the visit produces is, on the framework, not in tension with the framework. The permission is part of the framework's working logic.

What the framework does not permit is the assignment of the conditions to an external imposer the visitor can hold responsible. The conditions were not imposed by an external party. The conditions were selected by the visitor. The recognition is not, on the framework's working logic, a basis for self-blame — the visitor's selection at the time of selection was made under conditions and with information the visitor's current substrate does not have access to, and the standards of post-hoc accountability that apply to substrate-bound decisions do not straightforwardly apply to the selection. What the recognition is a basis for is the dissolution of the external-imposer posture. There is no external imposer to be angry at. There is the visitor who selected the conditions and the visitor who is now experiencing them. The two are continuous. The anger has, on the framework, nowhere to land.

What the framework requires, by way of positive orientation, is the stewardship of the visit. The visit was selected. The visit is now occurring. The visitor's responsibility, on the framework's working logic, is to be in the visit as fully as the visit's conditions permit — to engage the relationships, to do the work, to feel the feelings, to make the choices the visit's specific configuration was assembled around, and to do these things with the recognition that the visit is what the visitor came for. The recognition does not make the visit easier in the surface sense. It makes the visit clearer. The clarity is not consolation. The clarity is direction. The visitor knows, on the framework, what the visit is. The visit is not happening to the visitor. The visit is what the visitor is doing. The distinction is the central operational consequence of the framework, and it is the orientation from which all the subsequent chapters' practical content follows.

An analogy that the framework can sustain without strain is the analogy of the traveler who has booked a difficult journey. The traveler chose the destination, paid the fare, packed the bag, and arrived at the departure point in full knowledge that the route would be hard. Once the traveler is on the journey, the difficulties of the route are not impositions on the traveler. They are features of the journey the traveler selected. The traveler does not have, on the journey, the option of pretending the journey was assigned. The traveler does have the option of complaining about the difficulties, but the complaints have, given that the journey was chosen, the structure of complaints about features of a contract the complainant signed. The complaints are permitted. The complaints are also, in the deeper register, somewhat misplaced. The journey is what the traveler chose. The chosen journey is the journey the traveler is now on. The traveler's responsibility is the journey.

A second analogy that the framework can sustain captures the relationship between the pre-arrival visitor and the substrate-bound visitor more directly. Imagine a graduate student who designs a five-year research program, submits the program to the funding committee, secures the resources, enrolls in the program, and then proceeds through the work. The student in the third year of the program, encountering the specific difficulties the program's design entailed, does not relate to those difficulties as impositions by some external authority. The student designed the program. The student knew the difficulties would be encountered. The difficulties were factored into the design as features of the work the program was assembled to do. The student's third-year frustration with a specific obstacle is real and is permitted, but the frustration does not, on the working logic of the program, transfer to the program's designer as an external villain who imposed the obstacles. The student in the third year and the student who designed the program in the first year are the same student. The continuity is the operative point. The framework's pre-selection thesis is, on the second analogy's terms, the recognition that the visitor in the substrate now and the visitor who selected the conditions before arrival are the same visitor, with the temporary opacity introduced by the substrate's filtering machinery being the only barrier between the two registers of the same continuous identity.

The framework's working orientation can be summarized in four operational moves, each of which the chapter has implicitly developed but which are worth stating directly. The first move is the recognition that the conditions of arrival were selected by the visitor rather than imposed by an external authority. The second move is the willingness to experience the emotional content of the conditions fully — grief as grief, anger as anger, despair as despair — without using the framework's pre-selection thesis to short-circuit the experiencing. The third move is the dissolution of the external-imposer posture and the recovery of the agency the grievance posture had suspended. The fourth move is the active stewardship of the visit — the engagement with the relationships, the work, the difficulties, and the integrations that the visit's selected content was assembled around. The four moves are not sequential stages to be completed in order; they are interlocking operational positions to be held simultaneously, with each supporting the others and the absence of any one undermining the rest. The visitor who recognizes the selection but refuses the emotion has not yet entered the framework's working register. The visitor who feels the emotion but holds the external-imposer posture has not yet entered either. The framework's working integration requires all four moves together. The chapter has spent its length developing the conditions under which they become available.

From Recognition to Operation

The recognition that the life was chosen is not, by itself, the operational consequence of the framework. The recognition is the threshold. The operation is what follows. The next chapter examines the operational technology by which the visitor's accumulated pre-arrival content becomes accessible to the substrate while the substrate is still functional — the work the framework calls coherence, and that the cumulative scientific literature on attention, on autonomic regulation, on contemplative practice, and on the field of consciousness studies has begun to characterize as a measurable state of the system rather than an abstract spiritual condition. The operation of coherence is what makes the framework practically actionable. The chapter that follows develops it.

Before the operational chapters proceed, two points from the present chapter are worth marking for what follows. The first is that the framework's pre-selection thesis is not, on the working logic, a passive description of conditions the visitor must accept. The thesis is an active recognition that reorients the visitor's relationship to the conditions. The visitor who recognizes the conditions as chosen has a different relationship to the conditions than the visitor who treats them as impositions, and the different relationship makes different actions available. The grievance posture closes off certain actions and certain inner movements; the chosen-not-assigned recognition opens them. The opening is the operational point of the recognition. The recognition is for the sake of what it makes possible, not for the sake of the recognition itself.

The second is that the framework's working logic does not eliminate the difficulty of the visit. It re-locates the difficulty. The visit is still hard. The grief is still painful. The injustice is still real. The losses are still losses. The framework does not erase any of these. The framework changes the visitor's understanding of where the difficulties come from and what they are for. They come from the visit's selection rather than from external imposition. They are for the integration the visit was assembled to accomplish rather than for the punishment of a self the visit had no part in arranging. The change is not cosmetic. The change is operational. The same difficulties, met with the framework's working orientation, produce different inner movements than the same difficulties met with the grievance posture. The next chapter develops the technology by which the inner movements can be reliably produced. The technology has a name in the framework. The name is coherence.

✦ ✦ ✦

The boy on the ridge above Futami Ko harbor in the spring of 2010 had been doing, on the case file's accumulated documentation, the work the present chapter has been describing in the framework's operational register. The substrate had been carrying the content the visit had not closed. The trip to Chichi Jima had been the substrate's first opportunity to encounter the geography the unclosed content had been organized around. The encounter had produced the release the trauma-medicine literature would have predicted under any meaning-frame that integrated the substrate's content rather than holding it as an imposition the substrate merely endured. The boy on the ridge had not, on the case file's record, been operating from a framework he had read in a book. The boy had been operating from the framework the visit had brought with it. The framework had been the visit's own. The substrate had been the receiver. The release on the ridge had been the integration the visit had been organized around since the boy had been three. The visit's pre-selected content had become, in the morning on the ridge and on the boat over the wreck site, integrated into the substrate that had been carrying the content for the previous nine and a half years. The boy walked down from the ridge. The boy returned to Lafayette. The drawings shifted. The visit continued. The next chapter examines the substrate's technology by which any visit's accumulated content can become integrated into the substrate while the substrate is still functional, without requiring the geographic encounter that the boy's specific case had been organized around. The technology is available to every substrate. The chapter develops it.

Where the Argument Goes Next

The next chapter examines coherence as the operational technology of the framework. The chapter takes coherence in three registers. The first is the neuroscientific register, in which coherence refers to the measurable state of cross-system integration that the literatures on heart rate variability, autonomic regulation, and contemplative-practice neuroscience have established as a specific configuration of the nervous system. The second is the operational register, in which coherence refers to the alignment of the visitor's conscious orientation, embodied state, and acting choices with the visit's selected content. The third is the technical register the framework specifically develops, in which coherence is the substrate's capacity to receive the consciousness signal in higher fidelity than the substrate's ordinary configuration permits — the receiver model of Chapter 6 brought to its operational application.

The three registers are not three separate things. They are the same phenomenon described at three levels of analysis. The chapter develops the convergence and presents the operational consequences. The visit, on the framework, is what the visitor came for. Coherence, on the framework, is the technology by which the visitor can be increasingly present to the visit while the visit is happening. The chapter that follows builds the case.

CHAPTER 14

Coherence Is the Technology of Remembering

The structural magnetic resonance imaging suite at the Waisman Center on the western edge of the University of Wisconsin-Madison campus, on the floor that the Laboratory for Affective Neuroscience had been occupying since the 2001 reconfiguration of the building that had followed the original Lutz-Davidson collaborations with the Tibetan-Buddhist contemplative community the previous year, was a small windowless room on the second floor with the magnet enclosure at the center, the technician's console along the far wall, and a separate observation room behind the console wall in which the supervising investigator could watch the scan in progress through the viewing window. The Siemens 3T scanner in the magnet enclosure was the fourth-generation instrument the lab had acquired since the collaboration's beginning. The technician at the console at approximately ten in the morning on a Tuesday in May of 2016 was an experienced senior MRI technologist who had been doing the lab's contemplative-meditator scans for over a decade. The supervising investigator at the console next to the technician was Richard Davidson, who at the time of the scan was sixty-four years old, had been at Madison since 1985, and had been collaborating with Yongey Mingyur Rinpoche on the brain-imaging work the lab had been conducting on the long-term Tibetan-Buddhist meditators since September 2002, when Mingyur had first arrived in Madison with the delegation of monks the Mind and Life Institute had organized for the original gamma-coherence experiments the first chapter of this book opened with.

The other investigator in the observation room behind the console wall was Nagesh Adluru, a research scientist at the Waisman Center whose collaboration with the Davidson lab had been organized around the application of machine-learning methods to the structural imaging data the lab had been accumulating across the long-term-meditator project since the early 2000s. Adluru had developed, with the lab's broader team and in coordination with the Center for Predictive and Computational Phenotyping that the Waisman Center had established in 2014, the analytic pipeline that would, in the months and years that followed the scan on the morning in May of 2016, process the data the scanner was producing and submit the data to the Brain Age Gap Estimation framework — the BrainAGE algorithm — that had been developed in the international neuroimaging community since the early 2010s for estimating a subject's chronological age from the structural features of the gray matter the structural MRI captured.[202]

The subject in the scanner that morning was Mingyur Rinpoche, who at the time of the scan was forty-one years old, had been back from the four-and-a-half-year wandering retreat for approximately six months — the November 2015 return to Tergar Monastery in Bodh Gaya having concluded the retreat the previous chapter opened with — and was in Madison for the first follow-up scan the lab had been able to organize after his re-emergence. He was, on his arrival at the lab the previous afternoon, in the recovered configuration the family and the senior monks at Tergar had been observing since his return. The body was thinner than the body that had departed Bodh Gaya in June 2011. The face had the slight asymmetry on the left side that the food-poisoning episode at

[202]Adluru, N., Korponay, C. H., Norton, D. L., Goldman, R. I., and Davidson, R. J. (2020). "BrainAGE and regional volumetric analysis of a Buddhist monk: a longitudinal MRI case study." Neurocase 26(2): 79-90. The published paper presents the four-scan longitudinal series conducted on Yongey Mingyur Rinpoche over the period 2002 to 2016, with the comparison cohort of 105 demographically matched controls, and the BrainAGE analytic framework that produced the brain-age estimate. The fourth scan in the series, conducted in May 2016 when Mingyur was 41 chronological years old, produced the estimated brain age of 33.

Kushinagar in the first weeks of the retreat had left as a permanent trace. The hair, which had been growing during the wandering phase and which had been trimmed by the senior monks at Tergar in the weeks following his return, was at the moderate monastic length the order's standards permitted. The robes were the maroon-and-yellow robes of the order, in the configuration the lab's neuroimaging protocols had been documenting on him in every scan since 2002. He had arrived in Madison with two senior attendants from Tergar and with the small alms bowl from the wandering retreat that had been the only piece of equipment from the retreat he had brought with him on the flight from Kathmandu via Delhi and Frankfurt to Chicago.

He was lying on his back in the scanner cradle with the head coil positioned over the skull and the cushioning blocks at the sides of the head holding the position steady within the millimeter tolerance the structural protocol required. The acoustic noise of the scanner during the structural sequence was the loud rhythmic banging the gradient coils produced — the sequence the technician had configured for the high-resolution T1-weighted scan that would take approximately twelve minutes to complete. The padded headphones over the ears reduced the acoustic noise to a level Mingyur had described, in his recovered configuration, as comparable to the cycle of a Himalayan thunderstorm at moderate distance, which had been a sound he had become accustomed to in the third year of the retreat at the cave above the Dolpo valley where the storms had passed several times a week. He had asked the technician, before the scan had begun, to call out into the headphone microphone if any of the technical sequences he needed to extend would require him to remain in position longer than the protocol had indicated. The technician had agreed. He had then closed his eyes. He had begun the practice. The practice the technician had been observing on the console-monitor video feed of the scanner cradle, in the first ninety seconds after the scanning sequence had begun, was the practice the lab had been observing on Mingyur in every scan since 2002 — the slight slowing of the respiratory cycle, the slight settling of the small muscles around the eyes and the mouth, the cessation of the small adjusting movements of the head and shoulders that the cradle's tolerance permitted but that the protocol's reliability required to be minimized. The subject had entered the configuration the lab's protocols had been designed to capture.

The scan ran for the twelve minutes the protocol specified. The technician monitored the gradient outputs on the console screens, confirmed that the head position remained within tolerance across the duration of the sequence, and registered the standard internal quality-control checkpoints at the one-third and two-thirds positions of the run. Davidson stood at the back of the console area in the configuration he had taken at every scan of every long-term meditator the lab had ever processed — quiet, watching the screens, not commenting on the run in progress. Adluru in the observation room behind the console wall watched the live preview of the volumetric reconstruction as the slices accumulated. The brain in the scanner was, at first inspection of the slices the screen was rendering, a brain unremarkable in the immediate features the eye could resolve at low resolution. The gray-matter ribbon along the cortical surface was the gray-matter ribbon of a healthy adult human. The hippocampal volumes at the base of the temporal lobes were within the normal range for the chronological age the subject's birth certificate documented. The white-matter tracts beneath the cortex were the white-matter tracts of an undamaged adult brain. There was no superficial feature of the scan that would have indicated to a senior radiologist, looking at the slices alone, that the brain in the scanner belonged to a subject whose cumulative formal meditation practice time the lab's records had documented at over sixty-two thousand hours across the previous thirty-two years.[203]

---

[203]On the lab's documentation of Mingyur Rinpoche's cumulative meditation practice time at over 62,000 hours by the early 2010s, including the formal practice during the 1991-1994 three-year retreat at Sherab Ling, the multiple

The scan ended at the twelve-minute mark. The technician extracted the cradle from the magnet enclosure. The subject opened his eyes. The technician helped him sit up. He spoke briefly with the technician about the duration of the scan, which he had estimated, while in the cradle, at approximately twelve minutes. The technician confirmed that his estimate was within fifteen seconds of the actual duration the sequence had taken. He walked out of the magnet enclosure, accepted the cup of water the technician offered him, and sat down on the small bench against the wall while the technician completed the post-scan checks the protocol required. Davidson came out of the console area. They exchanged the brief greeting Davidson and Mingyur had developed over the fourteen years of the collaboration. Davidson asked, in the careful phrasing the lab had been using for the meditators across the long-term project, whether the scan had been comfortable. Mingyur said it had been comfortable. He asked, in the same careful phrasing the meditators had been using, whether the lab had captured what the lab had been trying to capture. Davidson said the data looked good on the preview screens and that the full analysis would take several months to complete. Mingyur thanked Davidson. He stood up. He walked with his two senior attendants toward the small dressing room at the end of the corridor where the lab had set aside changing space for the Tibetan delegations across the years of the collaboration.

The analysis took the months Davidson had estimated. The BrainAGE algorithm Adluru and his collaborators had configured for the long-term-meditator data set was, by mid-2017, returning the estimate the published paper would, three years later, report in the Neurocase volume that the team submitted to in the autumn of 2019. The estimate produced by the algorithm for the May 2016 scan,

---

subsequent confined retreats, and the cumulative hours of daily practice across his teaching career, see Goleman, D., and Davidson, R. J. (2017). Altered Traits: Science Reveals How Meditation Changes Your Mind, Brain, and Body. New York: Avery, pp. 245-247.

computed from the full set of cortical and subcortical gray-matter volume features the structural sequence had captured and compared against the 105-subject demographically matched comparison cohort the lab had assembled, was that the brain in the scanner that morning had the structural features of a brain whose chronological owner would have been thirty-three years old. The subject's chronological age at the time of the scan, computed from the birth-certificate record the Tergar administrative archive had provided, was forty-one. The brain in the scanner was, by the algorithm's analysis of the gray-matter inventory the structural sequence had captured, approximately eight years younger than the body that was carrying it. The substrate had been the platform of the four-and-a-half years of practice the previous chapter had opened with. The platform was carrying the measurable trace of what the practice had done to it. The substrate had been doing, on the available analysis, the work the framework of the present book describes the substrate as doing when the substrate is in the configuration the framework calls coherence.[204]

✦ ✦ ✦

The previous chapter completed the pivot from evidence to consequence. The framework has been earned. The grievance posture is incoherent under the framework's commitments. The four operational moves — recognition, emotional permission, dissolution of the external-imposer posture, and active stewardship of the visit

[204]Adluru et al. (2020), Tables 2-3 and Fig. 4. The brain-age estimate of 33 for the May 2016 scan was the most striking individual data point in the four-scan series, with the longitudinal trajectory across all four scans (2002, 2009, 2014, 2016) showing a consistently slower rate of brain aging relative to the matched controls. The paper explicitly noted that the structural differences were distributed across the gray matter rather than localized to specific regions, suggesting that the contemplative-practice effect operated at the level of coordinated whole-brain remodeling rather than as localized hypertrophy of specific functional regions.

— are the visitor's positions under the framework's working logic. What the previous chapter did not yet provide is the technology by which those positions are practically maintained. The recognition occurs. The orientation shifts. The visitor returns to ordinary life. The substrate-bound conditions reassert themselves. The recognition that had been clear in one moment becomes, hours or days later, faint. The grievance posture reappears. The work has to be redone. The chapter that follows examines what makes the work hold — what allows the framework's orientation to stabilize into a durable state of the substrate rather than recurring as a temporary insight that the substrate's ordinary configuration repeatedly absorbs and dissolves.

The framework's name for the stabilized state is coherence. The term is chosen deliberately to mark that the state in question is not abstract spiritual achievement but a specific configuration of the visitor's substrate that has, in recent decades, become measurable across multiple scientific registers. The substrate that produced the brain-age estimate of thirty-three at the chronological age of forty-one was a substrate whose coherent configuration had, across the cumulative practice time the lab's records had documented, become the substrate's default rather than the substrate's occasional achievement. The configuration was not a temporary state Mingyur was entering in the scanner and would leave on his return to the dressing room. The configuration was the substrate's stable baseline. The chapter takes coherence in three convergent registers and develops the conditions under which the stable baseline becomes available.

The first is the neuroscientific register, in which coherence refers to documented patterns of cross-system synchronization in the autonomic nervous system, the cardiac rhythm, the respiratory cycle, and the cortical electrical activity. The second is the operational register, in which coherence refers to the alignment of the visitor's conscious orientation, embodied state, and acting choices with the visit's selected content. The third is the technical register the framework specifically develops, in which coherence is the substrate's capacity to receive the consciousness signal at higher fidelity than the substrate's ordinary configuration permits. The three registers are not three separate phenomena. They are the same configuration described at three levels of analysis, with the empirical signature in each register tracking the empirical signature in the other two.

The chapter develops the case for coherence as the framework's operational technology by examining each register in turn, then assembling the convergence, then turning to the practical implications. The state is trainable. The training has documented neural, autonomic, and behavioral correlates. The substrate the framework's earlier chapters characterized as the receiver of the consciousness signal can, on the available evidence, be brought into configurations that receive the signal more clearly than the substrate's default state allows. The remembering that Chapter 1 promised the book would earn — the visitor's increased access to the pre-arrival content the substrate's ordinary filtering machinery occludes — is, on the framework's working logic, the experiential signature of the substrate's coherent configuration. The chapter develops the case by which this is so.

Coherence in the Neuroscientific Register

The single most thoroughly measured signature of physiological coherence in the contemporary literature is heart rate variability, the moment-to-moment variation in the intervals between successive heartbeats. The variability is not noise; it is the integrated output of the autonomic nervous system's two principal branches — the sympathetic, which accelerates the heart, and the parasympathetic vagal pathway, which slows it — operating against the rhythmic input of the respiratory cycle to produce a continuously fluctuating heart rate that, under conditions of physiological coherence, settles into a smooth sinusoidal pattern at approximately 0.1 Hz.[205] The 0.1 Hz frequency is not arbitrary; it is the resonant frequency of the baroreflex feedback loop that couples blood pressure regulation to cardiac rhythm. When breathing slows to approximately six breaths per minute, the respiratory cycle entrains the cardiac rhythm at the baroreflex resonance, producing the maximal possible heart rate variability and the smoothest possible sinusoidal pattern. The state is, in the technical literature, called cardiac coherence.

Cardiac coherence is not merely a curiosity of cardiovascular physiology. The clinical research on heart rate variability has established that the variability is among the most powerful single predictors of overall health, mortality, and psychological well-being in the contemporary medical literature. Reduced heart rate variability — the loss of the smooth sinusoidal pattern in favor of a flatter, more rigid heart rhythm — is associated with elevated cardiovascular mortality, depression, anxiety, post-traumatic stress disorder, and reduced cognitive performance. Elevated heart rate variability is associated with the opposite profile: cardiovascular longevity, emotional resilience, cognitive flexibility, and effective self-regulation under stress.[206] The neurovisceral integration

[205]Shaffer, F., and Ginsberg, J. P. (2017). "An overview of heart rate variability metrics and norms." Frontiers in Public Health 5: 258. Comprehensive synthesis: McCraty, R., Atkinson, M., Tomasino, D., and Bradley, R. T. (2009). "The coherent heart: Heart-brain interactions, psychophysiological coherence, and the emergence of system-wide order." Integral Review 5(2): 10-115.

[206]Thayer, J. F., Hansen, A. L., Saus-Rose, E., and Johnsen, B. H. (2009). "Heart rate variability, prefrontal neural function, and cognitive performance: The neurovisceral integration perspective on self-regulation, adaptation, and health." Annals of Behavioral Medicine 37(2): 141-153. Meta-analytic synthesis: Thayer, J. F., Åhs, F., Fredrikson, M., Sollers, J. J., and Wager, T. D. (2012). "A meta-analysis of heart rate variability and neuroimaging studies: Implications for heart rate variability as a marker of stress and health." Neuroscience and Biobehavioral

framework developed by the American psychophysiologist Julian Thayer and his collaborators specifies the mechanism: heart rate variability is the peripheral signature of the prefrontal cortex's regulatory influence over the autonomic nervous system, and the variability is therefore an index of the entire integrated regulatory hierarchy that connects cortical executive function, autonomic regulation, cardiovascular function, and the body's broader homeostatic systems. When the hierarchy is coherent, the variability is high. When the hierarchy is dysregulated, the variability is low. The state of cardiac coherence is the signature of a regulatory system operating in its integrated configuration.

The training of cardiac coherence has been the subject of substantial clinical literature. The most directly applicable technique is paced respiration at the baroreflex resonance frequency — slow breathing at approximately six breaths per minute, with equal inspiration and expiration phases, sustained for periods of five to twenty minutes. The technique is the basis of heart-rate-variability biofeedback as it is practiced in clinical settings. The clinical outcomes are documented: reductions in anxiety, depression, hypertension, asthma symptoms, and a range of stress-related conditions, with effect sizes in many cases comparable to or exceeding pharmacological interventions. The training does not require any spiritual framework to operate. The training operates at the physiological level, accessible to any visitor whose substrate has the autonomic system mammals share, and produces measurable outcomes on the standard clinical instruments.

The deeper consequence of cardiac coherence for the framework's case concerns the direction of information flow between heart and brain. The standard intuition is that the brain commands the heart through descending autonomic signals, with the heart acting as a peripheral organ that responds to central control. The neurocardiology literature of the last three decades has

---

Reviews 36(2): 747-756.

substantially revised this picture. The American physiologist Rollin McCraty and the Canadian researcher J. Andrew Armour, whose intrinsic-cardiac-nervous-system work was introduced in Chapter 10, have demonstrated that the volume of afferent neural traffic from the heart to the brain exceeds the volume of efferent traffic from brain to heart, and that the heart's afferent signals — transmitted through the vagus nerve, through the spinal cord, and through the bloodstream via hormonal release — exert measurable influence on cortical processing, on emotional regulation, and on cognitive function.[207] When the heart enters the smooth sinusoidal cardiac-coherence pattern, the afferent signals it sends to the brain change in character: they become more rhythmic, more synchronized with the respiratory cycle, and, in the available electroencephalographic measurements, they entrain cortical activity into states of cross-region synchronization that the substrate's ordinary cognitive states do not produce. The heart is, on this picture, not merely the pump the standard cardiovascular framework describes. The heart is an active partner in the substrate's overall regulatory and cognitive configuration, with the heart's coherent state being one of the principal inputs to the brain's capacity for the integrated processing that the framework's working orientation requires.

Coherence in the Contemplative-Practice Register

The second major body of evidence on coherence comes from the neuroscience of contemplative practice — the empirical investigation of what occurs in the brain and body during meditation, prayer, and related practices. The field has developed rapidly since

[207] McCraty, R., and Childre, D. (2010). "Coherence: Bridging personal, social, and global health." Alternative Therapies in Health and Medicine 16(4): 10-24. Earlier foundational paper: McCraty, R., Atkinson, M., Tiller, W. A., Rein, G., and Watkins, A. D. (1995). "The effects of emotions on short-term power spectrum analysis of heart rate variability." American Journal of Cardiology 76(14): 1089-1093.

the early 2000s, with several major research programs at universities including the University of Wisconsin under Richard Davidson, Massachusetts General Hospital under Sara Lazar, and Yale University under Judson Brewer. The convergent finding across these programs is that sustained contemplative practice produces specific, measurable changes in brain function and structure that correspond to changes in the practitioner's psychological state and behavioral patterns.[208]

The single most-cited finding in the long-term-meditator literature was published in 2004 by a collaboration led by Antoine Lutz in Davidson's laboratory, examining the electroencephalographic activity of eight long-term Tibetan Buddhist meditators with cumulative practice times ranging from 10,000 to 50,000 hours. During specific compassion-meditation states, the meditators produced sustained high-amplitude gamma-frequency oscillations — neural activity in the 25 to 70 Hz range — at amplitudes substantially exceeding anything in the published literature on ordinary human cognition. The gamma synchrony extended across widely separated cortical regions, indicating that the meditators were producing a state of cross-cortical neural coherence that the controls in the study could not match. The state was, by the standard neurophysiological measures, a configuration of the substrate that ordinary cognition does not typically achieve and that the contemplative training had made accessible to the practitioners. The subject who produced the most striking single trace in the Lutz 2004 paper was Mingyur Rinpoche, then twenty-eight years old, in the original September 2002 scanning sequence the first chapter of this book opened with.

[208]Lutz, A., Greischar, L. L., Rawlings, N. B., Ricard, M., and Davidson, R. J. (2004). "Long-term meditators self-induce high-amplitude gamma synchrony during mental practice." Proceedings of the National Academy of Sciences 101(46): 16369-16373. Comprehensive review: Tang, Y.-Y., Hölzel, B. K., and Posner, M. I. (2015). "The neuroscience of mindfulness meditation." Nature Reviews Neuroscience 16(4): 213-225.

The structural side of the contemplative-practice literature has been developed by Sara Lazar and her collaborators, who used magnetic resonance imaging to compare the cortical thickness of long-term meditators to age-matched controls. The findings, published in a series of papers beginning in 2005, established that long-term practitioners showed measurable increases in cortical thickness in specific brain regions — the prefrontal cortex, the right anterior insula, and the somatosensory cortex — that are associated with attention, interoception, and emotional regulation.[209] A subsequent longitudinal study, published in 2011, showed that an eight-week mindfulness program produced detectable increases in gray matter density in the same regions in previously untrained subjects. The substrate is not, on the available evidence, fixed in the configuration the substrate had at the beginning of the practice. The substrate is, within limits the field is continuing to characterize, modifiable by the practices the visitor undertakes.

The default mode network — the set of brain regions that activate when the brain is at rest, in the absence of a specific task — has been a particular focus of the contemplative-practice literature. The American psychiatrist Judson Brewer and his collaborators demonstrated in 2011 that experienced meditators show reduced activity in the default mode network during meditation relative to controls, and reduced default-mode activity even at rest outside of formal practice.[210] The default mode network is, in the broader

---

[209]Lazar, S. W., Kerr, C. E., Wasserman, R. H., Gray, J. R., Greve, D. N., Treadway, M. T., McGarvey, M., Quinn, B. T., Dusek, J. A., Benson, H., Rauch, S. L., Moore, C. I., and Fischl, B. (2005). "Meditation experience is associated with increased cortical thickness." NeuroReport 16(17): 1893-1897. Subsequent confirmation: Hölzel, B. K., Carmody, J., Vangel, M., Congleton, C., Yerramsetti, S. M., Gard, T., and Lazar, S. W. (2011). "Mindfulness practice leads to increases in regional brain gray matter density." Psychiatry Research: Neuroimaging 191(1): 36-43.

[210]Brewer, J. A., Worhunsky, P. D., Gray, J. R., Tang, Y.-Y., Weber, J., and Kober, H. (2011). "Meditation experience is

cognitive-neuroscience literature, associated with mind-wandering, self-referential thinking, and the construction of the narrative self the brain continuously generates. Reduced activity in this network corresponds, on the available correlations with subjective experience, to reduced rumination, reduced anxious self-monitoring, and increased capacity for present-moment attention. The contemplative training is, on the Brewer findings, producing a substrate whose default state is less occupied by the narrative-self construction and more available for the integrative attention the practice is cultivating. The pattern is precisely what the framework of this book would predict if the practice were producing a substrate more available to receive the consciousness signal in higher fidelity than the substrate's ordinary configuration permits.

A particularly consequential distinction in the long-term-meditator literature is the one Davidson and his collaborators have drawn between state effects and trait effects. State effects are the temporary configurations the substrate enters during active practice — the cardiac coherence during paced breathing, the gamma synchrony during compassion meditation, the reduced default-mode activity during sustained attention practices. Trait effects are the durable baseline changes the substrate undergoes when the state configurations are repeatedly produced across years of practice.[211] The trait-effect literature has documented that long-term contemplative practitioners exhibit, at rest, neurophysiological and

---

associated with differences in default mode network activity and connectivity." Proceedings of the National Academy of Sciences 108(50): 20254-20259. Foundational review: Raichle, M. E. (2015). "The brain's default mode network." Annual Review of Neuroscience 38: 433-447.

[211]Davidson, R. J., and Lutz, A. (2008). "Buddha's brain: Neuroplasticity and meditation." IEEE Signal Processing Magazine 25(1): 174-176. Comprehensive synthesis with the trait/state distinction developed in detail: Goleman, D., and Davidson, R. J. (2017). Altered Traits: Science Reveals How Meditation Changes Your Mind, Brain, and Body. New York: Avery.

structural features that resemble the state configurations of the practice, but as stable features of their default substrate. The brain-age finding the chapter opened with — the Adluru 2020 result that Mingyur Rinpoche's substrate, on the May 2016 scan, exhibited gray-matter features consistent with a substrate eight years younger than his chronological age — is a structural-MRI signature of the trait-effect phenomenon at its most consequential. The training had not, on the available analysis, produced a substrate that entered coherent configurations only during active practice. The training had produced a substrate whose default configuration was the coherent configuration the practice had cultivated. The visit's duration may, on this finding, be modifiable by the technology of coherence not only at the level of subjective well-being but at the level of the substrate's structural aging itself. The training is, in the strongest available reading, doing work the substrate would not, in its default operation, do on its own.

The Predictive Brain and the Stability of Identity

A complementary empirical phenomenon worth registering before the chapter turns to the predictive-coding framework is what the Hungarian-American psychologist Mihaly Csikszentmihalyi called flow — the state of complete absorption in an activity in which attention, action, and capacity align so fully that the ordinary self-monitoring of the prefrontal executive recedes and the activity proceeds with what practitioners across many domains describe as effortless concentration.[212] Csikszentmihalyi spent four decades studying the conditions under which flow occurs, the activities that most reliably produce it, and the well-being correlates of frequent experience of the state. The state has been studied across athletes,

[212]Csikszentmihalyi, M. (1990). Flow: The Psychology of Optimal Experience. New York: Harper & Row. Synthesis with the broader well-being literature: Csikszentmihalyi, M. (1997). Finding Flow: The Psychology of Engagement with Everyday Life. New York: Basic Books.

musicians, surgeons, programmers, chess players, and a range of contemplative practitioners, and the structural features that recur across the diverse populations include the matching of skill to challenge, the presence of immediate feedback on the activity's progress, the clarity of the activity's goals, and the visitor's full attention to the activity in the moment of its conduct. The neurophysiological correlates of flow have been investigated in recent years and include the same family of measurements that characterize the cardiac coherence and contemplative-practice findings: elevated heart rate variability, reduced default-mode-network activity, and the transient hypofrontality — reduced activity in specific prefrontal regions — that allows the executive's ordinary self-monitoring to step back while the integrated activity proceeds. The flow state is, on the available characterization, another mode of the substrate's coherent configuration, accessible through a different route than the contemplative practices but producing the same family of neurophysiological signatures and the same family of subjective markers — clarity, presence, and the experience of acting from a more integrated configuration than the substrate's ordinary divided attention permits.

The third major body of evidence relevant to coherence comes from the contemporary cognitive-neuroscience framework called predictive coding or, in its more general formulation by the British neuroscientist Karl Friston, the free-energy principle. The framework holds that the brain is fundamentally a prediction machine — that its primary computational function is to generate, continuously and hierarchically, predictions about the sensory inputs it will encounter, and to update those predictions when the actual inputs do not match the predicted ones.[213] The framework has

[213]Friston, K. (2010). "The free-energy principle: A unified brain theory?" Nature Reviews Neuroscience 11(2): 127-138. Accessible synthesis: Clark, A. (2013). "Whatever next? Predictive brains, situated agents, and the future of cognitive science." Behavioral and Brain Sciences 36(3): 181-204. Book-

accumulated substantial empirical support across multiple cognitive domains — perception, attention, motor control, emotion, social cognition — and is, as of the time of this writing, the closest available candidate for a unified computational theory of how the brain operates. The framework's central claim is that the brain does not, contrary to the older empiricist picture, passively register the sensory inputs it receives and construct its model of the world from those inputs. The brain generates, in advance, the model it expects to receive sensory input about, and treats the actual sensory input as feedback for updating the model when the model and the input diverge.

The framework has direct implications for how identity operates. The self that the visitor experiences as continuous through time is, on the predictive-coding framework, a high-level prediction the brain is continuously generating about its own state. The prediction is informed by the body's interoceptive signals, by the brain's memories of prior states, by the social and environmental conditions the brain is encountering, and by the brain's running model of what kind of agent the brain is. The continuous experience of being a particular self is, on this account, the brain's running prediction of itself, updated moment-to-moment but stabilized by the prediction's reliance on the same components moment after moment. Identity is, in the technical vocabulary of dynamical systems theory, an attractor — a stable configuration the system tends to settle into and to return to when displaced. The visitor's substrate-bound experience of being the visitor is the substrate's stable attractor in the predictive hierarchy.

This has consequences for the practical question of how the framework's operational orientation can be maintained against the substrate's ordinary configuration. The grievance posture is, on the predictive-coding framework, an attractor — a stable prediction the

---

length treatment: Hohwy, J. (2013). The Predictive Mind. Oxford: Oxford University Press.

brain has settled into about what kind of agent the visitor is and what kind of world the visitor is encountering. The framework's working orientation, in which the conditions of life are recognized as chosen and the visitor's stewardship of the visit is the operative posture, is a different attractor. The substrate cannot occupy both attractors simultaneously. The substrate has to settle into one or the other, and the settling is not a single decision but a continuous process of prediction-updating that depends on the conditions the substrate is encountering moment to moment. The technology of coherence is, in this register, the practice of repeatedly displacing the substrate from the grievance-posture attractor and stabilizing it in the framework's working-orientation attractor until the working-orientation attractor becomes the substrate's default. The practice is not, in this framing, mystical. It is the deliberate cultivation of a stable predictive configuration that produces the orientation the framework requires. The cultivation has documented neural, autonomic, and behavioral correlates, and the cultivation is what the contemplative-practice literature and the heart-rate-variability literature have, each in its own register, been characterizing.

The Substrate Is Modifiable

The empirical foundation for the practical optimism that the framework's operational technology can produce durable change in the substrate is the literature on neuroplasticity — the brain's documented capacity to modify its structure and function in response to experience throughout life. The doctrine that the adult brain was structurally fixed dominated mainstream neuroscience for most of the twentieth century. The doctrine has been progressively dismantled by experimental work that has accumulated for over four decades, with the contemporary consensus being that the brain remains structurally and functionally modifiable into late adulthood, with the modification operating through specific mechanisms that the field has characterized in considerable detail.[214] The mechanisms include synaptic strengthening and weakening at existing connections, the formation of new synaptic connections, the elimination of unused connections through synaptic pruning, the regulated production of new neurons in specific brain regions including the hippocampus, and the modification of myelination patterns that determine the timing and reliability of neural signaling. The mechanisms are documented in detail in the experimental literature and have been the subject of substantial clinical translation in fields including stroke rehabilitation, addiction treatment, and the management of chronic pain.

What the neuroplasticity literature contributes to the framework's operational case is the recognition that the substrate the visitor inhabits is not the same substrate from year to year. The substrate is, within the limits its underlying genetics and biology permit, continuously being reshaped by the experiences the visitor is having, by the practices the visitor is undertaking, by the orientations the visitor is sustaining, and by the relationships the visitor is in. The grievance posture, sustained for years, produces a substrate whose default configuration is shaped by the grievance posture. The framework's working orientation, sustained for years, produces a substrate whose default configuration is shaped by the working orientation. The substrate the visitor will inhabit in five years depends, in part, on the orientations the visitor is sustaining now. The substrate is not a fixed inheritance the visitor merely operates within. The substrate is an active partner in the visit, modifiable by the practices the visit is conducted through, and the practices the visitor

[214] Comprehensive synthesis of the contemporary neuroplasticity literature: Doidge, N. (2007). The Brain That Changes Itself: Stories of Personal Triumph from the Frontiers of Brain Science. New York: Viking. Foundational research synthesis: Pascual-Leone, A., Amedi, A., Fregni, F., and Merabet, L. B. (2005). "The plastic human brain cortex." Annual Review of Neuroscience 28: 377-401.

selects therefore have consequences not only for the immediate experience but for the substrate's longer-term configuration.

A particularly consequential finding from the neuroplasticity literature for the framework's operational case is the work of the Spanish-American neurologist Alvaro Pascual-Leone on the comparative effects of physical and mental practice. Pascual-Leone's collaborators at Harvard Medical School designed an experiment in the mid-1990s in which volunteers were trained to play a simple piano sequence over five days, with one group practicing the sequence physically on a keyboard for two hours daily and a second group performing only mental imagery of the same practice without physically touching the keyboard.[215] The motor-cortex changes that ordinarily occur in response to physical motor practice were measured by transcranial magnetic stimulation in both groups across the five days. The result was striking: the mental-practice group showed cortical changes at the same magnitude as the physical-practice group across the first four days, with the physical practice producing additional benefit only in the final phase of skill consolidation. The substrate's motor cortex was being modified, in the mental-practice group, by the visitor's deliberate attention to an imagined activity that the substrate was not actually performing. The finding has been replicated and extended in subsequent work and is now part of the standard literature on motor-skill acquisition. The implication for the framework's broader case is direct: the substrate is shaped not only by what the body does but by what the visitor attends to. The orientations the visitor sustains, even in the absence of corresponding physical action, are inputs to the substrate's ongoing reconfiguration. The visitor who deliberately attends to the framework's working orientation is, on the Pascual-Leone evidence,

---

[215]Pascual-Leone, A., Nguyet, D., Cohen, L. G., Brasil-Neto, J. P., Cammarota, A., and Hallett, M. (1995). "Modulation of muscle responses evoked by transcranial magnetic stimulation during the acquisition of new fine motor skills." Journal of Neurophysiology 74(3): 1037-1045.

doing physical work at the substrate's neural level even when the visitor's body is at rest. The internal work is not metaphorical. It is documented at the level of the motor cortex's measurable response to mental rehearsal alone.

The clinical translation of the contemplative-practice literature into a generally applicable therapeutic framework was developed most influentially by the American molecular biologist Jon Kabat-Zinn, whose Mindfulness-Based Stress Reduction program at the University of Massachusetts Medical Center has become, since its development in the late 1970s, the most widely studied secular adaptation of contemplative practice in mainstream medicine.[216] The MBSR program has been the subject of hundreds of clinical trials and multiple meta-analyses, with the consistent finding that the eight-week protocol produces significant reductions in anxiety, depression, chronic pain symptoms, and stress-related medical conditions, with effect sizes in many cases comparable to standard pharmacological interventions. The program's curriculum consists of formal contemplative practices, body-awareness exercises, and the deliberate cultivation of attention to the present moment, sustained over eight weeks with daily home practice. The clinical literature on the program is one of the most rigorous available demonstrations that the substrate the visitor inhabits is, on a timescale of weeks rather than years, responsive to systematic practice in ways that produce documented changes in psychological and physical health. The technology of coherence is not, on this body of evidence, an esoteric capacity available only to long-term

[216]Kabat-Zinn, J. (1990). Full Catastrophe Living: Using the Wisdom of Your Body and Mind to Face Stress, Pain, and Illness. New York: Delta. Meta-analytic synthesis: Goyal, M., Singh, S., Sibinga, E. M. S., Gould, N. F., Rowland-Seymour, A., Sharma, R., Berger, Z., Sleicher, D., Maron, D. D., Shihab, H. M., Ranasinghe, P. D., Linn, S., Saha, S., Bass, E. B., and Haythornthwaite, J. A. (2014). "Meditation programs for psychological stress and well-being: A systematic review and meta-analysis." JAMA Internal Medicine 174(3): 357-368.

contemplative specialists. The basic operations are accessible, the basic outcomes are documented, and the basic clinical translation has been in mainstream medical practice for over four decades.

The framework's technology of coherence is, on this consolidated picture, the cultivation of substrate configurations that increase the visitor's access to the visit's selected content. The cultivation operates through the autonomic regulation that heart rate variability measures, the cortical configuration that contemplative-practice neuroscience documents, the predictive attractor that the brain's hierarchical computational architecture maintains, and the underlying neuroplastic mechanisms that allow each of these to become stable features of the substrate over time. The framework's earlier chapters established that the substrate is the receiver of a signal the substrate does not generate. The framework's operational chapters establish that the receiver is tunable. The tuning is what the technology of coherence is.

The Three Registers Converge

Pulling the threads together. The neuroscientific register has documented, across the heart-rate-variability literature, the contemplative-practice neuroscience, the predictive-coding framework, and the neuroplasticity findings, that the substrate the visitor inhabits has multiple measurable configurations and that specific practices reliably produce specific configurations. The configurations include the smooth sinusoidal cardiac rhythm of physiological coherence, the cross-cortical gamma synchrony of long-term contemplative practitioners, the reduced default-mode activity that correlates with present-moment attention, and the stable predictive attractor that the visitor experiences as identity. The practices that produce these configurations are documented and trainable. The substrate is responsive to the practices over weeks, months, and years.

The operational register describes the same phenomenon from the visitor's first-person perspective. The visitor who has cultivated coherence experiences the visit differently than the visitor who has not. The visitor in the coherent configuration has access to a fuller range of emotional response, holds the visit's difficulties without being collapsed by them, makes choices that align with the visit's selected content rather than reflexively reproducing the grievance posture's predictable patterns, and maintains the recognition that the life was chosen as a stable orientation rather than an occasional insight that the substrate's default state repeatedly absorbs. The operational register is the substrate's neuroscientific configuration translated into the language of conscious experience and lived behavior. The two registers are descriptions of the same state at different levels of analysis.

The technical register, in the framework's specific vocabulary, names what the convergence accomplishes. The substrate in the coherent configuration is a substrate whose filtering machinery has been adjusted to permit the consciousness signal to come through at higher fidelity than the substrate's default configuration allows. The visitor's access to the pre-arrival content the substrate ordinarily occludes increases, in fragmentary and uneven ways, as the substrate's coherent configuration stabilizes. The increased access is what the framework calls remembering. The remembering is not, on the framework's working logic, the recovery of specific verbal facts about pre-arrival circumstances; the substrate's mature filtering machinery does not, in most cases, release such verbal facts even in coherent configurations. The remembering is, more accurately, the visitor's increasing recognition of the visit's selected content as the visit unfolds, the visitor's increasing capacity to act in alignment with the selection rather than against it, and the visitor's increasing sense that the life the visitor is living is the life the visitor came for. The recognition, the capacity, and the sense are the operational signatures of the remembering. The substrate's coherent configuration is the technical substrate of the remembering. The technology of coherence is, in the framework's working vocabulary, the technology of the remembering itself.

What This Establishes

The chapter has established three things. First, that coherence is a measurable state of the visitor's substrate with documented signatures across multiple scientific registers — autonomic regulation, cortical activity, predictive computation, and neural plasticity. The state is not abstract spiritual achievement. The state has neurophysiological correlates that the contemporary research literature is continuing to characterize at increasing precision. Second, that the state is trainable. The practices that produce coherence are documented, accessible, and effective at the level of measurable physiological and psychological outcomes. The training does not require the visitor to accept any specific metaphysical framework to operate, though the framework this book has developed provides a particular interpretive register in which the training's purpose can be understood at a deeper level. Third, that the framework of this book proposes a specific technical claim about what the coherent configuration accomplishes — that the configuration increases the substrate's capacity to receive the consciousness signal in higher fidelity than the substrate's default state allows — and that the claim, while not directly demonstrable by the available neuroscience, is consistent with what the neuroscience documents and is the framework's working interpretation of what the documented changes amount to.

The chapter is the operational counterpart to the evidence chapters that preceded it. The evidence chapters established the framework. The present chapter establishes the technology by which the framework is practically maintained. The training of coherence is the substrate's ongoing work, conducted across years and decades, that allows the framework's recognition to stabilize into the visitor's default orientation rather than recurring as occasional insight. The visitor who has done the work lives in a substrate whose default configuration is the framework's working orientation. The visitor who has not done the work lives in a substrate whose default configuration is whatever orientation the substrate had been settled into prior to the encounter with the framework — most commonly, the grievance posture the previous chapter identified as the framework's principal counter-position. The difference between the two visitors is not theoretical. The difference is the substrate the two visitors are inhabiting, and the substrate determines, more than the visitor's stated beliefs do, what the visitor's life actually consists of.

Two further observations close the chapter before the bridge to the next. The first is that the chapter's argument has held without invoking any practice or framework that is not, at present, accessible to a reader with no prior contemplative or somatic training. The cardiac-coherence breathing pattern can be acquired in minutes and produces measurable autonomic outcomes the first day it is practiced. The Mindfulness-Based Stress Reduction protocol is delivered in clinical settings worldwide and produces documented benefits at the eight-week mark. The mental-practice protocol the Pascual-Leone experiments employed requires only attention and consistency. The framework's operational technology is not gatekept by lineage, by initiation, or by the visitor's prior accomplishment. The chapter's case is that the technology is available, that the underlying mechanisms are documented, and that the visitor who begins the work begins at the appropriate place regardless of what the visitor's prior orientation has been. The second observation is that the chapter has spoken throughout of the substrate's coherent configuration as a state to be cultivated, but the chapter has also indicated that the configuration is, in the framework's working logic, the substrate's truer condition rather than a special achievement above the substrate's ordinary state. The substrate was assembled, on the framework's prior chapters, to receive the consciousness signal.

The default configuration of the modern substrate, distorted by stress, by chronic grievance, and by the standard conditions of contemporary life, is a configuration in which the substrate is doing the work of receiving inefficiently and under interference. Coherence is not, on this framing, an extraordinary spiritual achievement. Coherence is the substrate's recovery of its native operating configuration, the configuration the substrate was assembled to operate in. The technology of coherence is the technology of the substrate returning to itself.

✦ ✦ ✦

The subject who walked out of the Waisman Center on the Tuesday afternoon in May of 2016 with his two senior attendants from Tergar had been, on the analysis the lab would complete in the months that followed and would publish in 2020 in the Neurocase paper that established the case in the contemplative-neuroscience literature, the most clearly documented individual instance of what the chapter has been describing in the general register of the substrate's coherent configuration. The fifty-eight thousand hours of cumulative practice time the lab had recorded by 2002 had become, by the time of the May 2016 scan, over sixty-two thousand hours, with the four-and-a-half years of the wandering retreat the previous chapter had opened with constituting the most concentrated single period of practice in the cumulative total. The substrate the practice had been operating through had, on the structural-MRI evidence the scan had captured, become a substrate whose default configuration was the configuration the practice had cultivated. The brain that was eight years younger than the body carrying it was not a brain that had been temporarily young during the twelve minutes in the scanner. The brain was a brain whose default state was the state the practice had stabilized. The substrate had become what the visit's practice had been doing to it for the previous thirty-two years. The substrate was the technology the practice had been organized around. The substrate was the practice's accumulated trace.

The technology of coherence the chapter has been characterizing is the technology by which any substrate can, over the duration the substrate is available to the visit it is the platform for, become the substrate the visit is doing the work of becoming. Mingyur Rinpoche is one example. The framework does not require the reader to undertake the specific contemplative training the Tibetan-Buddhist tradition has organized around since the eleventh century, and the chapter has not argued that the specific training is the only available technology. The framework requires the reader to recognize that the substrate is modifiable, that the modification operates through identifiable practices, that the practices are accessible across multiple traditional and secular registers, and that the visitor who selects practices oriented toward the framework's working configuration is, over the years and decades available to the visit, doing physical work at the substrate's neural level that the substrate the visit ends with will measurably differ from the substrate the visit began with. The technology is available. The next chapter examines the integrated method the framework has been organized around.

Where the Argument Goes Next

The next chapter examines the integrated method that the framework's operational technology has been organized into. The method consists of specific practices, sequenced in specific ways, with specific outcomes documented at the level of the substrate's measurable configuration and the visitor's lived experience. The chapter examines the method's components, the rationale for the sequencing, the standards by which the method's progress is assessed, and the relationship of the method to the broader contemplative, somatic, and integrative traditions from which the method's components are drawn. The method is, in the framework's working vocabulary, the remembrance protocol — the systematic training by which the visitor's substrate becomes increasingly available to the visit's selected content while the visit is still in progress. The chapter that follows takes it up directly.

CHAPTER 15

The Method as Remembrance Protocol

The American evolutionary biologist Lee Alan Dugatkin arrived at the small airport in Novosibirsk on a Wednesday afternoon in January of 2014, on the connecting flight from Moscow that had taken approximately four hours across the Western Siberian Plain that lies between European Russia and the Ob River basin, with the temperature outside the aircraft at landing registering on the small monitor in the cabin at approximately twenty-eight degrees below zero on the Celsius scale. The driver who collected him from the arrivals hall was a young man from the Institute of Cytology and Genetics, who had been sent to drive him the thirty kilometers south of the city to the small scientific settlement of Akademgorodok, where the Institute had been located since the late 1950s and where the experimental fox farm that was the subject of his visit had been operating, in approximately the same location and under the supervision of approximately the same chief researcher, since the late summer of 1959. The road south from the airport to Akademgorodok was lined with the heavy snow of mid-January, in drifts the previous week's storms had pushed against the embankments to depths of nearly a meter, with the cleared roadway in the center cut at the width of two passing vehicles by the small fleet of municipal plows that the Soviet-era infrastructure had handed down to the post-Soviet town as one of the more reliable continuities the political dissolution of 1991 had not interrupted. The driver said little during the ride. Dugatkin watched the birch forests pass on either side of the road. The light, by the time the car arrived at the small house at the southern edge of the settlement where the chief researcher was waiting for him with the senior colleagues she had assembled for the first day of the interview sessions, was already beginning to fail. It was approximately three in the afternoon.

Lyudmila Nikolaevna Trut was eighty years old in January of 2014. She had been the lead researcher of the silver-fox domestication experiment since the death of Dmitri Belyaev in 1985, and she had been the senior researcher on the experiment, under Belyaev's overall supervision, since the spring of 1959, when Belyaev had selected her, then a recent graduate of the Department of Vertebrate Zoology at the Moscow State University, to run the program he had been developing in the preceding decade from his position at the Department of Fur Animal Breeding at the Central Research Laboratory in Moscow. The intervening fifty-four years had included her relocation from Moscow to Novosibirsk in 1959, her supervision of the original selection of the founding-generation foxes at the commercial fur farms in Estonia in the late 1950s, her construction of the experimental population at the Akademgorodok farm across the first decade, the systematic application of the tameness-selection protocol across the subsequent decades, the publication of the experiment's principal scientific findings in over two hundred papers in the Soviet and post-Soviet genetics literature, the survival of the experiment through the late-Lysenko-era political pressure of 1959 and 1960 that had threatened the closure of the entire Institute of Cytology and Genetics, the survival through the Brezhnev-era institutional stagnation, the survival through the catastrophic post-1991 collapse of Russian scientific funding that had nearly ended the experiment in the mid-1990s before international support had been organized to sustain it, and the supervision, through every one of the fifty-four years, of the daily breeding and behavioral selection at the farm that the experiment's continuity depended on.[217]

---

[217]On Lyudmila Trut's biography and her continuous fifty-four-year supervision of the silver-fox domestication experiment at the Institute of Cytology and Genetics in Novosibirsk, see Dugatkin, L. A., and Trut, L. N. (2017). How to Tame a Fox (and

She was, in the small front room of the wooden house at the southern edge of the settlement where the driver delivered Dugatkin that Wednesday afternoon, in the configuration the photographs from the period would subsequently record: a small woman in a heavy gray wool sweater and the warm felt slippers the senior researchers at the Institute had been wearing in their offices for the better part of fifty years, with her white hair drawn back in the simple style she had been wearing since at least the 1980s, with the steady direct gaze the family album photographs from the 1950s had already shown in her when she was twenty-five. The colleagues she had assembled for the interview sessions were the senior staff of the experiment's current administrative cohort — Anastasia Vladimirovna Kharlamova, who had been the experiment's deputy lead researcher since the late 1990s; Pavel Mikhailovich Borodin, who had been the Institute's principal collaborator on the broader behavioral-genetics work; and one of the senior fox-pen attendants who had been working at the farm since the early 1980s and whose family memory of the experiment extended back to the founding generation through her own mother, who had been one of the original caretakers. The samovar at the corner of the small table in the front room was active. The conversation was in Russian, with Trut's English — which she had been improving in the years preceding the book's collaboration, in preparation for the joint authorship the project required — providing the principal channel between her and the visitor.

Build a Dog): Visionary Scientists and a Siberian Tale of Jump-Started Evolution. Chicago: University of Chicago Press. The January 2014 interview sessions that produced the manuscript are documented in the book's acknowledgments (pp. 211-214); Dugatkin's three-week visit to Akademgorodok in mid-January 2014 was the principal data-collection trip for the project. The two-feet-of-fresh-snow morning at the fox pens is referenced in the book's photo gallery (Plate 14, captioned 'Domesticated foxes playing in the Siberian snow, January 2014').

The morning of the third day of Dugatkin's visit was the morning the snow had fallen most heavily the previous night, at a depth the senior colleagues had estimated at slightly over sixty centimeters, with the path from the small house at the southern edge of the settlement to the fox-pen complex at the eastern side of the farm having been cleared by the farm staff in the small hours before first light but the surrounding meadows and the access roads being still under the full depth of the fresh accumulation. Trut had announced at breakfast that she wanted to walk to the pens. The walk was approximately four hundred meters from the front door of the small house to the eastern gate of the complex, along the path that the experiment's daily staff had been walking, in approximately the same configuration, since the late 1960s. Dugatkin had asked whether the senior researchers would be joining the walk. Trut had said that the senior researchers would meet them at the pens, and that she would walk the path alone with the American author for the conversation the photographic record the publisher had requested would document at the pens themselves. She put on the heavy fur coat the Institute had issued her in the early 1990s. She put on the warm felt boots the Institute had issued her at approximately the same time. She picked up the small wooden walking stick she had been using since the previous winter when the icy patches on the path had begun to require additional support. She opened the front door. She stepped out into the snow.[218]

The snow that morning came up to mid-calf on Dugatkin's leg and to mid-thigh on Trut's, who had been a small woman in her

---

[218]Photographic documentation of Trut and Dugatkin at the experimental fox-pen complex on the morning of the third day of Dugatkin's January 2014 visit is preserved in Dugatkin and Trut (2017), Plates 13-15. The photographs include the path from the small wooden building at the southern edge of the settlement to the eastern gate of the fox-pen complex, the snow conditions on the morning of the visit, and the senior researcher in the heavy fur coat and felt boots that the Institute had issued to its senior staff in the post-Soviet period.

twenties and was a smaller woman at eighty. She walked the path through the snow in the slow steady cadence she had developed over the previous winters, lifting each foot above the accumulated depth, placing it forward in the soft fresh layer, transferring her weight onto the placed foot, lifting the trailing foot above the depth, placing it forward, transferring her weight, lifting again. The walking stick provided the third point of support the rhythm required. Her breath in the cold morning air formed the small clouds the photographs would subsequently record. Dugatkin walked beside her at the pace she set, which was slower than the pace either of them would have set alone but was the pace the conditions of the walk and the cumulative effect of fifty-five years of similar walks and the eighty years of the body that was conducting the walk and the small wooden stick at her right hand had determined was the pace the morning would proceed at. They did not speak for the first hundred meters. The fox-pen complex came into view at the bend in the path where the line of birches that bordered the farm's eastern boundary opened onto the cleared yard where the pens were arranged in their parallel rows. The foxes in the pens, on hearing the approach of the senior researcher whose footfall they had been hearing every morning of their lives, began the soft chorus of the welcoming vocalizations that the experiment's tameness-selection protocol had, across fifty-five generations of careful breeding, made the experiment's most striking single observable result.

She stopped at the eastern gate. She put her right hand on the wooden post the gate hung from, in the position that long habit had organized her body into when arriving at the gate after the walk from the house. The young woman who was the principal pen-attendant on the morning shift came out of the small heated kitchen at the western end of the complex with the bucket of the morning ration the foxes had been calibrated to expect at approximately the hour of Trut's arrival. The foxes' vocalizations intensified. Trut stood at the gate. She watched the pen-attendant move along the row toward the gate. She did not, in the moment, say anything to Dugatkin about the experiment, or about her own life, or about the framework she had inherited from Belyaev and had carried for the fifty-four years since his death. She watched the foxes. The foxes watched her. The wind from the northeast moved the small loose flakes the previous night's storm had left on the higher pen-frames into the soft cascade the photographs would record. The cold registered through the heavy fur coat at the level the eighty-year-old body had grown accustomed to across the fifty-five winters the body had conducted its work through. The eighty-one years that the body had been the platform of the life that had organized itself around the experiment had not, on the observable evidence, slowed the precision of the morning observation she was conducting at the gate. The substrate was eighty. The work the substrate was the platform of was the work the substrate had been organized around for the previous fifty-five years. The work continued. The substrate continued. The eighty was not, in the moment, the operative variable. The work was.[219]

She turned to Dugatkin after perhaps a minute of standing at the gate. She gestured at the row of pens with her free hand, the hand that was not on the wooden post that the long habit had her right hand resting on, and said, in the careful English she had been working on in the months preceding the visit, that the experiment had reached its fifty-fifth year of continuous selection that summer. She said that the fifty-six generations of foxes the selection protocol had produced were, at the moment they were both standing at the eastern gate, the

[219]On the formal classification of the silver-fox population at Akademgorodok by the early 2010s into the elite (most domesticated) class, the intermediate class, and the control aggressive class, and on the calibrated daily routine that the fox-pen staff had organized around the senior researcher's morning visit, see Trut, L. N., Plyusnina, I. Z., and Oskina, I. N. (2004). "An experiment on fox domestication and debatable issues of evolution of the dog." Russian Journal of Genetics 40(6): 644-655; and the more recent synthesis in Trut, L., and Dugatkin, L. A. (2017). "The early history of fox domestication." In Dugatkin and Trut (2017), Ch. 1.

result of approximately twenty thousand individual selection decisions she or her predecessors and successors had made across the fifty-five years — each year the staff selecting, from the cohort of pups born the previous spring, the small subset that would breed the subsequent generation, on the basis of the systematic tameness assessment the protocol had specified since 1959. She said that the assessment took several minutes per pup and was conducted on every pup born in every season. She said that the number of pups born in the experiment's history had passed sixty thousand approximately five years earlier, and that the cumulative observation hours the experiment had logged, across the fifty-five years and the multiple shifts of staff and the continuous attention to the breeding records, was a number she had not, in her own working memory, attempted to compute exactly but that, in her own estimate, exceeded one million observation hours by a substantial margin. She said that the experiment was, in her own assessment of what the experiment was, a method. The method had been operating in approximately the same form for fifty-five years. The method had produced what the method had been organized to produce. The foxes the pen-attendant was now distributing the morning ration to were the experiment's accumulated trace. The method was the technology by which the trace had been produced. She turned back to the pens. She did not say anything else for the next several minutes. She watched the foxes eat. Dugatkin stood beside her at the gate. The light continued to brighten. The work continued.

✦ ✦ ✦

The previous chapter established that coherence is the operational technology by which the framework's working orientation is practically maintained, and that the technology has documented neurophysiological and psychological signatures that the contemporary research literature has been characterizing at increasing precision. The chapter also established that the substrate the visitor inhabits is modifiable through specific practices, that the modifications stabilize over time as durable features of the substrate's default configuration, and that the cultivation of coherence is, in the framework's working vocabulary, the technology of the remembering itself. What the previous chapter did not yet provide is the systematic protocol by which the cultivation can be conducted in a sustained, sequenced, and assessable form across the years that the substrate's deeper reconfiguration requires.

The structural lesson of the woman at the eastern gate of the Akademgorodok fox-pen complex on the morning in January of 2014 is the lesson of sustained ordered practice with measurable outcomes at the scale of a single life. Fifty-five years. Twenty thousand selection decisions. Sixty thousand pups. Over one million cumulative observation hours. A method that had been operating in approximately the same form for fifty-five years and had produced what the method had been organized to produce. The chapter that follows is about the analogue of such a method at the scale of the visit the visitor is conducting. The visit is, on the framework's working logic, the visitor's own selection. The substrate the visit is being conducted through is, on the previous chapter's case, modifiable by the practices the visitor undertakes. The method by which the modification is organized — the sequenced program of practices that meet the structural requirements of substrate-level reconfiguration — is the subject of the present chapter.

A remembrance protocol, in the framework's working vocabulary, is a sequenced program of practices designed to cultivate coherence in its multiple registers, to stabilize the cultivation into trait-level features of the substrate, and to support the visitor's ongoing operational integration of the framework's recognitions into the conditions of the visit. The chapter develops the structural requirements such a protocol must meet, the tiered architecture by which the requirements are typically organized, the specific implementation that the broader research program of this book has developed under the name of the Gaia Method, and the limits the protocol cannot exceed regardless of how rigorously it is followed. The chapter is, in a sense the book's earlier chapters have not been, a chapter about what the reader can actually do with the framework the prior chapters have built. The pivot from argument to application that began in Chapter 13 reaches its operational specificity in the present chapter.

Before the chapter proceeds, two clarifying observations are worth registering. The first is that the framework of this book does not require the specific protocol the chapter will describe in order to be applied. The framework's working orientation can, in principle, be cultivated through any sufficient program of practices that meets the structural requirements the chapter develops. The contemplative traditions of multiple cultures and historical periods have produced such programs, and the contemporary integrative-medicine literature has documented a range of approaches that produce the coherent substrate configuration the previous chapter characterized. The protocol the present chapter describes is one implementation of the structural requirements. It is the implementation the broader research program of this book has developed and tested, and it has the practical advantages of integration and sequencing that ad hoc combinations of practices from disparate sources typically lack. But it is not the only available implementation, and the chapter does not assert that it is.

The second clarifying observation is that a remembrance protocol is not, on the framework's working logic, a means to an end the visitor could specify in advance. The protocol does not deliver predetermined experiences, recovered memories, or specific spiritual achievements. The protocol cultivates the substrate's coherent configuration. What the substrate's coherent configuration yields, for any particular visitor, is the visit that the visitor selected — which includes whatever specific content the visit was assembled around, whether or not that content matches the visitor's expectations at the time of beginning the protocol. The protocol is, in the framework's working vocabulary, the practice of becoming more available to the visit's content rather than the practice of acquiring specific contents the visitor wishes to possess. The distinction matters, because the spiritual-bypass pattern Chapter 13 identified appears reliably when the protocol is approached as an instrument for acquiring specific contents rather than as a discipline of substrate-level reconfiguration. The protocol is for what it is for. The visitor who begins it accepts that the visit is what the visit was selected to be, not what the visitor on day one of the protocol imagines the visit ought to become.

Structural Requirements of a Remembrance Protocol

Any protocol that meets the framework's operational requirements must address, in some sequenced form, five substrate registers that the previous chapters have identified as the principal sites of the coherence the framework cultivates. The five registers are interdependent, and a protocol that addresses fewer than all five will, on the available evidence, fail to produce the stable trait-level configuration the framework's working orientation requires. The chapter takes the five registers in the order in which a protocol typically addresses them, recognizing that the order is not strictly sequential and that mature protocol participants typically engage all five simultaneously at varying levels of emphasis.

The first register is autonomic. The substrate's autonomic nervous system, characterized in detail in the previous chapter under the name of cardiac coherence, is the foundation on which all the subsequent registers depend. A substrate whose autonomic regulation is chronically dysregulated — locked into sympathetic activation, locked into dorsal-vagal shutdown, or oscillating between these two without access to the integrated ventral-vagal state Porges's framework introduced in Chapter 10 — cannot conduct the higher-register practices the protocol's later phases require.[220] The substrate's executive function is, on the neurovisceral integration framework that Thayer's work has anchored, dependent on autonomic regulation in ways that make the autonomic substrate the necessary first focus of any serious protocol. Specific practices that address this register include paced respiration at the baroreflex resonance frequency, heart-rate-variability biofeedback, polyvagal-informed somatic practices, and the broader contemplative-traditions practices that work directly with breath and posture as autonomic regulators. The protocol's first phase consists, in most well-designed implementations, of dedicated work in this register until the substrate's autonomic regulation has stabilized into a baseline from which the subsequent work can proceed.

The second register is attentional. Once the autonomic substrate is stable enough to support sustained focus, the protocol's second phase typically engages the cortical-attentional substrate through practices that cultivate the specific kind of attention the contemplative-practice neuroscience literature has characterized — present-moment awareness, non-reactive observation of mental contents, sustained focus on a chosen object, and the open-monitoring states that the broader meditation traditions have developed.[221] The practices reduce default-mode-network activity,

[220]The dependency of higher cognitive function on autonomic regulation is formalized in the neurovisceral integration framework: Thayer, J. F., and Lane, R. D. (2009). "Claude Bernard and the heart-brain connection: Further elaboration of a model of neurovisceral integration." Neuroscience and Biobehavioral Reviews 33(2): 81-88. The clinical translation of the autonomic-first sequencing is developed in detail in Lehrer, P. M., and Gevirtz, R. (2014). "Heart rate variability biofeedback: How and why does it work?" Frontiers in Psychology 5: 756.

[221]The taxonomy of contemplative practices into focused-attention, open-monitoring, and non-dual styles is developed in Lutz, A., Slagter, H. A., Dunne, J. D., and Davidson, R. J. (2008). "Attention regulation and monitoring in meditation." Trends in Cognitive Sciences 12(4): 163-169. Comparative examination of

increase prefrontal regulatory capacity, and produce the cross-cortical synchronization that Lutz's work on long-term practitioners has documented. The Mindfulness-Based Stress Reduction protocol of Jon Kabat-Zinn, the insight-meditation lineages of the Theravāda Buddhist tradition, the dzogchen and mahāmudrā lineages of Tibetan Buddhism, and the contemplative-prayer practices of the Christian hesychast tradition each represent established systematic implementations of practice in this register. A remembrance protocol's attentional-register phase typically draws on one or more of these established lineages, adapted for the protocol's specific aims and integrated with the autonomic-register foundation the protocol's first phase has established.

The third register is somatic. The substrate's relationship to its embodied state — to the body's stored tensions, its musculoskeletal patterns, its interoceptive signals, its history of held trauma at the autonomic and tissue level — is the register the trauma-medicine literature of van der Kolk anchored in Chapter 10. A protocol that addresses only the autonomic and attentional registers without engaging the somatic substrate produces practitioners whose cognitive integration outpaces their embodied integration, with the predictable consequence that the body's stored material continues to operate as a counter-force to the substrate's coherent configuration.[222] The somatic-register practices that the contemporary integrative literature has documented include yoga in

---

contemplative protocols across traditions: Wallace, B. A. (2006). The Attention Revolution: Unlocking the Power of the Focused Mind. Boston: Wisdom Publications.

[222]The complementarity of cognitive and somatic registers in clinical practice is developed in Ogden, P., Minton, K., and Pain, C. (2006). Trauma and the Body: A Sensorimotor Approach to Psychotherapy. New York: W. W. Norton. The contemplative-traditions version of the same complementarity: Johnson, W. (2000). Aligned, Relaxed, Resilient: The Physical Foundations of Mindfulness. Boston: Shambhala. Somatic-experiencing foundational work: Levine, P. A. (1997). Waking the Tiger: Healing Trauma. Berkeley: North Atlantic Books.

its various lineages, somatic experiencing, body-oriented psychotherapy, the breathwork traditions that operate at the embodied level rather than the autonomic-regulation level alone, and the contemplative movement practices of multiple traditions. The protocol's somatic phase typically introduces these practices once the autonomic and attentional foundations have stabilized, and continues their integration alongside the protocol's subsequent phases.

The fourth register is cognitive. The substrate's predictive architecture, characterized in the previous chapter under the predictive-coding framework of Friston and Clark, is the register at which the visitor's working orientation either stabilizes into the substrate's default predictive attractor or fails to stabilize. The practices that address this register operate at the level of the visitor's explicit cognitive frame — the deliberate cultivation of the recognition that the life was chosen rather than assigned, the active practice of seeing the visit's difficulties as part of the visit's selected content rather than as external impositions, and the structured engagement with the framework's metaphysical commitments through study, reflection, and dialogue. The cognitive-register practices are, on the framework's working logic, what allow the framework itself to operate as a frame for the substrate's predictive hierarchy rather than as an abstract belief held alongside an unchanged predictive default. The cognitive-register practices include but are not limited to study of the underlying scientific and philosophical literature, structured journaling protocols that surface and engage the substrate's default predictive patterns, and dialogue with experienced practitioners who can identify and interrupt the spiritual-bypass patterns the framework is vulnerable to.

The fifth register is relational. The substrate the visitor inhabits is not, on any of the literatures the prior chapters have surveyed, an isolated unit. The substrate is embedded in the relational field of the other substrates the visit is conducted in proximity to, and the coherence of the visitor's substrate is, on the social-connection and mortality literatures surveyed in Chapter 9, structurally responsive to the coherence and connection of the substrates the visitor is in sustained contact with. A protocol that addresses only the individual substrate's internal registers, without engaging the relational substrate in which the individual substrate is operating, produces practitioners whose individual coherence is undermined every time they return to relational contexts that the protocol has not addressed. The relational-register practices include sustained participation in a community of practitioners engaged in compatible work, attention to the quality of the visitor's primary relationships and the conduct of the visit within them, and what Chapter 16 will examine in detail as the visitor's relationship to the broader symbiotic field the substrate's ascent through forms has assembled the visitor into. The relational register is, on the framework's working logic, the substrate's most consequential connection to the work the visit was selected to do.

The Tiered Architecture

Protocols that meet the five-register structural requirement are typically organized into tiered phases that sequence the engagement with the registers across the months and years the protocol is conducted over. The tiered architecture is not an arbitrary convenience; it reflects the substrate's actual sequence of receptivity, in which the higher-register practices depend on the foundations established by the lower-register practices in ways that make attempting the higher-register work prematurely unproductive at best and counterproductive at worst. The chapter describes the typical tiered architecture in general form, with the specific implementation the broader research program has developed described in the following section.

The foundation tier addresses the autonomic and attentional registers with primary emphasis, establishing the baseline substrate stability and present-moment attentional capacity on which the subsequent tiers depend. The foundation tier's work is, in the literature on contemplative training, the most extensively documented at the level of clinical outcomes, with the eight-week timeframe of programs like Mindfulness-Based Stress Reduction representing the well-supported minimum duration for the establishment of reliable baseline competence in this register. The foundation tier's practices are typically conducted daily, for moderate durations of fifteen to thirty minutes per session, with the cumulative time investment producing the substrate adaptations the literature has documented. The visitor who has completed the foundation tier has, on the available outcome measures, established the autonomic and attentional baseline from which the subsequent tiers' work becomes possible.

The integration tier addresses the somatic and cognitive registers, building on the foundation tier's substrate stability to engage the body's stored material and the visitor's predictive architecture in ways that the foundation tier has prepared the substrate to support. The integration tier's work is, in most well-designed protocols, the longest and most variable in its duration, because the somatic and cognitive material being engaged is specific to each visitor's particular history and predictive defaults. The visitor whose substrate carries substantial traumatic material will require more sustained somatic-register work than the visitor whose substrate does not; the visitor whose predictive defaults are particularly entrenched in the grievance posture will require more sustained cognitive-register work than the visitor whose defaults are not. The integration tier's typical duration spans months to years rather than weeks, and the tier's pacing is determined by the substrate's actual rate of integration rather than by externally imposed timelines. The visitor who has completed the integration tier has, on the framework's working logic, accomplished the substrate-level reconfiguration that allows the framework's working orientation to operate as the substrate's default predictive attractor rather than as an effortfully maintained alternative to a still-dominant grievance posture.

The embodiment tier addresses the durable stabilization of the prior tiers' work into trait-level features of the substrate, alongside the deepening of the relational register that the prior tiers have prepared the visitor to engage. The embodiment tier is, in some sense, the protocol's open-ended phase — there is no determinate endpoint, because the visitor's relationship to the visit's selected content continues to unfold across the duration of the visit, and the protocol's work continues to refine the substrate's receiving capacity at increasing precision. The visitor who has reached the embodiment tier has, on the framework's working logic, accomplished what the protocol was assembled to deliver: a substrate whose default configuration is the coherent receiving state, a visitor whose default orientation is the framework's working position, and a relational field whose conduct supports rather than undermines the substrate's ongoing operation. The remaining work of the embodiment tier is the ongoing service of the visit itself, conducted from the configured substrate the prior tiers have produced.

The Gaia Method as One Implementation

The protocol the broader research program of this book has developed under the name of the Gaia Method is one structured implementation of the architecture the previous sections have characterized. The Method's specific design draws on the contemplative traditions, the integrative-medicine literature, the heart-rate-variability and contemplative-practice neuroscience the previous chapter surveyed, and the framework's distinctive interpretive register in which the protocol's work is understood as the technology of the remembering the book's earlier chapters have argued for. The Method's curriculum is organized into the tiered architecture described above, with sequenced material at each tier and specific practices that address each of the five substrate registers in the proportions the tier requires.

The Method's foundation tier introduces the autonomic and attentional practices through guided audio sessions designed to be conducted daily over an initial eight-week period. The sessions address paced respiration at the baroreflex resonance frequency, the cultivation of present-moment attention, the introduction of the framework's conceptual material at a level appropriate to the foundation phase, and the establishment of the visitor's home practice as a reliable daily commitment. The foundation tier's outcomes are assessable by both subjective measures and, for visitors who choose to use them, objective measures including heart-rate-variability biofeedback equipment that documents the autonomic adaptations the practices are producing. The foundation tier is, in the Method's design, the bottleneck through which all subsequent work passes; visitors who do not establish reliable foundation-tier practice are not, on the Method's working logic, prepared to attempt the integration tier's more demanding work.

The Method's integration tier introduces the somatic and cognitive practices in extended sequences that engage the visitor's particular material across months of sustained work. The integration tier's content is more individualized than the foundation tier's, because the somatic and cognitive registers' work is responsive to the specific configuration of the visitor's substrate that the foundation tier's work has stabilized enough to permit examination. The integration tier draws on the body-oriented practices of yoga and somatic experiencing, on the structured journaling and dialogue protocols that engage the visitor's predictive defaults, and on the deeper exposition of the framework's conceptual material at the level the integration phase can sustain. The integration tier's duration is determined by the visitor's substrate rather than by externally imposed timelines, and the tier's completion is assessed by the substrate's demonstrated capacity to maintain the framework's working orientation through the conditions of ordinary life without effortful reassertion.

The Method's embodiment tier consists of the open-ended integration of the prior tiers' work into the visitor's ongoing conduct of the visit, conducted in the context of a community of fellow practitioners and supported by the continued availability of the prior tiers' practices as the visitor's substrate's needs change across the years. The embodiment tier introduces the relational register at its full weight, with sustained participation in the broader practitioner community being one of the principal supports for the substrate's continued coherent operation. The Method's embodiment-tier practitioners are, in the program's working logic, the operational expression of the framework — visitors whose substrate's coherent configuration has stabilized, whose orientation has settled into the framework's working position, and whose conduct of the visit is now informed by the recognition that the life is the visitor's own selected work.

What the Protocol Cannot Do

Honesty about the protocol's limits is, on the framework's working logic, more important than enthusiasm about the protocol's capacities. The chapter marks several specific limits explicitly because they are the points at which the protocol is most often misunderstood by visitors who begin it with mistaken expectations.

The protocol cannot bypass the visit's actual content. The visitor whose visit was selected to include specific difficulties will encounter those difficulties whether or not the protocol is followed, and the protocol does not, on the framework's working logic, remove the difficulties or make them easier in the surface sense. The protocol cultivates the substrate's capacity to meet the difficulties from a configured state rather than a reactive one. The grief is still grief. The injustice is still injustice. The losses are still losses. The protocol changes the visitor's relationship to the content, not the content itself. Visitors who begin the protocol expecting to be relieved of the visit's difficulties have, on this framing, misunderstood what the protocol is for, and the misunderstanding typically resolves either through the protocol's first encounter with content that does not yield to the practitioner's wishes or through the practitioner's eventual recognition that the protocol's actual outcomes are not what the practitioner was hoping to acquire.

The protocol cannot replace the relationships the visit was assembled around. The substrate's coherence is not, on the framework's working logic, an alternative to the visitor's engagement with the specific people, places, and conditions the visit was selected to include. The protocol's relational-register practices direct the visitor toward those engagements rather than away from them. Visitors who use the protocol as a route to a private spiritual achievement that allows them to disengage from their primary relationships have, again, misunderstood what the protocol is for, and the misunderstanding is one of the principal failure modes the broader integrative-traditions literature has documented across the contemplative paths.

The protocol cannot guarantee specific experiences. Visitors do, in the course of sustained protocol practice, sometimes have specific experiences that the protocol's framework can accommodate — moments of vivid presence, occasional fragments of what the framework would interpret as pre-arrival memory, dreams of unusual clarity, encounters with the visit's selected content at unexpected intensities. The protocol does not promise such experiences, does not produce them on schedule, and does not treat them as evidence of the practitioner's progress. The substrate's coherent configuration is the protocol's outcome; specific experiences are the substrate's incidental productions, valuable when they occur but neither necessary nor sufficient indicators of the protocol's working integration.

The protocol cannot substitute for the visitor's own active stewardship of the visit. The substrate's configuration is necessary but not sufficient for the framework's working orientation; what the configured substrate makes possible is the visitor's increasingly available engagement with the visit's selected content, and the engagement itself remains the visitor's own work. The substrate that is configured but uninhabited produces nothing. The visitor must show up. The protocol cultivates the substrate that allows the showing up to be conducted from the framework's working orientation. The showing up itself is the visit. The visit is what the visitor came for. The protocol is for the sake of the visit, not the other way around.

Failure Modes

Honest treatment of the protocol's working operation requires explicit attention to the specific patterns by which the protocol fails when it does fail. The patterns are not random. They cluster into recognizable categories, and visitors who recognize the categories in their own practice have the operational handles by which the failure modes can be interrupted before they become entrenched. The chapter takes the four most common failure modes in turn.

The first failure mode is spiritual bypassing, examined in detail in Chapter 13 in the work of Welwood and Masters. The bypassing pattern, in protocol terms, manifests as the deployment of the protocol's vocabulary and the protocol's conceptual framework to avoid the substrate-level material the protocol was assembled to engage. The visitor who is bypassing produces fluent use of the framework's language, attends regularly to the protocol's formal practices, and yet displays few of the behavioral and relational markers the genuine progress would produce. The bypassing visitor is, on the operational logic, using the protocol as a substitute for the work the protocol is supposed to make possible. The recognition of bypassing requires either the visitor's own honest self-examination, which is difficult because the bypassing pattern includes resistance to such examination, or the intervention of an experienced practitioner whose perception is not captured by the visitor's framework deployment. The Method's design includes specific structural features intended to interrupt bypassing — required somatic-register engagement, structured relational accountability, and assessment markers that are not solely cognitive — but the failure mode remains the most common one the protocol must contend with.

The second failure mode is premature advancement. The protocol's tiered architecture exists because the higher-register practices depend on foundations the lower-register practices establish. Visitors who push past the foundation tier without consolidating its outcomes attempt the integration tier's somatic and cognitive work on an autonomic substrate that has not yet stabilized to support it, with predictable consequences: the somatic material surfaces in volumes the substrate cannot integrate, the cognitive practice produces insights the substrate cannot maintain, and the relational practice runs into conflicts the substrate cannot regulate. The premature-advancement pattern is, in the broader contemplative-traditions literature, one of the principal failure modes long-recognized as requiring careful pedagogical management. The Method's tiered structure addresses the pattern by gating advancement on documented foundation-tier outcomes rather than on the visitor's wish to proceed.

The third failure mode is over-investment in measurement. The protocol's outcomes are partially measurable through the heart-rate-variability and related instruments the previous chapter examined, and the measurability has substantial pedagogical value because it makes the substrate's adaptations visible to the visitor in real time. The same measurability becomes a failure mode, however, when visitors become preoccupied with the measurements as ends in themselves rather than as indicators of the substrate-level configuration the protocol is cultivating. The visitor who is chasing higher heart-rate-variability numbers as a goal has substituted the indicator for the substrate. The protocol's working integration produces high variability as a side effect of the substrate's coherent configuration; the variability is not the configuration itself, and the substrate that has been optimized for variability without the broader configuration the variability is supposed to indicate produces an empty achievement that the visitor will eventually recognize as missing the protocol's actual point. The protocol's design accordingly de-emphasizes measurement at the level of competitive achievement and emphasizes measurement at the level of feedback for the practitioner's own self-understanding.

The fourth failure mode is isolation from the relational register. Visitors who conduct the protocol's first three registers in isolation from a community of fellow practitioners and from the relational field of the visitor's primary relationships produce, in many cases, substantial gains in the individual-substrate registers without corresponding gains in the relational register. The result is a practitioner whose internal coherence does not translate into the conduct of the visit in its actual relational conditions. The pattern is particularly common in contemporary settings where the protocol is conducted as an individual practice without the community structures that traditional contemplative paths assumed. The Method's design addresses the failure mode by integrating the relational register into the protocol from the foundation tier onward, with sustained community engagement being one of the protocol's required components rather than an optional supplement.

Markers of Progress

Visitors who undertake the protocol typically want to know how they will know whether the protocol is working. The chapter's response is structural: the protocol is working when specific markers, in specific registers, are documented in the visitor's substrate and lived experience. The markers are not arbitrary self-assessment. They are the same neurophysiological and psychological signatures the previous chapter characterized as the empirical signatures of the coherent configuration. The visitor whose protocol is working shows measurable changes in the markers across the protocol's duration; the visitor whose protocol is not working does not.

In the autonomic register, the principal marker is heart-rate-variability, measurable with the equipment that has become widely available at consumer-accessible price points in the last decade. The visitor whose foundation-tier work is producing the substrate adaptations the protocol intends shows progressive increases in baseline heart rate variability, increasing time spent in the cardiac-coherence pattern during practice, and progressively decreasing time required to enter the coherent pattern at the beginning of each practice session. In the attentional register, the principal markers are subjective: the visitor whose attentional practice is establishing the relevant capacities reports increasing periods of present-moment awareness, decreasing rumination, and increasing capacity to disengage from the substrate's default narrative-self production. In the somatic register, the markers include increases in the visitor's interoceptive awareness, reductions in chronic muscular tension, and the gradual surfacing and integration of stored bodily material that the foundation tier had been too unstable to permit. In the cognitive register, the markers include increasing fluency in the framework's vocabulary, increasing capacity to recognize the substrate's predictive defaults in real time, and the gradual stabilization of the framework's working orientation as the substrate's default rather than its effortfully maintained alternative. In the relational register, the markers include increasing quality of the visitor's primary relationships, increasing capacity to be present with others without reflexively defending or attacking, and the visitor's growing sense of being engaged with the broader symbiotic field that Chapter 16 will examine.

The markers across the five registers are not independent. They are interlocking and mutually reinforcing, with progress in any one register typically supporting progress in the others and stagnation in any one register typically inhibiting progress in the others. The protocol's design accounts for these interactions by sequencing the registers' engagement in the order that the substrate's actual receptivity supports, and by maintaining attention to all five registers throughout the protocol's duration rather than completing one register before beginning the next. The visitor who attends to the markers across all five registers across the months and years of the protocol's conduct has, in the cumulative record, the operational evidence of whether the substrate's reconfiguration is in fact occurring. The evidence is, in most well-designed implementations, unambiguous after twelve to eighteen months of sustained practice. The substrate that is going to respond does respond. The substrate that does not respond does not. The reasons for the difference, where the difference is observed, are the subject of the protocol's ongoing refinement and of the broader research program the Method participates in.

The long-term trait markers of the protocol's successful integration are different in character from the shorter-term state markers the prior paragraphs have characterized, and they are worth registering separately. The trait markers emerge not in the protocol's first months but across years of sustained practice, and they include features that the contemplative-traditions literature has long associated with mature practitioners across multiple traditions. The visitor whose protocol's trait integration has stabilized shows a default emotional regulation that does not require active management to maintain, a relational capacity that operates without effortful self-monitoring, a degree of equanimity in the face of the visit's difficulties that is not pseudo-acceptance imposed by the framework's vocabulary but the actual settled state of the configured substrate, and a sense of being in the right life rather than the imposed one — not as a cognitive insistence but as the substrate's lived experience. The trait markers are not pursued directly. They emerge as the cumulative consequence of the years of work the protocol's earlier phases have been preparing the substrate for. Visitors who attempt to bypass to the trait markers without the prior years of work produce, in the predictable failure mode, the pseudo-equanimity that the bypassing pattern characteristically generates and that the protocol's design has been at pains to forestall. The trait markers are real, they are reachable, and they are reachable only by the work the protocol has been organized to support.

✦ ✦ ✦

The eighty-year-old woman at the eastern gate of the Akademgorodok fox-pen complex on the January morning in 2014 had been, for the previous fifty-five years, the operating exemplar of what the chapter has been describing in the general register of a sequenced ordered method conducted across the duration of a life. She had not been conducting a remembrance protocol in the framework's specific vocabulary. She had been conducting a biological experiment, in the methodology of mainstream science, on the question of how an organism's behavioral and physical configuration responds to systematic selection across multiple generations. The experiment had been, by the available evidence, the most successful single contribution of the twentieth century to the understanding of how domestication actually works at the genetic, behavioral, and developmental levels of the relevant biology. The method by which the experiment had been conducted — daily attention, the documented selection criteria, the patient continuation across the political and institutional disruptions that the fifty-five years had included, the absence of premature claims about outcomes the work had not yet produced, the careful accumulation of the trace the work was assembled to produce — was the structural exemplar of what an integrated method conducted across the scale of a life looks like when it is conducted properly. The exemplar is not the visitor's specific method; the experiment is not a remembrance protocol; the foxes are not coherent substrates. The exemplar is the form of sustained ordered attention that, across the duration the substrate is available, produces a trace the substrate would not have produced on its own. The chapter has been characterizing such a method in the framework's specific vocabulary. The exemplar is the form.

The visitor who undertakes the Method, or any equivalent protocol that meets the five-register structural requirement and the tiered-architecture sequencing, is undertaking the same general form of work that the woman at the gate had been conducting for fifty-five years on the foxes the pen-attendant was now distributing the morning ration to. The work is daily. The work is sequenced. The work is gauged against measurable outcomes. The work is sustained across years. The work is, in its accumulated trace, the substrate's accumulated configuration. The next chapter examines the broader register that the substrate's accumulated configuration is, on the framework's working logic, the visitor's contribution to. The work is the visit's own work. The visit is the work the visitor came for. The next chapter takes up the symbiont obligation that the work, on the framework's structural commitments, is conducted within.

Where the Argument Goes Next

The closing chapter of the book examines the broader structural obligation that the framework's commitments imply — what the book has called, in its earlier framing of the substrate hierarchy, the symbiont obligation. The visit is not, on the framework's working logic, exhausted by the visitor's individual coherence. The visit is conducted in a relational field that includes other substrates, other forms, and the ascending hierarchy through which the substrate's broader purpose extends beyond the individual visit's contents. The closing chapter examines what the framework's commitments require by way of the visitor's relationship to that broader field — the form of love the framework specifically identifies as the operational expression of the symbiotic structure the earlier chapters of the book have characterized at the level of biological and physical evidence. The chapter brings the book's argument to its closing register, with the recognition that the visit's purpose is, on the cumulative weight of the framework's case, larger than the visit itself and larger than the visitor whose visit it is.

CHAPTER 16

The Symbiont Obligation

The small bedroom on the second floor of the apartment in the residential block at the southern edge of Akademgorodok where Lyudmila Nikolaevna Trut had been living since her daughter Aleksandra Dmitrievna had moved her in from the small wooden house at the experimental fox farm a few months earlier, when the daily walks the previous decade's mornings had been organized around had become more than the eighty-nine-year-old body the visit had been conducted through could reliably sustain across the increasingly variable footing of the Akademgorodok winters, was a room of approximately four meters by three at the eastern end of the apartment, with a single window looking out over the row of birches that bordered the residential block's parking area and the small park beyond. The bed was at the western wall. The bedside table on the right side of the bed held the small lamp the family had brought from her office, the framed photograph of Pushinka from the spring of 1974 that had been on the office desk for the previous fifty years, and the small wooden icon of Saint Nicholas that her mother had carried with her from the village in Estonia where the family had been raised. The bookshelf along the southern wall held the small library she had brought from the office when the move had been organized — Belyaev's papers in the bound collections the Institute had prepared in the years after his death, the manuscripts of her own principal publications, the copy of How to Tame a Fox that Dugatkin had inscribed for her in November of 2017, the small folder of correspondence with the international colleagues whose collaboration the experiment's later decades had been sustained by. The wardrobe opposite the bookshelf held the few clothes the move had retained — the heavy fur coat from the early 1990s, the warm felt boots that the previous winter's walks had been conducted through, a small selection of the simple dresses and sweaters she had been wearing since approximately the 1980s, and the dark blue suit she had been wearing for the formal academic photographs since approximately 1998 and that hung at the right end of the wardrobe in the position the family would, when the time came, retrieve it from.

It was the evening of Tuesday, the eighth of October, 2024. The temperature outside the window had dropped to approximately minus six degrees on the Celsius scale across the previous twelve hours, in the early-autumn cooling that the Novosibirsk climate produced in the second week of October each year. The light at the window had failed by six in the evening, in the way it failed in Akademgorodok in early October, with the long northern dusk holding for perhaps forty minutes before the full dark settled across the residential blocks and the parking area and the small park beyond. Aleksandra Dmitrievna had come into the bedroom at approximately seven to bring the small tray of dinner the family had prepared — the simple soup the family kitchen produced in the autumn, the small piece of brown bread, the cup of tea with lemon her mother had been taking with the evening meal for as long as Aleksandra had been able to remember. They had spoken briefly about the day. The Institute had sent a message earlier in the week confirming that the experimental fox population had been moved through the autumn transition without incident; the foxes were settled in their winter configuration; the staff at the farm were in their winter rotation. The senior researcher who had succeeded Anastasia Vladimirovna Kharlamova as the experiment's deputy lead the previous year had visited the apartment the previous weekend to update Lyudmila on the year's selection results, and Lyudmila had received the update with the same careful attention she had received every year's selection results since 1960. She had eaten the soup. She had eaten approximately half of the bread. She had drunk the tea. She had told Aleksandra that the evening had been a good one and that she would read for a small while and then sleep.[223]

Aleksandra had taken the tray. She had asked, in the way she had been asking each evening since the move, whether her mother needed anything else before the night. Her mother had said she did not. Aleksandra had kissed her mother on the forehead in the small careful way the family had been kissing her in the months since the move. She had left the small lamp on the bedside table on. She had closed the door of the bedroom most of the way, leaving the narrow gap the family kept open in case her mother called in the night. She had walked back down the corridor to the kitchen at the western end of the apartment, where her husband Mikhail was washing the dinner dishes at the sink with the small radio playing the evening news from the Novosibirsk regional station at the volume the family had calibrated to be audible but not intrusive. Aleksandra had set the tray on the counter. She had said to Mikhail that her mother had been steady at the meal. Mikhail had said that the foxes had been moved through the autumn transition without incident. Aleksandra had said that she would do the dishes if Mikhail would take the recycling down. Mikhail had said that he would do the recycling. The evening proceeded along the small careful track the family's evenings had been proceeding along since the move.

[223]On Lyudmila Nikolaevna Trut's final years and her peaceful death in her sleep on the night of October 8-9, 2024, at the age of ninety, just short of her ninety-first birthday on November 6, see the obituary by Dugatkin, L. A. (2024). "The daring Russian geneticist whose experiments on silver foxes explained domestication has died." Scientific American, October 24, 2024; and the obituary in The Daily Telegraph, October 23, 2024. The experimental silver-fox population at the Institute of Cytology and Genetics in Novosibirsk continues at the time of writing, under the supervision of the senior researchers Trut had been preparing as her successors through the previous decade.

Lyudmila Nikolaevna Trut read, in the bed in the small bedroom at the eastern end of the apartment, for approximately thirty minutes after Aleksandra had taken the tray. The book she was reading was the autobiography of Nikolai Ivanovich Vavilov, which she had been re-reading for the third or fourth time across the previous summer in the way she had been re-reading the books that had organized her professional formation since she was reading them as a graduate student in Moscow in the late 1950s. She read the chapter on the early Vavilov expeditions to Central Asia in the 1920s, which she had marked in her own annotations from the first reading. She read the chapter on the founding of the Vavilov Institute in Leningrad. She did not, on the family's reconstruction of the evening's sequence in the days that followed, finish the book. She placed the book on the bedside table beside the photograph of Pushinka and the small wooden icon of Saint Nicholas. She turned off the small lamp. She arranged the pillow under her head in the configuration she had been arranging the pillow in since approximately the 1960s. She closed her eyes.

She did not wake again. The morning of Wednesday, the ninth of October, was a morning of light frost on the parking area and on the lower branches of the birches outside the bedroom window, with the temperature having dropped slightly further across the night and the early sun being slow to clear the eastern horizon at the latitude Akademgorodok occupied. Aleksandra came to the bedroom door at approximately seven, in the way she had been coming to the bedroom door each morning since the move, with the small cup of warm water and the small spoon of honey that her mother had been taking at the start of each day since the move. She opened the door. The room was still in the configuration the evening had left it in. The small lamp on the bedside table was off. The book was on the bedside table beside the photograph of Pushinka. The pillow was in the position her mother had arranged it in. Her mother's face was in the small careful repose of a person who had been sleeping deeply, with no indication on the surface features of any disturbance the night had introduced. Aleksandra walked to the bed. She set the cup of warm water on the bedside table. She put her hand gently on her mother's shoulder in the small careful way the family had been waking her in the months since the move. The shoulder was cool. The pulse at the wrist she lifted to check was not, on her examination, present. The chest under the bedclothes was not, on her examination, moving. Her mother had passed across the substrate transition that ninety years and ten months and three days of the visit had been the platform for. The substrate had released the visitor at some point in the small hours of the night. The room remained. The book remained. The photograph remained. The visit was, at the moment her daughter's hand registered the cooling of the shoulder, the visit it had been the substrate for. The visit had completed what the visit had been organized to do.[224]

The news passed across the Institute that morning in the way the news of senior researchers' deaths typically passes — by phone to the deputy director, by email to the senior research staff, by quiet word at the fox farm where the morning shift had already begun the day's distribution at the pens, by formal notification to the Russian Academy of Sciences in Moscow, and to the international

[224]On the structural pattern of the integrated death — the visitor whose substrate releases the visit at the conclusion of the work the visit was assembled to do — and the contemplative-traditions descriptions of this pattern across multiple lineages, see Mingyur Rinpoche, Y., and Tworkov, H. (2019). In Love with the World: A Monk's Journey through the Bardos of Living and Dying. New York: Spiegel & Grau, Chs. 12-14; and Sogyal Rinpoche (1992). The Tibetan Book of Living and Dying. San Francisco: HarperSanFrancisco. The framework of this book takes no position on the specific phenomenology these traditional sources describe; it observes that the empirical pattern of peaceful integrated death in subjects whose lives have been organized around sustained meaningful work is consistent with the framework's working logic that the visit's completion of its assembled content is the substrate's signal that the substrate's local work is done.

collaborators in Louisville and Leipzig and Helsinki and the United States National Academy of Sciences in Washington, each of whom received the message at the time-zone-adjusted hours of October ninth that the news reached them. The foxes at the eastern gate of the experimental complex received the morning ration at approximately the hour Trut had been arriving at the gate for the previous fifty-five years. The senior pen-attendant who had been the family memory of the experiment back to the founding caretakers in the early 1960s distributed the ration in the configuration the protocol had specified since 1959. The chorus of welcoming vocalizations was the chorus of welcoming vocalizations of every morning of the previous fifty-five years. The wind from the northeast moved the loose autumn leaves on the higher pen-frames into the soft cascade the morning had organized around. The work continued. The substrate that had been the platform of the work for the previous sixty-five years remained. The visitor the substrate had been the platform for had completed the visit the visitor had come for. The visitor was now where visitors go when the substrate has released them. The chapter that closes the book takes up the structural register in which the going of any visitor is conducted.

✦ ✦ ✦

The book reaches its closing chapter. The framework has been built across fifteen chapters of cosmological, biological, neurological, clinical, archival, and physical evidence, and the framework's operational technology has been characterized at increasing precision through the chapters of Part Four. What remains is the question the framework's structural commitments imply but the prior chapters have only gestured toward: the question of what the visitor's relationship is to the broader field the visit is conducted within. The chapter that follows takes the question directly. The framework, on the closing chapter's argument, is not exhausted by what it says about the individual visit. The framework is, at its deepest commitment, a statement about the substrate the visitor came up through and the symbiotic field the substrate's ascent has integrated the visitor into. The visit is, on the closing chapter's framing, a visit in service of the field. The field is what the chapter examines.

The woman whose substrate released her in the small hours of the ninth of October in 2024 had been, for the previous sixty-five years of the visit's conducted work, an exemplar of what the closing chapter is about to characterize. She had served the experiment. The experiment had served the foxes. The foxes had served the broader scientific understanding of the substrate hierarchy the framework has been describing across the prior chapters. The understanding had served the international community of researchers whose ongoing work the experiment had contributed to. The visit had been, throughout, a visit conducted in service of a field larger than the visit. The closing chapter takes up the structural register in which any visit conducted in this configuration is to be understood.

The chapter develops the case in four movements. The first reprises the ascent thesis of Chapter 5 and develops the structural obligation the ascent implies — what the closing chapter will call, in the formal vocabulary the framework has been building toward, the symbiont obligation. The second anchors the obligation in the empirical evidence the contemporary biological literature has provided for the structural reality of symbiotic bonding, with the human-canine coevolutionary relationship serving as the most thoroughly documented example. The third names and develops the structural concept that this evidence supports, which the chapter calls Symbiotic Love and capitalizes throughout to mark its specifically structural rather than affective sense. The fourth extends the symbiotic field beyond the canine case to the broader biosphere — the ascending hierarchy of mutualistic relationships that the modern biological synthesis, anchored most consequentially by the American microbiologist Lynn Margulis, has established as the foundational mechanism of evolutionary novelty itself. The chapter closes with what the framework's full cumulative argument has been preparing the reader to receive.

A note on register before the chapter proceeds. The book has held a careful institutional voice across its sixteen chapters, allowing the scientific case to carry the argument's weight without rhetorical embellishment. The closing chapter sustains the same voice, but the closing chapter is also where the argument's full cumulative consequence is allowed, finally, to come into view. The reader who has held with the book to this point has earned the chapter's destination. The chapter trusts the reader accordingly.

The Ascent and the Obligation

Chapter 5 established that the human substrate is the integration point of an ascent through forms. The substrate carries, encoded in its developmental architecture and in the deeper substrates the developmental architecture is assembled from, the accumulated complexity of every prior form the ascent has passed through. The cell that the substrate is composed of recapitulates the cellular organization the eukaryotic line developed approximately two billion years ago. The body plan that the substrate is organized into recapitulates the chordate architecture the vertebrate line established approximately five hundred million years ago. The four-limbed structure that the substrate operates through recapitulates the tetrapod plan the lineage settled into approximately three hundred and eighty million years ago. The substrate is not, on this picture, a discrete creation that arrived independent of the forms below it. The substrate is the accumulated integration of those forms, with the human-specific features added at the top of a hierarchy that extends downward through every previous level of the ascent.

The locked architectural commitment of Chapter 5 was that the ascent is vertical in substrate complexity but horizontal in worth.

The forms that the ascent has passed through are not below the human form in any sense that authorizes the human substrate's instrumentalization of them. The forms are earlier in the substrate hierarchy, more limited in their integrative capacity, more specialized in their particular registers — but not lower in worth. The cellular substrate that the human body is composed of has the same status, on the framework's working logic, as the human substrate itself: the substrate is sacred because it is a substrate, not because of where it happens to sit in the ascent's complexity hierarchy. Every form the ascent has passed through is the visitor's prior substrate, and the visitor's relationship to the prior substrates is therefore not the relationship of a higher organism to inferior preceding stages but the relationship of an integrated configuration to the components that integration is composed of.

This has structural consequences for the visitor's obligations. The visitor cannot, on the framework's working logic, regard the forms below the human substrate as resources available for instrumental use without violating the same structural recognition that grounds the visitor's own status as a substrate worth recognizing. The cellular life of the body, the bacterial communities the body operates in symbiosis with, the animal life the visit is conducted in proximity to, the plant life the visit is sustained by, the soil and water and atmosphere that the substrate's continued existence depends on — each of these is the visitor's prior substrate, integrated into the configuration the visitor now experiences as the human form. The visitor's relationship to each of them is, on the framework's working logic, the relationship of an integrated whole to its constituting parts. The relationship is structural. The obligation that follows from the structure is what the closing chapter calls the symbiont obligation.

The symbiont obligation is not, on the framework's working logic, a moral imposition added to the framework from outside its evidentiary base. The obligation follows directly from the framework's commitments to the ascent's structure, to the substrate hierarchy's full extension below the human form, and to the receiver model's implication that the visitor's substrate is not the visitor but the configuration the visitor is locally received through. If the substrate the visitor inhabits is the accumulated integration of every prior form, then the visitor's care for the substrate cannot stop at the human form. The care must extend through the entire substrate hierarchy, because the entire substrate hierarchy is what the visitor's substrate is composed of. The obligation is not optional. It is structural.

The Human-Canine Coevolutionary Bond

The empirical evidence that the framework's structural commitments are not abstract speculation but documented features of the biological substrate is most extensively characterized in the literature on the human-canine coevolutionary relationship. The relationship between humans and the lineage of dogs is the most thoroughly studied case of cross-species symbiotic bonding in the contemporary biological literature, and the findings have established that the bond is not a cultural overlay imposed on two separately evolved species but the operational expression of mutual evolutionary adaptations the two lineages have produced in response to one another across approximately fifteen to forty thousand years of close association.

The molecular signature of the bond was documented most consequentially by the Japanese ethologist Takefumi Kikusui and his collaborators in a 2015 paper in Science that examined the role of oxytocin in mutual human-canine gaze.[225] Oxytocin is the peptide hormone most centrally associated with mammalian bonding behavior, with established roles in mother-infant attachment, pair-

[225]Nagasawa, M., Mitsui, S., En, S., Ohtani, N., Ohta, M., Sakuma, Y., Onaka, T., Mogi, K., and Kikusui, T. (2015). "Oxytocin-gaze positive loop and the coevolution of human-dog bonds." Science 348(6232): 333-336.

bonding in monogamous species, and the broader physiology of social affiliation. Nagasawa, Kikusui, and their collaborators demonstrated that mutual gaze between humans and dogs produces elevations in oxytocin in both species — a finding that parallels the oxytocin response in mother-infant gaze and that has not been observed in the gaze interactions between humans and wolves, despite wolves being the immediate evolutionary ancestor of the modern dog. The oxytocin response is, on the available data, a specifically domesticated-canine adaptation that emerged during the coevolutionary period and that has no analog in the canine lineage's wild ancestors. The bond is, in the strict biochemical sense, mutually felt at the molecular level by both species. The feeling is documented.

The morphological signature of the same coevolutionary adaptation was documented in 2019 by the British animal-behavior researcher Juliane Kaminski and her collaborators in a paper in the Proceedings of the National Academy of Sciences comparing the facial musculature of dogs to that of wolves.[226] The team established that dogs possess a specific facial muscle — the levator anguli oculi medialis — that allows the production of the so-called inner brow raise that produces the appearance of larger, more soulful eyes. The muscle is fully developed in domestic dogs, vestigial or absent in wolves, and there is no other plausible explanation for its evolution other than selection for the production of an expression that the human-canine bond reinforced across generations. The bond, on the morphological evidence, has literally shaped the dog's face. The face the dog presents to the human is, in the technical sense the Kaminski paper documented, the face the human-canine coevolutionary relationship produced for the purpose of being presented to humans.

[226]Kaminski, J., Waller, B. M., Diogo, R., Hartstone-Rose, A., and Burrows, A. M. (2019). "Evolution of facial muscle anatomy in dogs." Proceedings of the National Academy of Sciences 116(29): 14677-14681.

The communication channel between the two species is not a cultural construction. It is a piece of anatomy.

The autonomic signature of the bond at the longest timescales was documented in a 2019 paper by the Swedish ethologist Ann-Sofie Sundman and her collaborators, examining long-term cortisol levels in dogs and their human owners as measured by hair cortisol — an integrated measure of stress-hormone exposure across months of accumulated growth.[227] The finding was that the long-term cortisol profile of a dog and its owner are statistically synchronized — that the chronic stress level of the human and the chronic stress level of the dog track each other across months in ways that random pairing of dogs and humans does not produce. The synchronization was specifically influenced by the owner's personality and stress patterns rather than the dog's, indicating that the autonomic coupling runs principally from owner to dog rather than in the reverse direction. The two organisms, by the time they have lived together for sufficient duration, are operating as a single autonomic system across the longest timescales the body's stress regulation can integrate. The bond is, on this finding, not only molecular and morphological but autonomic. The two substrates share, in the most literal sense, the same regulatory state.

The deepest piece of evidence for the structural reality of the coevolutionary capacity is the silver-fox experiment initiated by the Soviet geneticist Dmitri Belyaev in 1959 at the Institute of Cytology and Genetics in Novosibirsk, and continued for over six decades under the direction of Belyaev's collaborator and successor Lyudmila Trut — whose own substrate's release the chapter opened with.[228] Belyaev's experiment selected, from each generation of

[227]Sundman, A.-S., Van Poucke, E., Holm, A. C. S., Faresjö, Å., Theodorsson, E., Jensen, P., and Roth, L. S. V. (2019). "Long-term stress levels are synchronized in dogs and their owners." Scientific Reports 9: 7391.
[228]Trut, L. N. (1999). "Early canid domestication: The farm-fox experiment." American Scientist 87(2): 160-169.

captive silver foxes, only those individuals that showed the least fearful and least aggressive response to human approach. Within approximately forty generations of selection on this single behavioral criterion, the resulting fox lineage exhibited the full suite of features that the broader biological literature calls the domestication syndrome: floppy ears, curly tails, piebald coat patterns, juvenile vocalizations persisting into adulthood, retention of juvenile facial proportions, and the characteristic seeking of human contact that distinguishes domestic dogs from wolves. The features had not been selected for. They emerged as a coherent package from the single selection criterion of tameness. The experiment established that the capacity for symbiotic association with humans has a specific genetic substrate, that the substrate can be isolated by selection on the bonding behavior alone, and that the substrate carries with it the entire suite of features that the canid lineage's domestication has produced over fifteen thousand years of association with the human form. The bond, on the Belyaev evidence, is not only the operational expression of two substrates encountering one another. The bond is what specific genetic substrates are configured to produce when the configurations are present.

Symbiotic Love

The convergence of these four bodies of evidence — molecular, morphological, autonomic, and genetic — establishes a specific empirical phenomenon for which the chapter now introduces the formal name. Symbiotic Love, capitalized throughout to mark its specifically structural rather than affective sense, is the framework's name for the operational expression of the bond that two substrates configured for mutual association produce when they

---

Comprehensive history: Dugatkin, L. A., and Trut, L. N. (2017). How to Tame a Fox (and Build a Dog): Visionary Scientists and a Siberian Tale of Jump-Started Evolution. Chicago: University of Chicago Press.

enter into sustained relationship. Symbiotic Love is not, in the framework's working vocabulary, primarily an emotion. It is a state of two substrates whose molecular, morphological, autonomic, and genetic configurations have been shaped, across evolutionary or developmental time, to produce the bond between them. The emotion the visitor experiences in the presence of a symbiotic partner is the substrate's local report on the bond's operation, not the bond itself. The bond is the substrate-level configuration. The emotion is what the configuration feels like from the inside.

The framing matters because Symbiotic Love, in the structural sense the chapter is developing, is the operational signature of what the framework's earlier chapters have characterized at multiple levels. It is the receiver model's interpersonal expression — two substrates each receiving the consciousness signal in higher fidelity in the presence of the other than either substrate can achieve alone. It is the coherent configuration's relational counterpart — the autonomic and cortical states that the previous chapters' practices cultivate within the individual substrate, here documented as the bonded state between two substrates whose nervous systems have, over sufficient duration, integrated into a single regulatory system. It is the symbiont obligation's lived expression — the substrate's recognition, at the autonomic and molecular level, of its continuity with another substrate the framework's structural commitments hold in the same regard.

Symbiotic Love is not, on the framework's working logic, restricted to the human-canine case. The canine relationship is the most thoroughly documented because the coevolutionary timescale has been sufficient to produce species-level adaptations on both sides, and the contemporary research has had access to instruments that make the molecular and autonomic signatures visible. The same structural pattern operates, with appropriate modifications, in every sustained symbiotic relationship the substrate enters into. The pair-bond between mated humans operates through the same oxytocin substrate, with the same autonomic synchronization, and with similar morphological adaptations on a developmental rather than evolutionary timescale. The parent-child bond operates through the same machinery. The bond between long-term human collaborators operating in close partnership operates through subset of the same mechanisms. The relationship between a human and a horse, between a human and the land the human has farmed for decades, between a human and the home the human has inhabited across years of attention — each of these is, on the framework's working extension, a relationship whose deeper substrate is the same machinery the canine literature has characterized. Symbiotic Love is, in this generalization, the operational signature of every relationship the substrate enters into with sustained mutual presence.

The operational implications for the visitor's daily conduct of the visit follow directly from the structural recognition. The visitor who has integrated the framework attends to the symbiotic relationships the visit is conducted in not as supplements to the work but as the central operational sites at which the work is conducted. The visitor's relationship to the visitor's primary partner is not, on this framing, an accessory to the visit's deeper purpose; the relationship is, in substantial part, what the deeper purpose consists of. The visitor's relationship to the visitor's children, to the visitor's parents, to the visitor's long-term collaborators, to the visitor's animal companions, to the visitor's land and home and food sources — each of these is a site at which Symbiotic Love operates as the framework's working orientation rather than as an abstract aspiration the visitor reaches for in occasional contemplative moments. The work is the relationships. The relationships are the visit. The recognition that the relationships are the operational sites at which the visit's broader purpose is conducted is the framework's most consequential practical implication, and it is the implication the closing chapter has been preparing the reader to recognize as the framework's full operational consequence.

The capitalization the chapter has adopted for the term is not affectation. It is the marker of a specifically structural concept that the framework treats as a load-bearing element rather than as a sentimental descriptor. Symbiotic Love is to the framework's interpersonal commitments what coherence is to the framework's intrapersonal commitments — the operational name for the configured state the framework's working orientation produces and the practical signature by which the framework's integration is recognizable. The visitor who has integrated the framework's commitments operates, in the visit's relational field, through Symbiotic Love. The visitor who has not is still operating through whatever default the substrate had been configured into prior to the framework's encounter. The difference, again, is not theoretical. The difference is the substrate the two visitors are inhabiting.

The Post-Human Re-Entry

A specific consequence of the framework that the closing chapter is now positioned to develop concerns the substrate hierarchy's symmetry. Chapter 5 established the ascent through forms — the structural sequence by which the substrate's integrative capacity increases across the levels of complexity the biological library has produced. The human form, on Chapter 5's argument, is the integration point at which the substrate's prior accumulations become available to the substrate's conscious operation. The closing chapter's framework permits, and on the locked thesis of Chapter 5 actively requires, that the framework's logic operate in the opposite direction as well. The realized visitor — the visitor whose substrate's configuration has integrated the framework's commitments at the trait level the prior chapter characterized — is not, on the framework's working logic, bound to continue arriving only into human substrates in subsequent visits. The substrate hierarchy is available in both directions. The visitor who has done the work the framework was assembled around may, in subsequent visits, deliberately return into substrates the integrative-capacity hierarchy classifies as below the human form, in service of the work the lower-substrate visits make possible.

The framework treats this possibility with appropriate care. The claim is not that every cross-species reincarnation case in the UVA archive — the small but documented subset that Chapter 11 examined — represents post-human re-entry by realized visitors. The cross-species cases are rare in the archive, and the framework's natural reading of the more common forward-direction cross-species cases is that they represent the lower-form perspectives the substrate hierarchy's ascent has not yet fully integrated. The claim the closing chapter does make is that the post-human re-entry possibility is structurally available in the framework, that it is consistent with the substrate hierarchy's commitments to the horizontal-in-worth thesis Chapter 5 locked, and that it is the framework's specific commitment to the recognition that the lower-substrate forms are not below the realized visitor in any sense that would make a return to them a regression rather than a choice. The framework's full deployment treats the substrate hierarchy as a library of available configurations rather than as a ladder whose lower rungs the visitor has finished with. The library remains. The visitor whose work calls for a particular configuration can, on the framework's commitments, return to it.

The closing implication of this commitment is that Symbiotic Love is not, on the framework's working logic, the visitor's relationship to a substrate hierarchy the visitor has transcended. Symbiotic Love is the visitor's relationship to the substrate hierarchy the visitor is, was, and remains continuously the same substrate hierarchy as. The dog the visitor loves is, in the framework's strongest reading, a substrate the visitor recognizes from the inside because the visitor's own substrate carries the canine substrate's accumulated complexity at the levels of the substrate hierarchy below the human form's integration point. The recognition is not metaphorical. The recognition is the substrate's actual continuity with the form being recognized. Symbiotic Love is the visitor's recognition of the visitor's own deeper substrate in the form the visitor is bonded to.

Symbiogenesis and the Living World

The deepest extension of Symbiotic Love into the broader biosphere is the framework's recognition that the substrate the visitor inhabits is itself the result of ancient symbiotic mergers, and that the principle of symbiosis is not only the framework's operational ideal but the documented mechanism by which the substrate the framework rests upon has come into existence. The principal scientific architect of this recognition is the American microbiologist Lynn Margulis, whose work across forty years established that the eukaryotic cell — the cell type that all multicellular life is composed of — is the product of a symbiotic merger between two previously independent prokaryotic lineages.[229]

The endosymbiotic theory Margulis developed holds that approximately two billion years ago, certain prokaryotic cells engulfed other prokaryotic cells without digesting them, and the engulfed cells continued to live inside their hosts, providing metabolic services in exchange for the protected environment the host provided. Over evolutionary time, the engulfed cells became

[229]Margulis, L. (1970). Origin of Eukaryotic Cells. New Haven: Yale University Press. Synthesis for general readers: Margulis, L. (1998). Symbiotic Planet: A New Look at Evolution. New York: Basic Books. The endosymbiotic theory Margulis developed has been confirmed across multiple lines of evidence — genetic, biochemical, and structural — and is now part of the standard high-school biology curriculum, though Margulis's broader claims about the centrality of symbiosis in evolutionary innovation remain more contested in the mainstream evolutionary literature.

permanent organelles within the host cell — the mitochondria that produce energy in every cell of the visitor's body today, and the chloroplasts that perform photosynthesis in every plant cell. The merger was not a curiosity. The merger produced the cell type without which all multicellular life — every animal, every plant, every fungus — would have been impossible. The eukaryotic cell, the substrate the visitor's body is composed of, is the symbiotic descendant of an ancient merger that two previously independent organisms entered into. The visitor's cells are, in the most literal possible sense, the visitor's prior symbiotic relationships made permanent.

The contemporary extension of Margulis's framework into the science of the human microbiome has, in the last two decades, made the symbiotic structure of the visitor's substrate visible at a level of detail Margulis's original work could only gesture toward. The Human Microbiome Project, an international research consortium funded by the United States National Institutes of Health beginning in 2007, has characterized the bacterial and archaeal communities that inhabit the human body across multiple anatomical sites — the gut, the skin, the mouth, the urogenital tract, the respiratory passages — at a level of taxonomic and functional precision that earlier eras of microbiology had no access to.[230] The findings establish that the human body, far from being a single organism, is more accurately described as a holobiont — a host plus the trillions of microbial cells that constitute the host's permanent microbial communities. The microbial cells outnumber the host's own cells, contribute essential functions to digestion, immunity, neurochemistry, and behavior, and have co-evolved with the host

[230]Human Microbiome Project Consortium (2012). "Structure, function and diversity of the healthy human microbiome." Nature 486(7402): 207-214. Synthesis of the broader holobiont framework: Bordenstein, S. R., and Theis, K. R. (2015). "Host biology in light of the microbiome: Ten principles of holobionts and hologenomes." PLOS Biology 13(8): e1002226.

lineage across timescales that make the host and the microbiome a single integrated unit rather than two separable populations. The visitor's substrate, on the available evidence, is not the visitor's body alone. The visitor's substrate is the integrated host-plus-microbiome configuration the body's continued operation depends on. Symbiotic Love, in this register, is the operational expression of a relationship the visitor is already in with trillions of bacterial and archaeal partners whose continued presence is the substrate's continued existence.

The reach of the framework extends further. The agricultural relationships through which the visitor is fed are themselves symbiotic — wheat and rice and the legumes that fix nitrogen for them, the soil microbiomes that mineralize nutrients for plant uptake, the pollinator species that the fruit and seed crops depend on, the rumen microbiomes that allow ungulates to convert cellulose into the protein the visitor consumes. The atmospheric oxygen the visitor breathes is the cumulative product of the photosynthetic activity of phytoplankton in the world's oceans and of the forests of the world's continents, sustained by the symbiotic relationships those producer organisms operate through with their own microbial partners. Every breath the visitor takes is the operational expression of symbiotic relationships extending across the entire biosphere. The visitor's own existence is, on the cumulative biological evidence, not the existence of a single organism in an external environment but the local expression of a planet-wide symbiotic system in which the visitor is one node among many. Symbiotic Love is not, on this scale, an aspiration. Symbiotic Love is what the visitor's substrate already is the local instance of.

The implications of Margulis's framework for the broader claim of this chapter are direct. Symbiotic Love is not, on the deepest biological reading, an unusual or specialized phenomenon. Symbiotic Love is the mechanism by which the substrates the visitor is composed of came into existence. The cells the visitor's body is built from are bonded mergers. The visitor's cells are bonded with the trillions of bacterial cells that constitute the visitor's microbiome and that the visitor's continued health depends on. The visitor's relationship to plants, to soil microbes, to the broader fungal networks of the ecosystem the visitor is sustained by — each of these is the same structural relationship in slightly modified registers.[231] The closing implication is that the visitor is not, on the deepest available biological understanding, an individual at all. The visitor is, in the formal sense the contemporary biology of symbiosis has been establishing, a community of substrates in sustained mutual association. The substrates are bonded. The bonds operate through the same molecular, autonomic, and structural mechanisms the chapter has characterized at the human-canine scale. Symbiotic Love is the operational expression of what the visitor's substrate already is. The visitor's recognition of Symbiotic Love as the framework's load-bearing concept is the visitor's recognition of the visitor's own deepest biological structure.

The planetary scale of this recognition was articulated most influentially by the British scientist James Lovelock and the same Lynn Margulis in their joint development of the Gaia hypothesis in the 1970s — the framework that holds that Earth's living and non-

[231]Comprehensive synthesis of the contemporary symbiosis-and-evolution literature: Gilbert, S. F., Sapp, J., and Tauber, A. I. (2012). "A symbiotic view of life: We have never been individuals." Quarterly Review of Biology 87(4): 325-341. The closing claim of the Gilbert paper — that the conceptual unit of biology cannot coherently be the individual organism because every organism is, on close inspection, a community of symbionts — is the formal statement of what the chapter's argument has been developing. On the specifically forest-symbiotic register of mycorrhizal networks and the inter-tree communication they support, see Simard, S. W., Perry, D. A., Jones, M. D., Myrold, D. D., Durall, D. M., and Molina, R. (1997). "Net transfer of carbon between ectomycorrhizal tree species in the field." Nature 388: 579-582; and the synthesis in Simard, S. W. (2021). Finding the Mother Tree: Discovering the Wisdom of the Forest. New York: Knopf.

living systems together constitute a self-regulating system that maintains the planetary conditions life depends on.[232] The Gaia hypothesis has had a complicated reception within mainstream science — the strong version, which would attribute purposive agency to the planetary system, is rejected by most working biologists, while the weak version, which holds that the planet operates as a self-regulating system through feedback loops between living and non-living components, is now part of the standard framework of Earth-system science. The contemporary discipline of Earth-system science treats the Gaia framing as a productive heuristic for understanding the integrated operation of the planet's atmospheric, oceanic, biological, and geological systems, even where the framing's stronger metaphysical claims remain contested. The framework of this book takes the integrated-system reading as sufficient for its purposes. The planet, on the integrated-system reading, operates as a vast symbiotic field in which the substrates that compose it — biological and otherwise — sustain conditions through their mutual regulation that no individual substrate could sustain on its own. The visitor's home is, in this framing, the largest symbiotic substrate the framework engages, and the visitor's relationship to that substrate is the largest scale at which Symbiotic Love operates.

✦ ✦ ✦

The book closes here. The framework has been built across the sixteen chapters of evidence, operational technology, and structural commitment that the closing chapter has now brought to its full extension. What remains is the closing recognition that the six

[232] Lovelock, J. E., and Margulis, L. (1974). "Atmospheric homeostasis by and for the biosphere: The Gaia hypothesis." Tellus 26(1-2): 2-10. Book-length development: Lovelock, J. E. (1979). Gaia: A New Look at Life on Earth. Oxford: Oxford University Press. Updated framework: Lovelock, J. E. (2006). The Revenge of Gaia: Earth's Climate in Crisis and the Fate of Humanity. New York: Basic Books.

substrates whose openings the book has been organized around constitute, together, the empirical signature of what the framework has been characterizing. The book's closing pages bring the six substrates into the present moment in which the reader is reading this sentence. Some of the six are still in the substrates that received their visits. Some have completed the visit those substrates were the platform for. The framework's working logic holds that the distinction does not exhaust the meaning of either. The presence of the visit, on the framework's commitments, extends across the substrate transition the body's matter is the platform for, into the relational field the visit was conducted in and the substrate hierarchy the visit was integrated into. The book closes with the six substrates' present-time presence in the field the visitor is one node within.

Yongey Mingyur Rinpoche, fifty-one years old in the spring of 2026, teaches at Tergar Monastery in Bodh Gaya, India, in the configuration his recovered post-retreat life has settled into across the decade since his November 2015 return from the wandering retreat that the four-and-a-half years before then had been organized around. The monastery is in the spring teaching season the international meditation community has been organizing its calendar around since the early 2000s. The students at the monastery this season include the long-term Western practitioners who have been following his work since the publication of his first English-language book in 2007, the Tibetan and Nepalese monastic community whose formation he has been responsible for since the founding of Tergar in the early 2000s, the small group of scientific collaborators from the Madison and Harvard labs whose ongoing imaging studies continue to use his substrate as the most extensively documented long-term-meditator reference in the contemporary contemplative-neuroscience literature, and the broader pilgrimage community that arrives at Bodh Gaya each spring in the season the historical Buddha's enlightenment is observed under the descendant Bodhi tree a kilometer up the road from the monastery's principal entrance. The substrate that produced the gamma-coherence measurements at twenty-eight in September 2002 and the brain-age estimate of thirty-three at forty-one in May 2016 is the substrate that is, at this moment, teaching at the monastery the substrate's life has been organized around. The visit continues. The substrate continues. The teaching is the visit's local expression.

Edda Audrey Kathleen Ruston, who lived in the substrate the world knew as Audrey Hepburn from 1929 to her death in January of 1993 at the small property at Tolochenaz on Lake Geneva, completed the visit that thirty-three years ago. The substrate had been released. The two sons the visit had produced — Sean Hepburn Ferrer and Luca Dotti — have, in the thirty-three years since the substrate's release, organized the Audrey Hepburn Children's Fund as the operational continuation of the UNICEF work the visit's final five years had been devoted to. The Fund, established the year of her death, has continued the field operations in the regions of severe child malnutrition the visit's final mission had begun in, with the work conducted by the staff and the local partners and the broader humanitarian community the original visits had built relationships with. The relationships the visit had assembled around — the children at the Ethiopian feeding station in March 1988, the children at the Somali camp at Mandera in 1992, the colleagues at UNICEF who had been working in those regions throughout the previous decades — continue, in the present moment of 2026, to be the operational sites at which the work the visit had been organized around is conducted by those whose visits the original visit's example had configured. The visit, on the framework's commitments, is not where the body that conducted it is. The visit is in the relational field the visit produced. The field continues.

Jim Springer and Jim Lewis, the identical twins the case archive of Thomas Bouchard's Minnesota Study had been organized around since their reunion in February of 1979, completed their visits at the substrate timescales the longevity literature would have predicted from the population statistics. Jim Springer's substrate released in September of 2021 at the age of eighty-five. Jim Lewis's substrate released in June of 2024 at the age of eighty-eight. The children and grandchildren the two visits had produced — Springer's three children in Dayton and the surrounding southwestern Ohio counties, Lewis's three children in Lima and the families they had built — carry the relational continuity the two visits had been organized around. The genetic substrate the two visits had shared from the single fertilized egg of February 1939 is, in the relational field of the children and grandchildren, the substrate that continues across the substrate transition the two original visits have completed. The Bouchard archive at Minnesota remains as the documentary record of what the two substrates accomplished when they were both still in the substrates that received them. The accomplishment was the demonstration of the substrate hierarchy's specific load-bearing role. The demonstration continues to operate in the scientific literature the broader behavioral-genetics community has been building on since 1979. The visits continue, on the framework's working logic, in the field the visits were conducted to support.

Lyudmila Nikolaevna Trut, whose substrate released in the small hours of the ninth of October in 2024 in the small bedroom at the eastern end of the apartment in Akademgorodok where the chapter opened, is, in the relational field she organized the experiment around for the previous sixty-five years, the present-time configuration the chapter has been building toward. The experiment continues. The senior researchers she had been preparing across the previous decade conduct the daily selection protocol the experiment has been organized around since 1959. The fifty-six generations of foxes the experiment has produced occupy the pens at the eastern side of the farm. The international collaborations — with Louisville, with Leipzig, with Helsinki, with the United States National Academy — continue to extend the experiment's data into the broader scientific literature on domestication, behavioral genetics, and the operational mechanisms of symbiotic bonding. Pushinka, the small ember-coated female Trut had lived with at the small house at the farm from 1974 to the substrate transition that fox completed in 1979, is in the framed photograph on the bedside table that the family did not, in the days after the morning of October ninth, move from the position it had occupied for the previous decade. The bond Trut and Pushinka had constituted is, in the framework's working vocabulary, the most thoroughly documented single instance of Symbiotic Love that the experimental record of the twentieth century produced. The bond did not end with the substrate transitions. The bond is in the relational field the experiment continues to extend across the substrate hierarchy the framework has been characterizing.

James Leininger, who would turn twenty-eight on the tenth of April in 2026, is in the configuration his post-childhood substrate-bound life has organized around since the conclusion of the access window the Stevenson-Tucker case archive had documented as the typical closure pattern for cases like his. The drawings shifted after the Chichi Jima trip of 2010. The nightmares stopped. The verbal statements the case file had been built on faded into the substrate's mature filtering machinery in the way the literature had predicted. The young man who arrived at college in the late 2010s and at his early career configuration in the early 2020s does not, on the available documentation of his current life, carry the verbal access to the prior personality's content that the four-year-old had carried with such striking clarity in the early 2000s. What the young man carries is the substrate that was the platform of the visit James M. Huston Jr. had begun in 1923 and that had ended at the harbor at Chichi Jima in 1945, integrated through the substrate transition that the gap between the death of one body and the conception of the next had been the substrate's local interval. The visit continues in the substrate the young man is conducting it through. The substrate carries, on the framework's working logic, the accumulated content of both the prior visit and the present visit, integrated at the level the mature filtering machinery permits. The young man is the present-time configuration of a continuity the framework's case archive has been building the evidence for across the previous six decades. The visit continues.

Chanai Choomalaiwong, who turned fifty-eight in October of 2025 at the village in the central plains of Thailand where he had been born in October of 1967, is in the configuration his post-childhood substrate-bound life has organized around since the visit to the family of Bua Kai Lawnak that the boy of three had organized the family's attention around in the early 1970s. The two birthmarks the midwife had pointed out to his mother on the morning of his birth — the small round one at the back of the head, the larger irregular one above the left eye — remain as the visible trace of the substrate transition the visit had been organized through. The schoolteacher's family in the neighboring village remains in the relational continuity the boy's recognition of them in 1972 had reopened. The substrate that received the visit James M. Huston Jr. had been the substrate of in a different hemisphere of the planet and a different decade of the war is the substrate that received the visit Bua Kai Lawnak had been the substrate of in the village in the central plains of Thailand five years before the boy's birth. The substrate hierarchy operates, on the framework's commitments, in every register and on every continent. The continuity is the continuity. The trace on the substrate the body is carrying is the trace on the substrate. The visit, in the configuration the man is now conducting it through, is the same visit the boy of three had been conducting in the early 1970s and the schoolteacher had been conducting in the 1950s. The visit, on the framework's strongest reading, has been the same visit throughout.

✦ ✦ ✦

The book reaches its closing turn. The framework has been built across the sixteen chapters of evidence, operational technology, and structural commitment that the closing chapter has now brought to its full extension. The case has been made. The recognition has been offered. The visitor who has held with the book to this point has received what the book had to deliver — the documented evidence that the visit was chosen, the operational technology by which the choice can be lived from a configured substrate, the integrated method that one principled implementation of the technology has produced, and the structural obligation that the framework's commitments imply at every scale from the cellular to the planetary.

What follows from the book's argument is not, on the framework's working logic, anything the book itself can specify. The visit is the visitor's. The conditions of the visit are the visitor's selected content. The relationships the visit is conducted in are the visitor's specific relational field. The work the visit was assembled to do is the work this visit and no other was assembled to do, and the work is therefore not generalizable across visitors and not specifiable in any concrete program the book could prescribe. What the book has done is establish the framework within which the work becomes recognizable as the work, and the operational technology by which the work can be conducted from the substrate's configured state rather than from the grievance posture's reactive default. The work itself is the visitor's. The work awaits the visitor's engagement. The visitor's engagement is the visit.

The book has used, throughout its sixteen chapters, the vocabulary of the visit, the visitor, and the substrate to describe what other vocabularies would have named differently. The choice was deliberate. The vocabularies that emphasize the soul, the spirit, the higher self, the divine portion within — each of these vocabularies carries cultural and historical accretions that complicate the framework's working operation in ways the book wanted to avoid. The visit-vocabulary is intentionally austere, intentionally institutional in tone, intentionally close to what the empirical literatures the book has surveyed actually say. The vocabulary's modesty is its virtue. The framework named in this vocabulary is what the cumulative evidence actually supports, without the embellishments the older vocabularies have accumulated and without the metaphysical commitments those vocabularies require beyond what the evidence will sustain.

The reader who has integrated the framework to the point of the closing chapter does not, on the framework's working logic, need the framework's vocabulary to remain conscious in daily life. The vocabulary is a scaffolding. The scaffolding is for the substrate's reconfiguration during the period when the reconfiguration is still being accomplished. The substrate that has been reconfigured operates from the framework's working orientation without needing to remind itself of the framework's vocabulary at each moment of operation. The visitor whose substrate has integrated the framework moves through ordinary life, conducts ordinary relationships, makes ordinary choices, and meets ordinary difficulties — but does so from the configured state the framework's earlier chapters have characterized. The framework, fully integrated, becomes invisible. The visit, fully integrated, becomes the visitor's life.

The closing chapter has been the book's longest commitment to the affirmative. The argument has been built from rigorously documented empirical sources, with the institutional voice held throughout, with the hostile-witness logic invoked at every register where the framework could be supported by the testimony of researchers who would not have chosen the framework's interpretation themselves. The closing chapter has, with the framework's case fully made, allowed itself the recognition that the case adds up to. The visitor's life is the visitor's own work. The work is sacred, in the sense that the structural commitments the framework has documented support. The work extends beyond the visitor to the symbiotic field the visit is conducted in. The field is what the visitor came to participate in, and the participation is what the visit consists of.

The substrate awaits the visitor's full inhabitation of it. The configured state awaits the visitor's daily practice of it. The relationships await the visitor's presence to them. The symbiotic field — biological, social, planetary — awaits the visitor's recognition of it as the larger substrate the visitor's substrate is one node within. The visit awaits the visitor's choice to live it from the configured state rather than from the default. The choice is available now, in the substrate the reader is reading from, in the moment the reader is finishing this sentence. The book has done what the book set out to do. The rest is what awaits.

www.ingramcontent.com/pod-product-compliance
Lightning Source LLC
LaVergne TN
LVHW100248210926
848480LV00003B/863